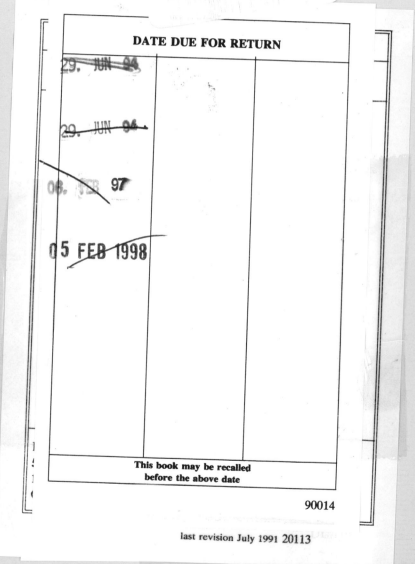

DATE DUE FOR RETURN

29. JUN 94		
29. JUN 94		
06. FEB 97		
05 FEB 1998		

This book may be recalled
before the above date

90014

last revision July 1991 20113

NITROGEN FIXATION

NITROGEN FIXATION

Volume 1   *Ecology*

Volume 2   *Rhizobium*

Volume 3   *Legumes*

Volume 4   *Molecular biology*

See the Appendix in this volume for contents lists of Volumes 1–3.

NITROGEN FIXATION

Volume 4
Molecular biology

Edited by

W. J. BROUGHTON
Université de Genève

and
A. PÜHLER
Universität Bielefeld

CLARENDON PRESS · OXFORD

1986

Oxford University Press, Walton Street, Oxford OX2 6DP

Oxford New York Toronto
Delhi Bombay Calcutta Madras Karachi
Petaling Jaya Singapore Hong Kong Tokyo
Nairobi Dar es Salaam Cape Town
Melbourne Auckland

and associated companies in
Beirut Berlin Ibadan Nicosia

Oxford is a trade mark of Oxford University Press

Published in the United States
by Oxford University Press, New York

British Library Cataloguing in Publication Data

Nitrogen fixation.
Vol. 4: Molecular biology
1. Nitrogen—Fixation
I. Broughton, W. J.
546′ 71159   QR89.7
ISBN 0-19-854575-4

Library of Congress Cataloging in Publication Data
(Revised for volume 4)
Main entry under title:
Nitrogen fixation.
Includes bibliographies and indexes.
Contents: v. 1. Ecology.—v. 2. Rhizobium.—[etc.]—
v. 4. Molecular biology.
1. Nitrogen—Fixation—Collected works. 2. Micro-
organisms, Nitrogen-fixing—Collected works.
I. Broughton, W. J.
QR89.7.N55 1981   589.9′5′04133   80-40612
ISBN 0-19-854540-1 (v. 1)
ISBN 0-19-854575-4 (v.4)

Set by DMB (Typesetting), Oxford
Printed in Great Britain
at the University Printing House, Oxford
by David Stanford
Printer to the University

Contents

List of contributions   vii

1. ENZYMOLOGY IN FREE-LIVING
   DIAZOTROPHS
   Robert R. Eady   1

2. BIOCHEMICAL PHYSIOLOGY OF
   Rhizobium DINITROGEN FIXATION
   Robert A. Ludwig and Gert E. de Vries   50

3. ELECTRON TRANSPORT TO NITROGENASE
   IN DIAZOTROPHS
   H. Haaker   70

4. TRANSCRIPTIONAL ANALYSIS OF THE
   NITROGEN FIXATION REGION
   (*nif*-REGION) OF *Klebsiella pneumoniae* IN *Escherichia coli*
   W. Klipp and A. Pühler   95

5. AZOSPIRILLUM
   Claudine Elmerich   106

6. AZOTOBACTER BIOLOGY, BIOCHEMISTRY,
   AND MOLECULAR BIOLOGY
   Betty E. Terzaghi and Eric Terzaghi   127

7. CYANOBACTERIAL NITROGEN FIXATION
   Robert Haselkorn   168

8. GENETICS AND MOLECULAR
   BIOLOGY OF SYMBIOTIC NITROGEN
   FIXATION BY *Rhizobium* sp. AND *R. japonicum*
   C. E. Pankhurst   194

9. PLASMID MOLECULAR GENETICS OF
   Rhizobium leguminosarum, Rhizobium trifolii,
   AND *Rhizobium phaseoli*
   R. J. M. van Veen, R. J. H. Okker, P. J. J. Hooykaas,
   and R. A. Schilperoort   224

10. MOLECULAR BIOLOGY OF SYMBIOTIC
NITROGEN FIXATION
BY *Rhizobium meliloti*
Adam Kondorosi   245

11. HOST-SPECIFIC GENE EXPRESSION IN
LEGUME ROOT NODULES
T. Bisseling, R. C. van den Bos, and A. van Kammen   280

APPENDIX: CONTENTS OF VOLUMES 1–3   313

INDEX   315

Contributors

T. Bisseling
Laboratory of Molecular Biology
Agricultural University
De Dreijen 11
7603 BC Wageningen
The Netherlands

R. C. van den Bos
Laboratory of Molecular Biology
Agricultural University
De Dreijen 11
7603 BC Wageningen
The Netherlands

W. J. Broughton
Université de Genève
1, ch. de l'Impératrice
CH-1292 Chambésy/GENÈVE

Robert R. Eady
ARC Unit of Nitrogen Fixation
University of Sussex
Brighton BN1 9RQ
UK

Claudine Elmerich
Unité de Physiologie Cellulaire
Départment de Biochimie et
Génétique Moléculaire
Institut Pasteur
F-75 724 Paris Cedex
France

H. Haaker
Biochemistry Department
Agricultural University
Wageningen
The Netherlands

Robert Haselkorn
Department of Biophysics and Theoretical Biology
University of Chicago
Chicago,
Illinois 60637
USA

P. J. J. Hooykaas
Department of Plant Molecular Biology
MOLBAS Research Group
University of Leiden
Wassenaarseweg 64
2333 AL Leiden
The Netherlands

A. van Kammen
Laboratory of Molecular Biology
Agricultural University
De Dreijen 11
7603 BC Wageningen
The Netherlands

W. Klipp
Lehrstuhl für Genetik
Biologie VI
Universität Bielefeld
Postfach 8640
D-4800 Bielefeld
West Germany

Adam Kondorosi
Institute of Genetics
Biological Research Center
Hungarian Academy of Sciences
PO Box 521
H-6701 Szeged
Hungary

Robert A. Ludwig
Thimann Laboratories
University of California
Santa Cruz
California 95064
USA

R. J. H. Okker
Department of Plant Molecular Biology
MOLBAS Research Group
University of Leiden
Wassenaarseweg 64
2333 AL Leiden
The Netherlands

C. E. Pankhurst
Applied Biochemistry Division
DSIR
Palmerston North
New Zealand

A. Pühler
Lehrstuhl für Genetik
Biologie VI
Universität Bielefeld
Postfach 8640
D-4800 Bielefeld
West Germany

R. A. Schilper
Department of Plant Molecular Biology
MOLBAS Research Group
University of Leiden
Wassenaarseweg 64
2333 AL Leiden
The Netherlands

Betty E. Terzaghi
Tissue Culture and Genetic Engineering Group
Plant Physiology Division
DSIR
Palmerston North
New Zealand

Eric Terzaghi
Tissue Culture and Genetic Engineering Group
Plant Physiology Division
DSIR
Palmerston North
New Zealand

R. J. M. van Veen
Department of Plant Molecular Biology
MOLBAS Research Group
University of Leiden
Wassenaarseweg 64
2333 AL Leiden
The Netherlands

Gert E. de Vries
Division of Natural Sciences
University of California
Santa Cruz
California 95064
USA

1 Enzymology in free-living diazotrophs

Robert R. Eady

1.1 INTRODUCTION

Nitrogenase from all organisms studied to date can be separated into two O_2-sensitive redox proteins, one of which contains molybdenum and iron (MoFe protein), and the other of which contains iron (Fe protein). Neither protein has any catalytic activity towards dinitrogen alone, but when recombined, in the presence of MgATP and a source of low potential reducing equivalents, catalyse the reaction:

$$N_2 + 8H^+ + 8e^- + 16MgATP \rightarrow 2NH_3 + H_2 + 16MgADP + 16P_i.$$

This reaction is inhibited by MgADP, but surprisingly, in view of its high MgATP requirement, not by NH_3 or glutamine, the immediate assimilation product.

The enzyme is both structurally and mechanistically complex. There have been many recent reviews on specific aspects of the molecular enzymology of this system (see Eady and Smith 1979; Mortenson and Thorneley 1979; Lowe et al. 1980; Emerich et al. 1981; and references therein). In this chapter my intention is to review new developments and provide a broader perspective of the peripheral biochemistry of N_2-fixation.

The component proteins of nitrogenase have been purified to homogeneity from a number of diazotrophs with quite different physiology. Many of these proteins have been extensively characterized and irrespective of the source from which they are purified show only minor differences from each other, and often form an active nitrogenase when recombined with the complementary protein from another organism. An exception to this generalization is the nitrogenase of phototrophic bacteria where, as discussed below, the Fe protein has been isolated in two forms; one of which is covalently modified and is inactive. Details of the anaerobic techniques required for the successful purification have been reviewed (Emerich and Burris 1979; Eady 1980a,b). More recently, the purification of the MoFe protein from *Anabaena cylindrica*

1

(Hallenbeck *et al.* 1979), large-scale purification of Av1 and Av2 from *A. vinelandii* (Burgess *et al.* 1980) and the active and inactive forms of Fe protein from *R. rubrum* (Gotto and Yoch 1982*a*; Ludden *et al.* 1982*a*), and *R. capsulata* (Hallenbeck *et al.* 1982) have been reported.

1.2 THE MoFe PROTEINS

1.2.1 Structure

The larger of the component proteins of nitrogenase, the MoFe proteins, are tetramers with an $\alpha_2\beta_2$ structure and relative molecular masses near 220 000. It is generally assumed that fully-active MoFe proteins contain 2Mo and 33 ± 3 Fe atoms per tetramer with somewhat less S^{2-} than Fe. The data in Table 1.1 indicate that the actual metal contents vary considerably, probably due to contamination of preparations with material lacking metal. Based on the activity of 300 nmole C_2H_2 reduced min^{-1} ng atom Mo^{-1} Smith (1983) calculated, an extrapolated maximum activity of 2700 nmol C_2H_2 reduced min^{-1} mg protein^{-1}. In practice, preparations of this specific activity have only occasionally been reported, and then not characterized extensively.

1.2.2 Subunit structure

Table 1.1 shows the range of relative molecular masses and the subunit structure of various purified MoFe proteins. The α-subunit (faster migrating on sodium dodecyl sulphate (SDS) electrophoresis, encoded by *nifD* in *K. pneumoniae*) has an approximate relative molecular mass of 51 000 and the β-subunit (encoded by *nifK*) of 60 000. The amino-acid sequence of the complete *nifD* gene product of *A. cylindrica* 7180, *C. pasteurianum*, *Parasporia rhizobium* sp ANU289, *Cowpea rhizobium* IRc78, and *A. chroococcum* are compared with the partial data available for *R. meliloti*, *R. trifolii*, *A. vinelandii*, and *K. pneumoniae* as shown in Fig. 1.1. There are five invariant Cys residues which occur in highly conserved regions and form obvious candidates for ligands to the FeS clusters these proteins are thought to contain. *NifD* of *C. pasteurianum* shows a markedly lower degree of homology within the generally conserved regions, in particular a region of 49 residues towards the C terminus which have no counterpart in any of the other four polypeptides with sequences spanning this region. In the case of Av1, Ac, and Kp1 five of the six Cys residues in the D-subunit are conserved. The Cys_{88} x x x x Cys_{93} grouping observed in Cp1 (characteristic of simple FeS proteins) is not conserved in Kp1 or Av1.

Complete sequence data for the *nifK* product are available for *A. cylindrica* 7180 and *Parasponia rhizobium* ANU289; these, together with partial sequence data for *C. pasteurianum*, *A. vinelandii*, and *A. chroococcum* are shown in Fig. 1.2. There is considerable sequence homology

TABLE 1.1
Metal and acid-labile sulphide content of MoFe proteins

	Relative molecular mass native protein	subunits*	Molybdenum g-atom mol^{-1}	Iron g-atom mol^{-1}	S^{2-} g-atom mol^{-1}	Specific activity nmol C_2H_2 reduced min^{-1} mg MoFe protein^{-1}
Av1	234 000	54 000 / 60 000	2	30–32	—	2106
Ac1	223 000	60 000	1.9	22	20	2000
Xa1	232 000	57 500 / 61 000	2.2	23	20	1070
Rj1	197 000	58 000 / 61 000	1.3	29	26	1000
Rl1	211 000	57 000		30–36		
Kp1	219 000	51 000 / 59 600	2		—	2150
Bp1	215 000	— / —	2	33	21	2750
Cp1	203 000	50 200 / 59 500	2	24	24	2250
Acy1	223 000	55 000 / 52 800	2.2	20	20	1200
Rc1	230 000	55 000 / 59 000	1.3	28	26.1	1800

These data were obtained with highly-purified apparently homogeneous preparations.
* The variability encounted with SDS electrophoresis gives approximate values only (see Kennedy *et al.* 1976); see Eady (1980) for references except Av1 (Burgess *et al.* 1980); Bp1 (Emerich and Burris 1978); Acy1 (Hallen-beck *et al.* 1979) and Rc1 (Hallenbeck *et al.* 1982).

between these polypeptides, but the Cys distribution is more variable than is the case with *nifD* and *nifH* products. There are only three conserved Cys residues, clustered near the N terminus. Within this cluster Pr, Av, and Ac have another conserved Cys and Ab has a Cys within two residues of this position. All of these Cys residues lie with a region of highly conserved sequence. The *nifK* product, but not the *nifD* product has a region of amino-acid sequence which is similar to two proteins of *E. coli* which hydrolyse ATP, the RecA protein, and the β-subunit of ATPase (Robson 1984) (see section 1.2.8). Assuming conventional ligation to the four 4Fe–4S clusters present in MoFe proteins, both the α- and β-subunits could each ligate only one cluster.

Although no obvious sequence homology is apparent between the α-and β-subunits of MoFe proteins, X-ray diffraction data of crystals of Cp1 allow the calculation of a 6 Å rotation function which indicates that there is structural homology between the α- and β-subunits (Yamane *et al.* 1982). Optical diffraction studies of electron micrographs of Av1 polymers are consistent with a tetrahedral arrangement of the subunits (Voordouw *et al.* 1983).

The sequence data for *C. pasteurianum* indicate that the slower migrating subunit of Cp1, observed during SDS electrophoresis, corresponds to the faster migrating subunit of Acy1, Kp1, and Av1 (Hase *et al.* 1981) emphasizing the unreliability of this method when applied to MoFe proteins (see Kennedy *et al.* 1976). The distribution of Mo and Fe atoms between the subunits is not known since their separation requires denaturation and consequent loss of metals from the protein. Recently, Av1 has been resolved into three spots on two-dimensional electrophoresis in urea, and peptide mapping shows the presence of a modified α-subunit (Harker and Wollstein 1981).

1.2.3 Redox centres

A variety of spectroscopic and physical techniques have been used to investigate the nature of the redox centres of nitrogenase proteins. Spectroscopic techniques include electron paramagnetic resonance (EPR); [57]Fe Mössbauer spectroscopy, Mo and Fe X-ray absorption spectroscopy (EXAFS), [96]Mo and ^{1}H electron nuclear double resonance (ENDOR), linear electric field effect (LEFE), and magnetic circular dichroism (MCD). These techniques have provided information as to the environment of the Mo and Fe nuclei and their interaction with the unpaired spin of electrons in paramagnetic species of these proteins. Comparison of spectral properties of nitrogenase proteins with simpler, better-understood FeS proteins has often been used diagnostically.

In addition, cluster extrusion of FeS centres has allowed the number and types of clusters present to be determined. In this technique, the protein is denatured by an organic solvent in the presence of a thiolate ligand. Under the appropriate conditions, FeS clusters undergo ligand exchange and are extruded as 2Fe–2S or 4Fe–4S clusters which can then be identified by comparison of their spectral properties with authentic FeS clusters.

1.2.4 FeMo clusters

There is circumstantial evidence that molybdenum forms part of the active site at which N_2 is reduced (see Smith 1977; Richards 1980; and below). Mo EXAFS studies of $S_2O_4^{2-}$-reduced Cp1 and Av1 indicate that molybdenum is in a sulphur environment 4–5 S atoms at 2.37 Å with 3–4 Fe atoms at 2.67 Å with possibly 2 O, N, or C atoms at about 2.1 Å (Cramer *et al.* 1978; Conradson, personal communication). Mo = O is only observed in O_2-inactivated protein.

All S_2O_4-reduced MoFe proteins exhibit, below 40 K a unique EPR spectrum with g values near 4.3, 3.7, and 2.01, which integrates to one-electron/Mo atom (Rawlings *et al.* 1978) and has been assigned to transitions in the $S = 1/2$ ground state of a spin $S = 3/2$ system. [57]Fe, but not [95]Mo, substituted proteins show a broadened EPR spectrum indicating spin interaction with iron; however, electron spin echo decay curves of Cp1 showed weak interactions with [95]Mo and [1]H (Orme-Johnson *et al.* 1981). Analysis of the Mössbauer effect of Kp1, Cp1, and Av1 indicates that 6Fe atoms are associated with the $S = 3/2$ centre.

The molybdenum and approximately 50 per cent of the iron of MoFe proteins can be extracted from precipitated protein by N-methyl-formamide (NMF) as an iron-molybdenum cofactor (FeMoco) (see Shah and Brill 1977; Smith 1980; Burgess *et al.* 1980). The cofactor which has a composition of $MoFe_{6-8}S_{4-9}$ has a broadened EPR spectrum with g values of 4.3, 3.7, and 2.01 and is capable of restoring activity to the inactive apoprotein synthesized by WO_4^--derepressed *A. vinelandii*, or by some mutant strains of *K. pneumoniae* and *A. vinelandii*. Preparations of FeMoco are extremely O_2 sensitive and do not contain amino acids; there is no general agreement on the sulphur content. Depending on the method used to precipitate the MoFe protein, two different forms of FeMoco, both active, but with different apparent relative molecular masses can be isolated (Smith 1980). The similarity of FeMoco from different sources is shown by the finding that on a molybdenum basis cofactors isolated from Kp1, Bp1, and Rr1 are as effective as that from Av1 in reconstituting activity to extracts of the mutant *A. vinelandii* UW45.

Mo and Fe EXAFS of FeMoco indicate that the environment of the molybdenum is changed only slightly by extrusion of FeMoco, and that the arrangement of Fe atoms is asymmetric in this cluster (Antonio *et al.* 1982). This interpretation is in accord with Mössbauer and

Protein sequence alignment (rotated 90°). Nine aligned sequences labelled Acy, Cp, Rm, Rt, Pr, cR, Kp, Av, Ac, shown in four blocks.

Block 1

```
Acy:  M---TPPENKNLVDENKELIQEVLKAYPEKSRKKREKHLNVH-EENKSDCG-------VKSNI
Cp :  -----SENL--KDEILEKYIPKTKTRGHIVIKTEETPNPE-----------IAI------ANT
Rm :  MSL---DYENDNELHARLIDEVLSQYPDKTARRRAKKHLSYAKNKQETAEEGQVVSECDVKSNI
Rt :  MSL---DYENDGDLHARLIDEVLSQYPDKTAKRRAKKHLSYAKNKGREALEQGSDALCETGVKSNI
Pr :  MSLATTQSIAEIRARNKELIEQEVLKVYPEKTAKRRAKHLNVHQAGKSDCG-------VKSNI
cR :  MSLASTQSIAEIRARNKELIEQEVLKVYPEKTAKRRAKHLNVHQAGKSDCG-------VKSNI
Kp :  M----MTNATGERNLALIQEVLKVFPPETARKERRKHMMVSDPKMKSVGK-------CIISNR
Av :  -----TGMSREEVESLIQEVLEVYPEKARKDRA----------------------CIISNK
Ac :  -----SGMSREEVESLIQEVLEVYPEKARKDRAKHLAVNDQSVTQSKK--------CITSNK
```

Block 2

```
Acy:  KSVPGVMTSDFQERDIVFGGCAYAGSKGVVWGPIKDMIHISHGPVCGYYYVG--NYYVGTGINSPGTM
Cp :  RTVPGIITARGCAYAG-CKGVVWGPIKDMVHITHGP-IGCYFYTWGGRRP--KSKPED-GLTTGV-PNEY
Rt :  KSIPGVMTSDFQRKDIVFGGCAYAGSKGVVWGPIKDMIHISHGPVCGYHYSWGSQRR--NYYVGLTTGVDAFVTL
Pr :  KSIPGVMTSDFDFRGKDIVFGGCAYAGSKGVVWGPIKDMVHISHGP-GCYYVGSRR--NYYVGTTGIDSFVTL
cR :  KSQPGVMTVRGCAYAGSKGVVFGPIKDMVHISHGPAGCYVGAGCR--NYYTG--SGVDSPGTL
Av :  KSQPGLMTIRGCAYAGSKGVVWGPIKDMIHISHGPVCG-YSS--RPVSN-YYIGTTGVNAFVTM
```

Block 3

```
Acy:  HFTSDFQELDVHELTKLIEELDVALTKLIEEDAIHEAYEMF-FPLNRGVSIQSECPIGSIGDDIEAVAKKTSKQIG-
Cp :  VFSTDMQESELDV--LKDAIHEAYEMF-FPLNRGVATCPVLIGDDILAVA-SKEIG-
Rt :  QFTSDFQRREKDV--IKKLSLPPLNNGITISECPIGDDIEAVSRSKKEYGG
Pr :  QFTFDFR...
cR :  QFTSDFQR...
Kp :  NFSDFQEKDV--SAGHHIANT...SVSKVVKGAEQTP
```

Block 4

```
Acy:  KPVVPLRCEGFRGVSQSLGHHIANDAIRDWIFPEYDKLKKETRLDFEPSPYDVALIGDEYNIGG
Cp :  IPVHAFSCEGYKGVSQSLGHHIANNTVMTDDIGKGNKEEKK--YSINVLGEYNIGG
Pr :  KTIVPVRCEGFRGVSQSLGHHIANDAVRDWIFDQVREADGKPK--PYDVAIIGDYNIGG
cR :  KTIVPVRCEGFRGVSQSLGHHIANDVVRDWI
Kp :  KPVIPVRCEGFRGR
Av :  CEGFR
```

6

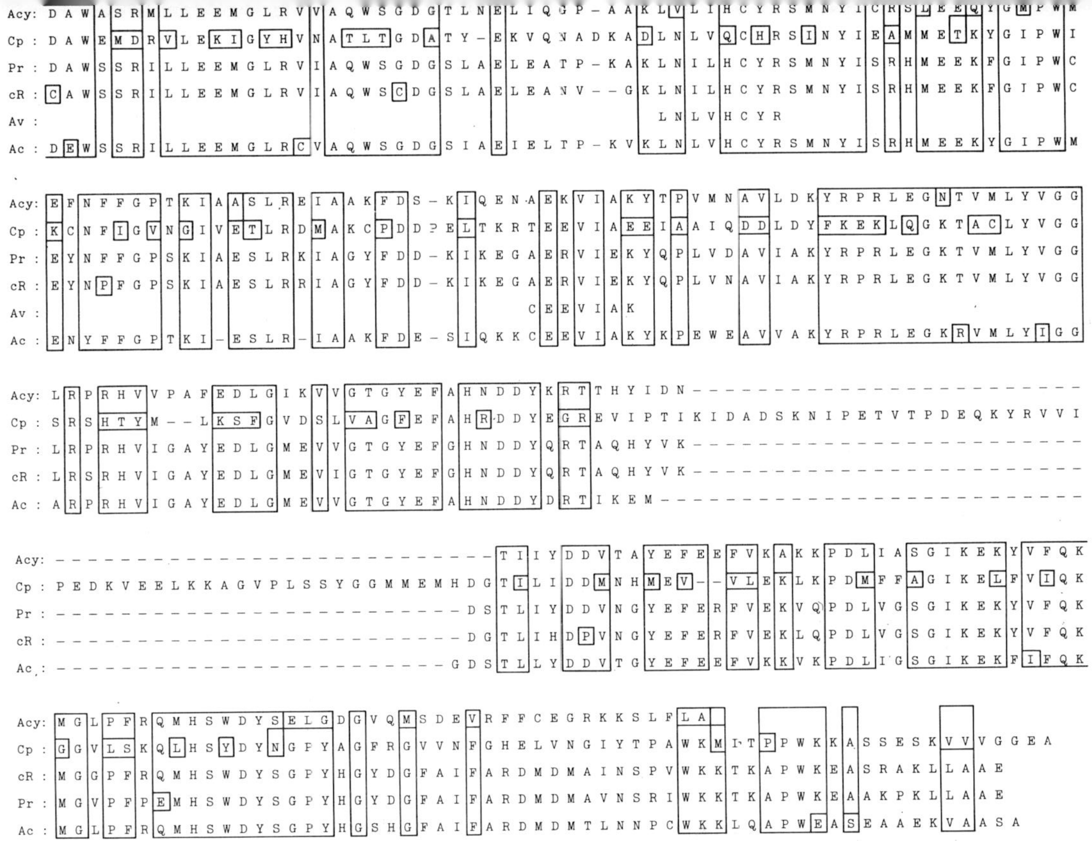

Fig. 1.1. Comparison of the amino-acid sequences encoded by *nifD*. Data for Acy1 (Lammers and Haselkorn 1983); Cp1 (Hase *et al.* 1984). RM1 (Török and Kondorosi 1981); Rt1 (Scott *et al* 1983); Pr1 (ANU289)(Weinmann *et al.* 1984); Rc1 (Yun and Szalay 1984); Kp1 (Scott *et al.* 1981); Av1 (Lundell and Howard 1981); and Ac1 (Jones 1984). Sequences have been aligned to maximize homology and the boxed areas enclose sequences which are either invariant or in which only one residue is not conserved. No allowance has been made for conservative substitution and the invariant Cys residues are marked with an asterisk.

```
Ac :  T P A D G Q F R M Y A G G T Q E E M K D A P N A L S T V L L H P W Q L E K T K K L V E G T W N D E V P K

Acy: - L R P F G V K G T D E F L T A V S E L T G K A I P E E L E M E R G R L V D A I T D S Y A W I H G K K F A I
Cp : - - T P I G V S A T D E F I M A L S E A T G K E V P A S I E E B E R G Q L I D L M I D A Q Q Y L G Q K K V A L
Pr : - - N Y P V G V S A T D D L L V A L S R I S G K E I P A Q L A R E R G R L V D A I A D S S A H I H G K K F A I
Av :
Ac : L N I P M G L D W T D E F L M K V S E I S G H R I P A S L T K E R G V L V D H M T T A H T W L H G N R F A L

Acy: - Y G P D L I I S I T S F L L E M G A E P V H I L C N N Q D D T F K K E M E A I L A A S P F G K E A K V W I
Cp : - L G P D E A I L K L G A I P K Y V T G T P G M K F E K E I D A M L A E A G I - E G S K V K V
Pr : - Y G P D L C Y G L A A F L L E L G A E P T H V L S T N G N N V A C - E N A T L F A G S F F G - E L P A Y P
Av : F L L E L G C E P V H I L C H D G N K R
Ac : W G D P D F V M G V M K F L L E L G C E P V H I L C N N G S K R W K K A V D A I L A S S P D G N N A T V Y V

Acy: Q K D F W H F R S L L F T E P V D F F I G N S Y G K Y L W R D T S I P M V R I G Y P L F D R H H L H R Y S
Cp : E G D F F D V H Q W I K N E G V D L L I S N T Y G K F I A - E E N I P F V R E G F P I M D R Y G H Y Y N P
Pr : G R D L W H M R S L L F T E P V D F L I G N T H G K Y L E R D T G T P L I R I G F P I F D R H H H H R P V
Av :
Ac : G K D L W H L R S L I F T D

Acy: T L G Y Q G G L N I L N W V N T L L D E M D R S T N L J G K T D I S F D L I R
Cp : K V G Y K G A I R L V E E I T N V I L D K I E R - - - E C T E E D F E V V R
Pr : V W G Y Q G G L N V L V K I L D K - F D E I D K K T S V L G K T D Y S F D I I R
```

FIG. 1.2. Comparison of the amino-acid sequences encoded by *nifK*. Data for Acy1 (Mazur and Chui 1982); Cp1 (Hase *et al.* 1984); Pr1 (ANU289) (Weinman *et al.* 1984); Av1 (Lundell and Howard 1981); and Ac1 (Jones 1984). The sequences have been aligned to maximize homology and the boxed areas enclose sequences which are either invariant; or in which only one residue is not conserved. No allowance has been made for conservative substitution, and the invariant Cys residues are indicated by an asterisk. *Nitrogenase nomenclature.* Nitrogenase component proteins are indicated by the number 1 or 2 for the MoFe protein and the Fe protein respectively. Acy is *Anabaena cylindrica*; Ac is *Azotobacter chroococcum*; Av is *Azotobacter vinelandii*; Bp is *Bacillus polymyxa*; Cp is *Clostridium pasteurianum*; Cv is *Chromatium vinosum*; Kp is *Klebsiella pneumoniae*; Pr is *Paspalum rhizobium*; Rc is *Rhodopseudomonas capsulata*; Rr is *Rhodospirillum rubrum*; Rl is *Rhizobium lupini* bacteriods; Rj is *Rhizobium japonicum* bacteroids; Rm is *Rhizobium meliloti*; Rt is *Rhizobium trifolii*; Xa is *Xanthobacter autotrophicus*.

ENDOR (Hoffman *et al.* 1982) data on the environment of the six magnetic Fe atoms.

Isolated FeMoco is inactivated by mercurials; the addition of stoichiometric amounts of thiophenol sharpens the EPR spectrum; EDTA and o-phenanthroline bleach the EPR signal which is restored by the addition of Zn or Fe. Activity is retained after these treatments.

A different Mo- and Fe-containing cluster can be isolated by extraction of precipitated protein with methylethylketone (Shah and Brill 1981). This cluster, which does not have FeMoco activity, has the formulation $MoFe_6$, and an intense EPR spectrum with g values of 2.05 and 2.01. The addition of NMF transforms this EPR species to that with the $g = 4.3, 3.7$, and 2.01 EPR signal. The lack of FeMoco activity of this cluster suggests that the two additional Fe atoms present in FeMoco are essential for activity.

Both FeMoco and the MoFe clusters reduce C_2H_2 in the presence of sodium borohydride, a reaction which is inhibited by CO (Shah *et al.* 1978). Cyclopropene is also reduced in the FeMoco system but gives a different stoichiometry of products from the nitrogenase-catalysed reduction (McKenna *et al.* 1979) (see Section 1.4.2). Therefore the relevance of the borohydride reduction of these compounds in the presence of the clusters to the enzyme-catalysed reduction is questionable.

The molybdenum cofactor which has been isolated from other molybdenum-containing enzymes by extraction with NMF has been shown to be a different entity from FeMoco (Pienkos *et al.* 1977).

1.2.5 FeS centres (P clusters)

The Mössbauer parameters of the non-cofactor Fe in Kp1, Av1, and Cp1 are very unusual and have no counterpart in other FeS proteins (Smith and Lang 1974; Münck *et al.* 1975; Huynh *et al.* 1980). The Fe atoms are all ferrous and are in two different environments. Cluster extrusion of $S_2O_4^{2-}$-reduced Cp1 in the presence of o-xylyl-α, α' -dithiol quantitatively extruded non-cofactor Fe as four 4Fe–4S clusters. MCD studies on Kp1 and Av1 indicate that 4Fe–4S clusters in the oxidation states $+1$, $+2$, or $+3$ found in other FeS proteins are not present (Stephens *et al.* 1979), and a probable oxidation level of [4Fe–4S]^O in the $S_2O_4^{2-}$-reduced proteins has been assigned. Consistent with this, a transient EPR signal with $g_{av} = 1.93$ arising from P clusters was observed in rapidly frozen samples of Kp1 during ferricyanide oxidation (Smith *et al.* 1983). Oxidized P clusters are EPR silent, but have a very complex magnetic Mössbauer spectrum at 4.2 K which indicates that they are in different environments. Detailed analysis indicated that the spin state of oxidized P clusters lay between 3/2 and 9/2. MCD data are consistent with $S = 5/2$ or $7/2$, for Kp1 (Johnson *et al.* 1981), and preliminary NMR shift requirements indicate $S_2O_4^{2-}$-reduced Kp1 to have two $S = 3/2$ cluster/mol and dye-oxidized protein to have four $S = 5/2$ clusters/mol (Smith, J. *et al.* 1982).

The spectral properties of the P clusters are unique for 4Fe–4S centres; there are no data to support the presence of 3Fe–3S centres in MoFe proteins.

1.2.6 Redox properties

The midpoint potential of dye-oxidized $S_2O_4^{2-}$-reduced FeMoco centres have been determined for Kp1, Bp1, Av1, Ac1, Cp1, and Cv1 from EPR measurements. The midpoints are species-dependent and range from -260 mV for Cv1 to 0 mV for Cp1 (see O'Donnell and Smith 1978). With the exception of Cv1 where two one-electron (1e) waves were determined ($E_{m\,7.5} = -60$ mV and -260 mV) the other MoFe proteins showed a single one-electron wave. Mössbauer studies with Kp1 gave E_m values of -216 mV for the FeMoco and -340 mV for the P centres (Smith *et al.* 1980). Corresponding values for Av1 using amperometric techniques were -280 mV and -480 mV for FeMoco and P centres respectively (Watt *et al.* 1980). For unknown reasons, the value for the FeMoco centres obtained by this method is 250 mV more negative than the value obtained from EPR measurements; however, it is apparent that the P centres have lower redox potentials than the FeMoco centres.

Oxidative titrations of Cp1 (Orme-Johnson *et al.* 1981) and Kp1 (Smith *et al.* 1980) indicate that six electrons are given up monotonically. In contrast, the FeMoco and P centres of Av1 undergo differential oxidations, but there is no general agreement as to the number of electrons associated with each process (Smith 1983 for a discussion). In one report, four electrons were removed from the P centres, leaving the FeMoco centres unaffected. During such an oxidation a new EPR signal with g values of 2.05, 1.94, and 1.81 characteristic of $[4Fe-4S]^{+1}$ clusters is observed, which increases to a maximum intensity as two electrons are removed and then is bleached as a further two electrons are removed (Orme-Johnson *et al.* 1981). These data have been interpreted as the P centres being paired such that when one member of each pair is oxidized they give rise to this EPR signal and when both are oxidized they are EPR silent but detectable by Mössbauer spectroscopy. A similar EPR signal is observed as a transient in rapid-freeze samples when Kp1 is oxidized with ferricyanide (Smith *et al.* 1983). This signal which is at a maximum intensity after 50 ms and absent after 1 s reaction, has been assigned to P centres in the $[4Fe-4S]^{1+}$ oxidation level (P^{+1}) on the basis of the effect [57]Fe-labelled P or FeMoco centres on the EPR linewidth. This assignment is supported by the observation that when Kp1 in which only the P centres had been previously oxidized is reacted with ferricyanide the signal is not observed.

Multiple sequence alignment (two blocks). Species labels: Acy, Cp, Rj, Rm, Rt, Pr, Kp, Av, Ac.

Block 1 (positions ~1–75):

```
Acy:  M T D E N I R Q I A F Y G K G G I G K S T T S Q N T L A A M A E M G Q R I M I V G C D P K A D S T R L M L H S K A Q T T V L H L A A E R G A V E D L E L
Cp :  - - M A S L R Q V A I Y G K G G I G K S T T S Q N L A A L S G L H A K T I M V V G C D P K A D S T R L L L G L A Q K S D T I L S L H L R E E G - E D V E L
Rj :  - M A A L R Q I A F Y G K G G I G K S T T S Q N T L A A L V D I G K I L A Q D T I L S L H L A A S A G S V E D L E L
Rm :  M A A L R Q I A F Y G K G G I G K S T T S Q N T L A A L V E L G K K I L H A K T K G S A A T E G S V E D L E L
Rt :  M A A L R Q I A F Y G K G G I G K S T T S Q N T L A A L V D L G K K I L N S K A Q D T I L D L H L A A K G S V E D L E L
Pr :  M S S L R Q I A F Y G K G G I G K S T T S Q N L V A A L V E M G K K V M I V G C D P K A D S T R L L L H A K A Q D T I M E M A A E B V G S V B D L E L
Kp :  T M R Q C A I Y G K G G I G K S T T T Q N L V A A L A E M G K K V M I V G C D P K A D S T R L I L H S K A Q N T I M E M A A E A G T V E D L E L
Av :  A M R Q C A I Y G K G G I G K S T T S Q N L V A A L A E M G K K V M I V G C D P K A D S T R L I L H A K A Q N T I M E M A A E A G T V E D L E L
Ac :  A M R Q C A I Y G K G G I G K S T T S Q N L V A A L V E M G K K V M I V G C D P K A D S T R L I L H S K A Q N T I M E M A A E A G T V E D L E L
```

                                                                    100•

Block 2 (positions ~76–150):

```
Acy:  H E V M L T G R F G V K C V E S G G P E P G V G C A G R G I T A I N F L E E N G A Y Q D L D F V S Y D V L G D V V G R G N G A V K G G F A M P I R E G K A Q E I Y
Cp :  D S I L K E G Y G G I R C V E S G G P E P G V G C A G R G I T S I N M L E Q L G A Y T D D L D Y V F Y D V L G D V V G R G N K A V K A Q E I Y
Rj :  E D V M K V G Y Q D I R C V E S G G P E P G V G C A G R G V I T S I N F L E E N G A Y - E N I D D Y V S Y D V L G D V V G R G N K A V K A Q E I Y
Rm :  E D V L K V G Y R G I K C V E S G G P E P G V G C A G R G V I T S I N F L E E N G A Y - D D Y V S Y D V L G D V V G R G N K A V K A Q E I Y
Rt :  G D V L K T G Y G G I K C V E S G G P E P G V G C A G R G V I T S I N F L E E N G A Y - D D Y V S Y D V L G D V V G R G N K A V K A Q E I Y
Pr :  E D V M K V G Y K D I R C V E S G G P E P G V G C A G R G V I T A I N F L E E N G A Y E N I D D Y V S Y D V L G D V V G R G N K A V K A Q E I Y
Kp :  E D V L Q I G Y G D V R C A E S G G P E P G V G C A G R G V I T A I N F L E E D G A Y E D D L D F V F Y D V L G D V V G R G N K A V K A Q E I Y
Av :  E D V L K A G Y G V K C V E S G G P E P G V G C A G R G V I T A I N F L E E N G A Y E D D L D F V F Y D V L G D V V G R G N K A V K A Q E I Y
Ac :  . . . . . . . . . . . . . . . . . . . . . . . . . . . . . . . . . . . . . . . . . . . . . . . . . . . . . . . . . . . . . . . . . . . . . . .
```

                                                                    150

12

```
Acy:  I V I S G E M M A M Y A A N N I A R G I L K Y A H S G G V R L G G L I C N R K V D R E D E L I M N L A E R L N T Q M I H F V P R D N I V Q H A E L R R
Cp :  I V A S G E M M A L Y A A N N I S K G I Q K Y A K S G G V R L G G L I C N E R K V A N E Y E L L D A F A K E L G S Q L I H F V P R S P M V T K A E I N K
Rj :  I V M S G E M M A M Y A A N N I S K G I L K Y A N S G G V R L G G L I C N E R O T D K E L A E L A K L G T Q L I Y F V P R D N V V Q H A E L R R
Rm :  I V M S G E M M A L Y A A N N I A K G I L K Y A S A G S V R L G G L I C N E R H T D R E L D L A E A L A A K L A A K L I H F V P R D N I V Q H A E L R R
Rt :  I V M S G E M M A L Y A A N N I A R G I L K Y A N S G G V R L G G L I C N E R Q T D R E L D L A E A L A A K L A A K L G T Q L I Y F V P R D N I V Q H A E L R R
Pr :  I V M S G E M M A M Y A A N N I S K G I L K Y A N S G G V R L G G L I C N E R Q T D R E D E L I I A L A E K L G T Q M I H F V P R D N I V Q R A E I R R
Kp :  I V C S G E M M A M Y A A N N I S K G I V K Y A K S G K V R L G G L I C N S R Q T D R E D E L I I A L A N K L G T Q M I H F V P R D N V V Q R A E I R R
Av :  I V C S G E M M A M Y A A N N I S K G I V K Y A N S G S V R L G G L I C N S R N T D R E D E L I I A L A A K L G T Q M L H F V P R D N V V Q R A E I R R
Ac :  I V C S G E M M A M Y A A N N I S K G I V K Y A N S G S V R L G G L I C N S R N T D R E D E L I I A L A A K L G T Q M L H F V P R D N V V Q R A E I R R

                                                250
Acy:  M T V N E Y A P D S N Q G E Y R A L A K K I - N N D - K L T I P T P M E M D E L E A L K I E Y G L L D D T K H S E I I G K P A E A T N R S C R N
Cp :  Q T V I E Y D P T C E Q A E E Y R E L A K V D A N E L F - V I P K P M T E Q R H L E E I L M Q Y G L M D L I K H S E I I G K T A A E L A A S
Rj :  M T V L E Y A P D S K Q A D H Y R K L A A K V H N N G G K G I I P T P I S W D E L E D M L M E H G I I K A V D E S - I I G K T A A E L A A S
Rm :  M T V I Q Y A P R S K Q A G E Y R A L A E K I H A N S G R G T V P T P I T W E E L E D M L L D F G I M K S - D E Q M L A E L H A K E A K V I A P H
Rt :  M T V I Q Y A P R S K Q A A E Y R W L A E K I H S N S G K G I I P T P I T W E E L E D M L L D F G I M K S - D E Q M L E E L L A K E V Q A A V A P
Pr :  M T V L E Y A P E S Q Q A D H Y R N L A T K V H N N G G K G I I P T P I S M D E L E D M L M E H G I M K P V D E S - I V G K T A A E L A A S
Kp :  M T V I E Y D P A C K Q A N E Y R T L A Q K I V N N I M K - V V P T P C T M D E L E S L L M E E F G I M E E E D T S - I I G K T A A E E N A A
Av :  M T V I E Y D P F K A K Q A D E Y R A L A R K V V D N K L L - V I P N P I T M D E L E E L L M E F G I M E E V E D E S - I V G K T A - E E V
Ac :  M T V I E Y D P T A K Q A D E Y R I L A R K V V E N K M L - I I P N P I T M D E L E A L L M E F G V M E E E D E S - I V G K G A A E E
```

FIG. 1.3. Comparison of the amino-acid sequences encoded by *nifH*. Data for Acy2 (Mevarech *et al.* 1980); Cp2 (Tanaka *et al.* 1977); (Fuhrmann and Hennecke 1984); Rm2 (Török and Kondorose 1981); Rt2 (Scott *et al.* 1983*a*); Pr2 (ANU289) (Scott *et al.* 1983*b*); Kp2 Sundaresan and Ausubel 1981; Scott *et al.* 1981; Av2 (Hausinger and Howard 1982); Ac2 (Jones 1984). The sequences have been aligned to maximize homology and the boxed areas enclose sequences which are either invariant, or in which only one residue is not conserved. No allowance has been made for conservative substitution and the invariant Cys residues are indicated by an asterisk.

13

The behaviour of the transient EPR signal as a function of Kp1 and ferricyanide concentration indicated that an intramolecular process was involved, and that the initial rapid phase of oxidation of Kp1 was associated with oxidation of the P centres with the FeMoco centre remaining EPR active. The disappearance of the EPR signal was attributed to the spin-state change of $S = 1/2$ to the EPR silent $S = 5/2$ state of the P centres in the $+1$ oxidation level. Further oxidation of the P centres resulted in the bleaching of the 4.3, 3.7, and 2.01 EPR signal of the FeMoco centre at the same rate as the bleaching of the transient $\tau = 4.1 \pm 0.8$ s^{-1} (23 °C), and was attributed to the intramolecular transfer of an electron from FeMoco to the oxidized (P^{2+}) P centres before spin-state change could occur. The similarity of the τ for this reaction with that for the rate-limiting step in nitrogenase turnover may indicate that both processes are linked to the same conformational change in the MoFe protein.

When dye-oxidized Av1 is re-reduced with $S_2O_4^{2-}$ only the P centres are reduced (Zimmermann *et al.* 1978) whereas in Kp1 both the P centres and the FeMoco centres are re-reduced (Smith and Lang 1974). These data suggest that isolated Av1 does not undergo the conformational change which allows the internal electron transfer to occur.

The differential exposure of FeMoco and P centres of Cp1 has indicated that the P centres are responsible for the circular dichroism (CD) at 400 nm (Orme-Johnson *et al.* 1981). This has enabled the demonstration that, like Av1, re-reduction of dye-oxidized Cp1 with $S_2O_4^{2-}$ reduces only the FeMoco and not the P centres. The oxidized P centres were re-reduced by Cp2 + MgATP with a first-order rate constant of 7 s^{-1}; this is sufficiently fast to allow more oxidized states of some of these clusters (which appear as EPR transients in the steady-state under perturbed conditions (Lowe *et al.* 1978) to be kinetically competent to be involved in turnover.

Further reduction of $S_2O_4^{2-}$-reduced MoFe proteins is only achieved by Fe protein and MgATP at potentials lower than -440 mV, and under these conditions the FeMoco centres become EPR silent. A preliminary report indicated that when Av1 in this state is rapidly separated from Av2 and MgATP the EPR silent form relaxes with a time constant of 600 s (Orme-Johnson *et al.* 1981). This has been interpreted as being due to intramolecular equilibration between the two FeMoco centres of an $\alpha_2\beta_2$ tetramer. If this assignment is correct it indicates that the FeMoco centres act independently in electron transfer during substrate reduction.

1.2.7 Interaction with substrates and inhibitors

A role for the MoFe protein in substrate binding was suggested by the observation that the EPR-observable pK_a of isolated Kp1 was displaced in the presence of C_2H_2 (Smith *et al.* 1973). Similar effects have been reported for Bp1 and Xa1. Subsequently, it was shown that during turnover in the presence of CO (which inhibits the reduction of all substrates except H^+), or under conditions of limited electron flux in the presence of C_2H_2 or added C_2H_4, a number of EPR signals arising from the MoFe protein could be detected (Davis *et al.* 1979; Lowe *et al.* 1978). Both tight and weak CO binding sites on Cp1 and Kp1 have been reported, and in the case of Kp1 an ethylene-binding site ($K = 1.3$ mM) giving rise to an EPR signal with *g* values 2.125, 2.000, 2.000 also bound C_2H_2 ($K = 8$ μM). This signal, indicative of binding of substrate and product at a single site, has now been demonstrated to arise from FeMoco centre of Kp1 (Hawkes *et al.* 1983). However, since ^{13}C-substituted C_2H_2, C_2H_4, and CO produced no detectable linewidth broadening of the EPR signals there is no evidence for *direct* binding of these compounds to the EPR active centres.

It has recently been established that Av1 has an uptake-hydrogenase activity in the presence of suitable oxidants (Wang and Watt 1984). This reaction which is catalysed in the absence of Fe protein and MgATP was detected in preparations of Av1 which were not contaminated by conventional hydrogenase. The rate of H_2-uptake was maximal with dichlorophenolindophenol as oxidant, with a 4e-oxidized state of Av1.

More direct evidence implicating FeMoco as a site of substrate-binding has been provided by a study of the properties of Kp1 isolated from a *nifV* mutant (see Section 1.5.1 (g)). This protein has altered properties which result in the enzyme *NifV⁻* Kp1 Kp2 being a poor nitrogenase but an effective C_2H_2 reductase (McLean and Dixon 1981). In addition, CO inhibits H_2 evolution, in contrast to the behaviour of Kp1 Kp2 isolated from wild-type. When FeMoco, isolated from the purified *NifV⁻* Kp1 is used to reconstitute activity of *NifB⁻* Kp1 it confers the *nifV* phenotype on the reconstituted enzyme (Hawkes and Smith 1983), i.e. altered substrate reduction patterns and response to inhibitors is associated with a modified FeMoco.

Additional data implicating FeMoco in N_2-binding is the finding that the maximum concentration of the enzyme-bound dinitrogen hydride intermediate involved in N_2 reduction shows a direct correlation with the molybdenum content of the Kp1 present in the quenched mixture (Lowe *et al.* 1983).

1.2.8 Binding of nucleotides

Kp1 and Ac1 freed of contaminating adenylylate kinase activity catalyse the slow exchange of $^{32}PO_4$ into the terminal P of both ATP and ADP (Miller *et al.* 1980). This exchange reaction is not observed with Kp2 or Ac2. It is unlikely to be due to contaminating activities since the column equilibration technique was used to demonstrate that Kp1 bound four ^{14}C-ATP molecules per tetramer with $K_m = 600 \pm 100$ μM (Miller *et al.* 1980). This binding is weaker than is observed with the Fe proteins and may account for the failure of Tso and Burris (1973), using the small-scale gel exclusion technique, to observe binding to Cp1. More recently, Braaksma *et al.* (1982) failed to detect binding to Av1. These differences may be due to different conformational states of the isolated MoFe proteins being involved in binding MgATP.

Proton NMR relaxation studies using the water proton NMR relaxation enhancement due to the paramagnetic ion Mn^{2+}, has been used to study metal ion and ATP interaction with Kp1 (Kimber *et al.* 1982). The specific activity of Kp1 was found to be directly related to its ability to bind Mn^{2+} being a maximum of four Mn^{2+} sites at a specific activity of 2900 (the highest reported activity for Kp1; Miller *et al.* 1980). Since Mg^{2+} can completely displace Mn^{2+} it was inferred that Mg^{2+} can bind to these four sites. Data were also obtained indicating the formation of the ternary complex ATP-metal-Kp1, and have been interpreted as fully active Kp1 having four ATP binding sites (Kimber *et al.* 1982). Additional evidence suggesting that MoFe proteins bind nucleotides is provided by the conservation of an amino-acid sequence found in other ATP-hydrolysing proteins (Robson 1984) (see Section 1.2.2).

1.3 Fe PROTEINS

1.3.1 Introduction

Fe proteins from eight organisms have been obtained as homogeneous protein preparations: Av2, Ac2, Xa2, R12, Kp2, Bp2, Cp2, and Rc2. As with the MoFe proteins, the specific activity shows considerable variation (Table 1.2). The extreme O_2-sensitivity ($t_{1/2} = 45$ s) and their general instability make it uncertain to what extent these apparently homogeneous preparations are contaminated by inactive material, although the ATP-enhanced reactivity of iron towards the chelator bathophenanthroline sulphonate has been used as a measure of this (Ljones and Burris 1978). The Fe proteins of phototrophic bacteria are subject to covalent modification which, in some cases, results in two bands being observed on SDS electrophoresis (see Section 1.3.6). These proteins have relative molecular masses in the range 57 000–72 000, are dimers of the structure γ_2 (encoded by *nifH* in *K. pneumoniae*), and contain approximately four atoms and four S^{2-} atoms per dimer (see Table 1.2). As discussed below (Section 1.3.3) the present consensus is that these proteins contain a single 4Fe–4S cluster which traverses the oxidation levels $+1$ and $+2$. Dithionite-reduced

TABLE 1.2
Iron and acid-labile sulphide contents of Fe proteins

	Relative molecular mass	Fe g-atom mol^{-1}	S^{2-} g-atom mol-1	Specific Activity nmol C$_2$H$_2$ reduced min^{-1} mg Fe protein^{-1}
Av2	64 000	3.7	—	2000
Ac2	65 400	4	3.9	2000
Xa2	72 600	3.8	2.4	1260
Rl2	65 800	3.1	—	434
Kp2	66 800	4	3.8	980
Bp2	55 500	3.2	3.6	2521
Cp2	57 600	4	4	2709
Rc2	63 000	4.1	5	600–800

These data were obtained with highly-purified apparently homogeneous preparations; see Eady (1980) for references except Av2 (Burgess *et al.* 1980); Xa2 (Berndt *et al.* 1978); Bp2 (Emerich and Burris 1978); and Rc2 (Hallenbeck *et al.* 1982).

Av2 has been crystallized and preliminary X-ray diffraction data show the space group to be P2, 2, 2 (Rees and Howard 1983).

1.3.2 Primary structure of the Fe proteins (*nifH* product)

Nucleotide sequencing of cloned *nif* regions which include *nifH* provided data from which the amino-acid sequences of Kp2, Acy2, Rj2, Rt2, Rm2, Pr2, and Ac2 were deduced. Figure 1.3 compares their sequence with those of Cp2 and Av2 determined from conventional amino-acid sequencing procedures. It is apparent that there is a high overall degree of conservation of the amino-acid sequence from different organisms, despite the divergence of the nucleotide sequences. However, sufficient DNA homology exists to enable cloned *nifH* DNA from *K. pneumoniae* to be used to probe for the presence of analogous DNA in other organisms, including symbiotic bacteria and blue-green algae (Nuti *et al.* 1979; Mazur *et al.* 1980; Ruvkun and Ausubel 1980).

There are five invariant Cys residues, which occur in regions of a high degree of conservation and are likely candidates for ligands to the 4Fe-4S cluster which these proteins are thought to contain, and chemical modification experiments of Av2 have indicated that Cys$_{97}$ and Cys$_{132}$ are involved (Hausinger and Howard 1983). Nevertheless, there is no apparent homology with the position of Cys residues in the MoFe proteins or with other FeS proteins. If the FeS cluster in the Fe protein bridges the two subunits, then it is reasonable to assume that two of the five conserved Cys residues act as ligands to the cluster. Cysteine residues have also been implicated in the binding of ATP to Cp2

(Section 1.3.5). In Kp2 at least, there is no evidence to support disulphide bridging within or between the subunits (Eady *et al.* 1972).

If aromatic residues are involved in redox reactions of the Fe proteins then the data of Fig. 1.2 show that $Tyr_{8, 114, 124, 148, 159, 171, 230,}$ and $_{240,}$ and also His_{209} are possible candidates, since they are invariant and occur within highly conserved regions of the primary structure. The assignment of a particular amino-acid residue to a direct role in electron transfer processes in FeS proteins is fraught with difficulties. In the case of HiPIP from *Chromatium vinosum*, X-ray crystallographic studies have suggested that Tyr_{19} which is in contact with the 4 Fe–4S cluster has an active role in redox reactions (Carter *et al.* 1974). However, kinetic studies and chemical modification of His_{42} of HiPIP with diethyl-pyrocarbonate have shown that both deprotonation or modification of this residue perturb the polypeptide structure to destabilize the reduced 4Fe–4S cluster, and so facilitate its oxidation by ferricyanide (Nettesheim *et al.* 1980). These studies indicate that direct interaction with the cluster is not necessarily a prerequisite for facilitation of electron transfer by a particular residue since the protein can mediate the geometry of the cluster and thus the properties of the 4Fe–4S cluster. From the available sequence data and the formation of active or partially-active heterologous nitrogenases, it is apparent that the structure of the Fe proteins as a class is highly conserved. The primary structure of Rm2, Rj2, Pr2, Ac2, Av2, Kp2, and Acy2 have an overall homology of 67–71 per cent; Cp2 is of lower homology, 58–67 per cent, consistent with its behaviour in forming mostly inactive heterologous nitrogenases with MoFe proteins from other organisms.

In general, the rate of accumulation of acceptable mutations is slowest among residues forming the active site and those surfaces involved in subunit contact or protein–protein interaction. In the case of Fe proteins these constraints on the evolution of the protein structure arise from the requirements for maintaining ATP, ADP binding sites, the cysteinyl geometry for the 4FeS cluster, and contact surfaces between the subunits, the electron donors flavodoxin or ferredoxins, in addition to the MoFe proteins. Such functionally important regions are likely to fall in the very highly conserved sequences which share 90–95 per cent homology.

1.3.3 FeS clusters

The Fe and S^{2-} atoms present in the Fe proteins have been assigned to a single 4Fe–4S cluster on the basis of cluster extrusion data and by comparison of various spectral parameters with FeS proteins containing characterized 2Fe–2S and 4Fe–4S centres. These data are summarized below:

(a) Cluster extrusion of Cp2 with thiophenol resulted in the near quantitative removal of the Fe as a 4Fe–4S cluster (Orme-Johnson and Davis 1977; Gillum *et al.* 1977).

(b) The large effect of an applied linear electric field (LEFE) on the *g* values of the EPR spectrum of Cp2 is consistent with the presence of a 4Fe–4S centre since this effect is an order of magnitude larger for 4Fe compared with 2Fe centres (Orme-Johnson *et al.* 1977).

(c) The Mössbauer spectrum of ^{57}Fe Kp2 (Smith and Lang 1974) is a doublet at 195 K which collapses, as the temperature is decreased to 4.2 K to a magnetic multiplet allowing the EPR signal of $S_2O_4^{2-}$-reduced Kp2 to be assigned to clusters of Fe atoms. The single doublet observed at 195 K is characteristic of 4Fe–4S centres in contrast to 2Fe–2S centres where the Fe atoms are not equivalent in the reduced state.

(d) Cp2 and Kp2 have features in their CD and MCD spectra which arise from metal centres and are similar to those of 4Fe–4S proteins (Stephens *et al.* 1979). Oxidation–reduction properties were consistent with 4Fe–4S clusters in the +2 and +1 states respectively. Quantitation of the MCD spectra indicated that the low integration of the EPR signal was not due to incomplete reduction of the centre. In addition, no optical effects due to the presence of additional chromophores were observed.

More recently, preparations of Av2 have been isolated which have a higher Fe and S content (Braaksma *et al.* 1983). The reason for this difference is not clear at present, but it may be due to the physiological status of the organisms at the time of harvest (Haaker, personal communication).

As isolated, in the presence of $S_2O_4^{2-}$ all Fe proteins exhibit a rhombic EPR signal with $g_{av} = 1.95$ detectable below 30 K. The signal has very anisotropic linewidths and integrates to fractions of a spin per mol; typical values are: Cp2, 0.2; Ac2, 0.24; Ca2, 0.28; and Kp2, 0.45. The low integration could be accounted for by interaction with an additional rapidly-relaxing paramagnetic centre (Lowe 1978), but as yet this centre has not been identified.

1.3.4 Redox properties

The $g = 1.95$ EPR signal of $S_2O_4^{2-}$-reduced Fe proteins is bleached on oxidation with O_2, dyes, or by enzymic oxidation in assays with limiting reductant and low concentrations of MoFe protein. Under the latter conditions it is assumed that only catalytically competent Fe protein is measured. More recently, the $SO_2^{\cdot-}/SO_3^{2-}$ redox couple has been used to poise the potential during redox experiments (Braaksma *et al.* 1982). The E_m for Cp2 was pH-independent indicating that proton transfer is not involved in the redox process.

The presence of MgATP decreases the E_m by approximately 110 mV for Cp2 and by -42 mV for Av2; MgADP has a similar effect resulting in a decrease of -80 mV for both proteins. Curve-fitting these data for Av2 indicated redox processes with $n = 2$ for the protein alone and in the presence of MgATP, but $n = 1$ in the presence of MgADP (Braaksma *et al.* 1982). These data contrast with Cp2 where $n = 1$ was obtained in the presence and absence of MgATP (Zumft *et al.* 1974).

The number of electrons involved in these processes is not resolved. Dye-oxidized Ac2 showed two phases of absorbance changes on reduction with SO_2^-. The first, very rapid phase restored the EPR spectrum and accounted for one electron; the second was slower and accounted for a futher electron, and was only associated with changes in absorbance. MgADP inhibited the rapid phase and MgATP was without effect(Thorneley *et al.* 1976).

Dye-oxidized Cp2 required 1.4–2 electrons/mol for re-reduction as measured by absorbance changes (Walker and Mortenson 1973) but when enzymically oxidized only 1e/mol (Ljones and Burris 1978). Further work will be required to resolve this problem since in the latter case MgADP would be present and may possibly result in an $n = 1$ behaviour from Cp2.

An additional redox pattern which is not understood is the observation that in the presence of MgATP or MgADP both Kp2 and Av2 catalyse the decompostion of $S_2O_4^{2-}$ and become oxidized (Stephens *et al.* 1982).

1.3.5 Nucleotide binding

The binding of MgATP results in a gross conformational change in Fe proteins. The accessibility of Fe towards chelating agents is increased, as is the reactivity with DTNB; the symmetry of the EPR signal changes from rhombic to axial and O_2 sensitivity is enhanced (see Eady and Smith 1979; Mortenson and Thorneley 1979, for review).

There is general agreement that Av2, Cp2, Kp2, and Bp2 bind two ATP per molecule (Table 1.3). However, the recent reports that Av2 and Kp2 in the presence of MgATP become oxidized as excess $S_2O_4^{2-}$ is decomposed and that $S_2O_4^{2-}$-reduced Av2 binds only one MgATP in contrast to $Av2_{ox}$ which binds two MgATP may require re-evaluation of the number of binding sites. The flow dialysis method used by Cordewener *et al.* (1983) allows rapid measurements of binding before oxidation can occur. In this context it should be noted that MgATP does not always result in a change in the EPR spectrum and that CD measurements do not detect ATP binding to $S_2O_4^{2-}$-reduced proteins.

Cysteine has been implicated in the binding of MgATP, since at saturating ATP levels, two Cys residues of Cp2 are protected from

TABLE 1.3
ATP binding to various Fe proteins

Fe protein	Number of binding sites	K_D (μM)	Method used to measure binding	Reference
Bp2	1.4	104	[14C] ATP gel-exclusion	Emerich and Burris (1978)
Kp2	—	400	EPR	Smith et al. (1973)
Av2	2	400 and 200	BPS	Hageman et al. (1980)
Av2	1.3±0.3	560±110	[32P] ATP flow dialysis	Cordewener et al. (1983)
Av2$_{ox}$	2.2±0.4	290±50	CD	
Av2$_{ox}$	2	75		Stephens et al. (1981)
Cp2	1.7	53	[14C] ATP gel exclusion	Emerich et al. (1978)
Cp2	1.7	16.7	EPR	Tso and Burris (1974)
Cp2	2	tight	CD	Zumft et al. (1974)
Cp2$_{ox}$	2	75		Stephens et al. (1981)
Cp2	2	tight	[31P] ATP NMR	Mortenson and Upchurch (1981)
Cp2	2	85	BPS	Ljones and Burris (1978)

carboxymethylation by iodoacetic acid (Mortenson 1978; see also Likhtenstein 1980). In Av2 the conserved Cys_{85} has been implicated as forming part of the nucleotide binding site since MgATP and to a lesser extent MgADP protect this residue from chemical modification (Hausinger and Howard 1983). Three regions of conserved amino acid sequences in the Fe proteins which are similar to nucleotide-binding proteins have been identified (see Robson 1984). The difference between Kp2 where MgATP enhances reactivity with DTNB (Thorneley and Eady 1974) may be due to reaction with S^{2-} or to the $S_2O_4^{2-}$-free Kp2 having become oxidized.

MgADP has been reported to inhibit MgATP enhancement of the reactivity of Cp2 with Fe chelators and a binding constant of 5.3 μM to a single site has been reported from ^{14}C studies (Tso and Burris 1973). The analysis of BPS reactivity data in terms of two identical (Ljones and Burris 1978) or two dissimilar (Hageman *et al.* 1980) MgATP binding sites has been questioned by Cordewener *et al.* (1983) who fitted the data to a single-site model.

1.3.6 Regulation of nitrogenase activity by covalent modification of the Fe protein

R. rubrum and other non-sulphur purple bacteria possess a control system for regulating nitrogenase activity in which the Fe protein is inactivated *in vivo* by covalent modification. Extracts containing inactive nitrogenase are stimulated by the addition of Mn^{2+} and show non-linear substrate reduction kinetics with time. This behaviour, which is attributed to the presence of an Mn^{2+}-dependent activating enzyme, has been observed in *R. rubrum, R. palustris, R. vannelli, R. capsulata*, and in the non-phototroph *A. lipoferum* (see Eady 1981; Zumft *et al.* 1981). Among the phototrophs, *R. sphaeroidies* is atypical since although there is evidence for reversible activation/inactivation *in vivo*, Mn^{2+} does not stimulate activity in extracts which also show linear rates of substrate reduction (Haaker *et al.* 1982). Mn^{2+} is required for N_2-dependent growth of *R. rubrum* and *R. capsulata* except under nitrogen starvation (Yoch 1979).

In *R. rubrum*, inactive Rr2 is the predominant form in extracts of N_2 or glutamate-grown cells, and also in N_2-limited and cells grown on glutamate plus N_2 (Sweet and Burris 1982). The unmodified active Rr2 was first identified in NH_4^+ limited batch cultures (Carithers *et al.* 1979). Purified preparations of both active and inactive Rr2 have been compared (Ludden *et al.* 1982a; Gotto and Yoch 1982a). Inactive Rr2 contains covalently-bound phosphate, ribose, and an adenine-like entity (Ludden and Burris 1978). It was originally proposed that inactive protein contained two modifying groups but recent work (Ludden *et al.* 1982a; Gotto and Yoch 1982a) is more consistent with only one being present. EPR measurements indicate that it is unable to transfer electrons to Rr1 (Ludden *et al.* 1982*b*).

SDS electrophoresis resolves inactive Rr2 into two bands of apparent relative molecular masses of 30 000 and 31 500. Active Rr2 migrates as a single band corresponding with the faster-running band of the doublet, and ^{32}P-labelling experiments showed the inactive species (upper band) contained phosphate (Ludden *et al.* 1982*a*; Gotto and Yoch 1982*a*). This difference in migration pattern allows the extent of interconversion of the two forms of Rr2 to be determined as a result of changes in physiological conditions (Preston and Ludden 1982; Gotto and Yoch 1982*a*). When purified from *R. capsulata*, the inactive and active forms of Rc2 do not resolve on SDS electrophoresis (Hallenbeck *et al.* 1982). This difference may not reflect a real difference between the regulated forms of the Fe proteins of photoheterotrophs since it is possible that, as has been shown for MoFe proteins (Kennedy *et al.* 1976), the commercial source of the SDS used in electrode buffers used in electrophoresis can alter the banding pattern observed.

Inactive Fe protein can be partially reactivated *in vitro* by a membrane-bound protein in a reaction which requires Mn^{2+} and MgATP (Ludden and Burris 1976, 1979; Nordlund *et al.* 1977; Gotto and Yoch 1982*b*). During activation the doublet observed on SDS gels becomes partially converted to the single band pattern characteristic of fully active Rr2 (Preston and Ludden 1982; Gotto and Yoch 1982*a*). It was initially reported that during reactivation only the adenine-like entity was removed (Ludden and Burris 1978); however, activation experiments with ^{32}P-labelled protein (Preston and Ludden 1982; Gotto and Yoch 1982*a*) have shown that most of the phosphate is released from the upper band during reactivation and formation of the lower band. Since the linkage of the modifying group is proposed to be protein–P–ribose–Ad* these data are consistent with the removal of the entire group during activation by activating enzyme. The requirement of ATP for activation explains the non-linearity of substrate reduction with time and the stimulation of activity by Mn^{2+} in extracts. However, as noted above, extracts of *R. spheroidies* which show reversible inactivation of nitrogenase *in vivo* do not show either of these effects (Haaker *et al.* 1982). ATP is not hydrolysed during the activation reaction and is not required for reactivation of oxidized Rr2. Its role may be in producing a conformational change in Rr2 to allow reactivation by the activating enzyme–Mn complex. Rr2 has also been shown to undergo non-enzymic activation on heating at 47 °C (Dowling *et al.* 1982).

Activating enzyme can be solubilized from the chromatophore fraction with 0.5 M NaCl and has been purified from *R. rubrum* (Nordlund *et al.* 1977; Gotto and Yoch 1982*b*) and *S. lipoferum* (Ludden and Burris

1978). Preparations are unstable but can be stabilized by Mn^{2+} (0.2 mM); are O_2-sensitive; and have an approximate relative molecular mass of 25 000 (Zumft and Nordlund 1981; Gotto and Yoch 1982*b*). Activating enzyme from *R. rubrum* will activate nitrogenase from *A. lipoferum*. Borate specifically inhibits activation of Rr2 (Ludden 1981), as does O_2-treated activating enzyme. In *R. rubrum*, the synthesis of activating enzyme is not under *nif* control since it is present in NH_4^+-grown or aerobic culture conditions under which repression of Rr2 occurs (Triplett *et al.* 1982).

The factors which result in the covalent inactivation of the Fe protein *in vivo* are not well-understood. In *R. rubrum*, nitrogenase activity of N_2-and glutamate-grown cells is high, but during harvesting and preparation of extracts, the Fe protein is spontaneously inactivated. This does not occur with NH_4^+-limited cells unless they are subjected to carbon-starvation or light limitation for some time before harvesting (Yoch and Cantu 1980). Under some conditions the addition of NH_4^+, glutamine, or aspartate results in the loss of nitrogenase activity *in vivo* and the appearance of inactive Fe proteins in extracts.

The effect of added NH_4^+ in producing an inhibition of nitrogenase activity *in vivo* has been extensively investigated. Under some conditions the addition of low concentrations of NH_4^+ results in a rapid but transient inhibition of nitrogenase-catalysed H_2 evolution or C_2H_2 reduction which is restored as the external NH_4^+ is depleted. The time required for the relaxation of inhibition is proportional to the NH_4^+ added. The extent and speed of the onset of inhibition depends on the nutritional status of the cultures. NH_4^+-starved organisms show a rapid and complete inhibition of activity, in contrast with glutamate or N_2-grown organisms which are unaffected by added NH_4^+ despite the fact that it is taken up by the cells. Susceptible cultures of *R. rubrum* (Yoch and Gotto 1982*a*) and *R. capsulata* (Hallenbeck *et al.* 1982) are more rapidly inhibited at low light intensities, or at high light intensities in the presence of uncouplers. The effect of light intensity on the effectiveness of NH_4^+ as an inhibitor of nitrogenase *in vivo* may well account for conflicting reports in the literature as to whether inhibition occurs or not in particular instances.

In vivo labelling studies have shown that during NH_4^+-shock, *R. rubrum* incorporates $^{32}PO_4$ into Rr2 (Preston and Ludden 1982) and that *R. capsulata,* when made permeable by toluene treatment, incorporates ^{14}C-ATP specifically into GS and Rc2 as nitrogenase activity is inhibited (Michalski *et al.* 1982). In the presence of the analogue 8-Br-Ad 5′P, incorporation of label into Rc2 but not GS is decreased. Since treatment with phosphodiesterase removes the label from GS and not Rc2, binding *via* adenine, rather than phosphate as in GS-AMP, is indicated.

Steady-state labelling of glutamate-grown *R. rubrum* with 8 ^{3}H-adenine or 2 ^{3}H-adenine led to incorporation of radioactivity into the upper band of Rr2 which was lost after reactivation (Nordlund and Ludden 1983).

Assimilation of the added NH_4^+ appears to be required for inhibition of nitrogenase activity. When glutamine synthetase is inhibited by the addition of methionine sulphoximine (MSX), NH_4^+ no longer inhibits nitrogenase activity in *R. rubrum* (Sweet and Burris 1981), *R. sphaeroidies* (Jones and Monty 1979; Haaker *et al.* 1982), *R. capsulata* (Hillmer and Fahlbusch 1979; Yoch 1980), and *R. palustris* (Alef and Zumft 1981); no inhibition by NH_4^+ is observed in glutamine auxotrophs of *R. capsulata* (Wall and Gest 1979).

In *R. capsulata* GS is subject to adenylylation–de-adenylylation control of activity (Johansson and Gest 1977). The adenylylation state depends not only on the nitrogen status, but also on the light intensity, being highly adenylylated (low activity) at high light intensities. Rapid changes occur in the adenylylation level on shift down of illumination. In *R. capsulata* and *R. palustris* it has been shown that in response to NH_4^+ shock, and during subsequent recovery, nitrogenase activity in extracts correlates with the state of adenylylation of GS (Hillmer and Fahlbusch 1979; Alef and Zumft 1981). However, in *R. rubrum* under these conditions the GS activity does not correlate with its degree of adenylylation (Falk *et al.* 1982). Under conditions of carbon-starvation or in nitrogen-deficient cultures the correlation of adenylylation of GS and nitrogenase activity is also not observed (Alef and Zumft 1981; Yoch and Cantu 1980).

In *R. sphaeroidies* it has been established that the inhibitory effect of NH_4^+ is not due to perturbation of the ATP/ADP ratio nor to the collapse of the $\Delta\Psi$ component of the membrane potential (see Haaker, this volume). Extracts prepared up to 8 min after the addition of NH_4^+ (sufficient to completely inhibit nitrogenase activity *in vivo*) had fully-active nitrogenase in extracts (Haaker *et al.* 1982), indicating that under these conditions the loss of activity was not due to covalent modification of the Fe protein.

It is apparent that NH_4^+ has more than one locus at which it acts to inhibit nitrogenase activity in phototrophs, and its effect appears to require the presence of functional glutamine synthetase.

1.4 STUDIES ON NITROGENASE DURING TURNOVER

1.4.1 Introduction

In the early 1970s the path of electron transfer from Fe protein to MoFe protein was established by the application of rapid-freeze EPR spectroscopy and Mössbauer spectroscopy. The subsequent use of stopped-flow absorption spectroscopy to monitor oxidation changes in the 4Fe–4S centre of the Fe proteins has facilitated studies on the electron-transfer reactions. Electron transfer from the Fe protein is ATP-dependent and too fast to be rate-limiting in turnover. More recently, quench-flow studies have (as described below) established that ATP hydrolysis is coupled to the electron-transfer reaction, allowed the detection of an enzyme-bound intermediate in N_2 reduction, and allowed the study of pre-steady-state product formation.

Hageman and Burris (1978, 1979) perturbed Av nitrogenase by using a hundred-fold excess of Av1 over Av2 which resulted in a temporal separation of substrate reduction from ATP hydrolysis. They observed a 4 min lag before H_2 evolution became linear but no lag for ATP hydrolysis. During the lag period, the EPR signal of Av1 became 50 per cent bleached. These data are consistent with an involvement of ATP in electron transfer between the two proteins before substrate reduction occurs.

Eady *et al.* (1978) with Kp nitrogenase and Hageman *et al.* (1980) with Av nitrogenase used rapid-quench techniques to observe a pre-steady-state burst of ATP hydrolysis before the linear steady-state rate of hydrolysis was attained. In the case of Kp nitrogenase it was shown that the τ for ATP hydrolysis of 44 ± 4 ms at 10 °C was indistinguishable from that for the rate of electron transfer from Kp2 to Kp1 determined by stopped-flow spectrophotometry. These observations indicate that ATP hydrolysis and electron transfer to the MoFe proteins are tightly coupled processes and occur rapidly relative to the turnover time of nitrogenase. No evidence for enzyme-bound phosphate or ADP was obtained in acid-quenched samples of Av nitrogenase.

It has been suggested that ATP has an additional role in nitrogenase function other than its involvement in electron transfer (see Eady *et al.* 1980), although this view has been criticized (Hageman *et al.* 1980). In principle, it is possible to resolve this question by quantifying the number of electrons transferred and the ATP molecules hydrolysed in the pre-steady-state burst and comparing this with the steady-state ratio of ATP/H_2. However, uncertainties as to the amount of fully-active proteins present and the observation that the EPR signals of the MoFe proteins are not fully bleached in the approach to the steady-state complicate such calculations. Bearing in mind these reservations, an approximate ATP/2e of 4 has been calculated (Eady *et al.* 1980; Hageman *et al.* 1980) which compares with the minimum value of 4 observed under optimal conditions in the steady-state (Ljones and Burris 1972; Watt *et al.* 1975; Imam and Eady 1980).

1.4.2 Substrate specificity

In addition to N_2, nitrogenase will reduce a number of substrate analogues (e.g. C_2H_2, CN, N_3^-, CH_3NC), and in the absence of an added reducible substrate, protons are reduced to H_2 (for a detailed review see Hardy 1979). All reductions require MgATP; CO inhibits the reduction of all substrates except H^+; and H_2 specifically and competitively inhibits N_2 reduction. During the reduction of N_2 a dinitrogen-hydride intermediate is formed which can be trapped as hydrazine (N_2H_4) following quenching with acid or alkali.

More recently, cyclopropene ($\triangle$) has been shown to be a substrate giving the products propene and cyclopropane in a 2 : 1 ratio (McKenna *et al.* 1976). This substrate is comparable with N_2 and C_2H_2 in effectiveness as a substrate. The 2 : 1 ratio of products occurs over a wide range of experimental conditions and *in vivo* in *Azotobacter*. This ratio is not maintained in the BH_4-FeMoco model system for the enzyme-catalysed reaction (McKenna and Huang 1979; McKenna *et al.* 1979). The fluoro analogue, 3, 3-difluorcyclopropene, and diazirine (a ring-strained diazene analogue) are also substrates (McKenna *et al.* 1981). N_2H_4 has been shown to be a poor substrate for Av nitrogenase; half-saturation occurs at hydrazine concentrations of 30 mM (Burgess *et al.* 1981). A re-investigation of azide reduction (Dilworth and Thorneley 1981) by Kp nitrogenase has revealed that, in addition to NH_3, N_2H_4 is a reaction product. ^{14}N–^{15}N-labelling data and product balances indicate that azide is reduced as:

$$N_3^- + 3H^+ + 2e \rightarrow N_2 + NH_3$$
$$N_3^- + 7H^+ + 6e \rightarrow N_2H_4 + NH_3$$
$$N_3^- + 9H^+ + 8e \rightarrow 3NH_3.$$

Pre-steady-state kinetic data suggest that N_2H_4 is formed by the cleavage of the α-β N-bond of bound azide to leave a nitride ($\equiv N$) intermediate which subsequently yields NH_3. During azide reduction, added $^{15}N_2H_4$ does not contribute ^{15}N to NH_4, consistent with the low affinity of N_2ase for N_2H_4 and the lack of equlibration of enzyme-bound intermediates giving rise to N_2H_4 with N_2H_4 in solution.

A re-investigation of cyanide as a substrate (Li *et al.* 1982) has shown that the substrate is HCN ($K_m = 4.5$ mM) and that CN^- is an inhibitor of electron flux ($K_i = 27$ μM) to HCN and H^+. Inhibition data suggest that HCN, which can completely suppress H_2 evolution, binds to a relatively oxidized enzyme species.

At low rates of electron flux through nitrogenase, protons are reduced in preference to C_2H_2 or N_2. In addition, mutual inhibition patterns of alternative substrates and inhibition by CO suggested that modified or multiple sites exist for different reducible substrates (Rivera-Ortiz and Burris 1975). This model was supported by the time course of substrate reduction by the heterologous nitrogenase formed by Cp2 and Kp1 which showed no detectable lag for H_2 evolution, but a 10 min lag for C_2H_2 and a 35 min lag for N_2 reduction (Smith *et al.* 1976). A similar temporal separation of substrate specificity for H^+ and C_2H_2 was

observed with Kp nitrogenase at 10 °C or high Kp1/Kp2 ratios Thorneley and Eady 1978), while at short reaction times C_2H_2 was not reduced.
These lag periods are of too long a duration to be pre-steady-state lags
and have been attributed to the slow attainment of enzyme species
capable of reduction of different substrates under these perturbed conditions (Lowe *et al.* 1978).

Of quite a different nature are the lags and bursts observed in pre-
steady-state studies of H_2 evolution under argon, H_2 evolution concomitant with N_2 reduction, dinitrogen hydride intermediate, and NH_4
formation which have provided most insight to the mechanism of N_2
reduction by nitrogenase as discussed in Section 1.4.4.

In *K. pneumoniae* the substrate specificity of nitrogenase is additionally
controlled by the *nifV* product, see Section 1.5.

1.4.3 Interaction of nitrogenase with H_2 and HD formation

Because it is the only known inhibitor which specifically and competitively inhibits N_2 reduction, H_2 and its interaction with nitrogenase has
been studied intensively. In the absence of an added reducible substrate the electron flux through nitrogenase results in the reduction of
protons to H_2. When increasing concentrations of N_2 are added, H_2
evolution is inhibited to an extrapolated minimum of 25 per cent of the
electron flux. The reaction catalysed is then:

$$N_2 + 8H^+ + 8e^- + 16ATP \rightarrow 2NH_3 + H_2 + 16ADP + 16P_i.$$

The evolution of H_2 under these conditions is thought to be
mechanistically important in N_2 reduction rather than an inherent inefficiency of nitrogenase. Since H_2 evolution still occurs under 50 atmos
pressure of N_2 (Simpson and Burris 1984).

The formation of HD by nitrogenase in mixtures of D_2/H_2O of
H_2/D_2O has attracted considerable attention because of the requirement for, or the stimulation of, this reaction by N_2. Early studies were
made with partially-purified nitrogenase proteins which made it uncertain as to the extent to which uptake hydrogenase was a contaminant.
More recently (Wherland *et al.* 1981), studies with highly-purified Av1
and Av2 free of conventional hydrogenase activity have shown that H_2
evolution and HD formation show a marked dependence on the
Av1/Av2 ratio. However, the proportion of e^- flux diverted into HD is
constant in the absence, but not the presence, of N_2. This clearly
establishes that HD formation is a property of nitrogenase and that
there are two separate routes for HD formation, only the major route
being N_2 dependent; the existence of an N_2-independent route has
recently been questioned (Li and Burris 1983; Guth and Burris 1983).
Electron balance data indicate that HD formation requires reductant
and has a stoichiometry of $1e^-$/HD formed, and in addition, $^3H_2$ is not

significantly exchanged into H_2O. These observations establish that HD is *not* formed by a simple reversible exchange reaction, of the type

$$D_2 + H_2O \rightleftharpoons HD + HOD.$$

In the absence of N_2 (50 per cent D_2 50 per cent Ar), only 6 per cent of the e^- flux (81.9 nmol HD $min^{-1}mg$ protein) is utilized for HD formation. However, under 50 per cent D_2, 10 per cent Ar, and 40 per cent N_2, 50 per cent of the e^- flux is diverted from NH_3 formation to HD formation. This is consistent with the suggestion that H_2 inhibits N_2 reduction by the formation of HD depriving nitrogenase of reductant, and also with the observation that when the contribution of N_2-independent HD formation is subtracted from the total, the apparent K_m for N_2 reduction is the same as that for HD formation $(0.12 \pm 0.1$ atm) (Burgess *et al.* 1981). This shows that HD formation and H_2 inhibition of N_2 reduction are manifestations of the same phenomenon. The mechanism of HD formation has been suggested as occurring by the reaction of D_2 with a diazene-level enzyme-bound intermediate (NH=NH) on the route to NH_3 (Burgess *et al.* 1981) or as occurring as a hydride-deuteride exchange on a metal centre, facilitated by the displacement of a metal-bound N_2 atom (Chatt 1980), see Section 1.4.4.

1.4.4 A model for the enzymic reduction of N_2

There is strong circumstantial evidence (discussed above) that N_2 is bound to the MoFe protein, and comparison with the chemistry of characterized N_2 complexes suggests that the metal (most likely molybdenum) must be in a low oxidation state. During the reduction of N_2 an enzyme-bound intermediate, which gives rise to N_2H_4 on quenching the functioning enzyme in acid or alkali, has been detected (Thorneley *et al.* 1978). Scheme 1.1 shows the methodology used to trap the intermediate. The concentration of the intermediate is highest at the maximal rate of NH_3 formation and correlates with the molybdenum content of the Kp1 used, providing strong evidence for an intimate involvement of molybdenum in catalysis.

An analysis of the pre-steady-state kinetics of the rates of formation of the dinitrogen-hydride intermediate, H_2 evolution under N_2 compared with H_2 evolution under Ar, and the formation of NH_3 have provided great insight to events occurring during the reduction of N_2 by Kp nitrogenase (Thorneley and Lowe 1981, 1982, 1984*a, b*; Lowe *et al.* 1983; Lowe and Thorneley 1984*a, b*). The kinetics are characterized by lag and burst phases before the rate of product formation becomes linear and the time course for the intermediate yielding shows an overshoot and oscillation as the steady-state is approached (Table 1.4). Computer-modelling of these complex kinetics has led to the

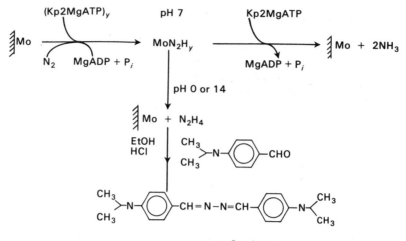

$$\varepsilon_{max} 458 \text{ nm} = 7 \times 10^5 \text{M}^{-1} \text{cm}^{-1}$$

SCHEME 1.1

TABLE 1.4
*Pre-steady-state data for the formation of the dinitrogen-hydride
intermediate and products by Kp nitrogenase*

Substate	Product	Lag (ms)	Burst (ms)
H$^+$ under Ar	H$_2$	100	Linear
H$^+$ in presence of N$_2$	H$_2$	100	1000
N$_2$	N$_2$H$_4$	250	1500
	(derived from dinitrogen hydride intermediate)		25% overshoot in approach to steady state
N$_2$	NH$_3$	400	1500

Data were obtained at 23 °C pH 7.4 using the rapid-quench apparatus to stop the reaction at various times. The quenched reaction mixtures and gas headspace were then analysed for NH$_3$, N$_2$H$_4$ (see Scheme 1.1), and H$_2$. The duration of the lag period is derived from a linear extrapolation to the time axis and the burst from the time when the steady state was activated. Data from Thorneley and Lowe (1981, 1982), see also Lowe *et al.* (1983). These data can be rationalized in terms of the mechanism outlined in Scheme 1.2.

development of a proposed mechanism (Scheme 1.2) in which the rapid sequential transfer of each electron is followed by a rate-limiting step ($k = 6.4$ s^{-1}). The slow step is probably dissociation of Kp2$_{ox}$ from reduced Kp1 since stopped flow studies indicate this occurs with $k = 6.4 \pm 0.8$ s^{-1}. The point in Scheme 1.2 at which H$_2$ is evolved, the N$_2$H$_4$-intermediate and NH$_3$ are formed are the number of electrons

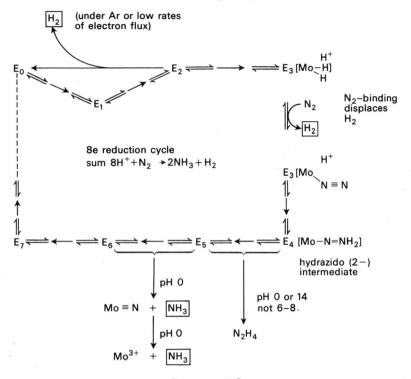

images/scheme_1_2

SCHEME 1.2

A proposed mechanism for N_2 reduction by nitrogenase

E_0 represents $S_2O_4^{2-}$-reduced MoFe protein. In the species E_1, E_2, etc. the subscript denotes the number of electrons donated from the Fe protein, and also the number of slow protein dissociations. (Each step in the cycle is comprised of rapid complex formation, rapid electron transfer, and slow protein dissociation.) The release of $2NH_3$ from E_5 and E_6 may be a consequence of the acid conditions of the quench. If NH_3 is released at this point then E_6 and E_7 represent oxidized species relative to the $S_2O_4^{2-}$-reduced protein. (From Thorneley and Lowe 1981, 1982, 1984a, b; Lowe and Thorneley 1984a, b).

transferred which best fit the experimental data. In this mechanism, N_2 binds to the enzyme by displacement of H_2, thereby accounting for the minimal stoichiometry of 25 per cent of electrons giving rise to H_2 evolution during N_2 reduction. The binding of $Kp2_{ox}$ to the hydridic species E_2 and preventing H_2 evolution has been proposed as a critical role for the Fe protein (Thorneley and Lowe 1982). The concentration of species E_2, which is capable of wastefully evolving H_2 and relaxing to E_0, is minimized since dissociation of $Kp2_{ox}$ is slow ($k = 5$ s^{-1}), and further reduction to the N_2-binding species fast since reduced Kp2 is complexed at the diffusion-controlled rate.

The initial assignment of the dinitrogen hydride intermediate as a hydrazido $(2-)$ species (Thorneley *et al.* 1978) was based on comparison of the reactivity of Mo-complexes towards acid–base hydrolysis. This assignment is consistent with the appearance of the intermediate after the four slow steps as shown in Scheme 1.2.

HD formation in the absence of N_2 is suggested to occur *via* the dihydride species E_3:

$$\text{Mo}\overset{\displaystyle H^+}{\underset{\displaystyle H}{\diagup\!\!-\!H}} + D_2 \rightleftharpoons \text{Mo}\overset{\displaystyle H^+}{\underset{\displaystyle D}{\diagup\!\!-\!D}} + H_2;$$

or in the presence of N_2 by displacement of N_2:

$$\text{Mo}\overset{\displaystyle H^+}{\underset{\displaystyle N}{\diagup\!\!-\!N}} + D_2 \rightleftharpoons \text{Mo}\overset{\displaystyle H^+}{\underset{\displaystyle D}{\diagup\!\!-\!D}} + N_2.$$

An alternative model for HD formation involving reaction of D_2 with a diazene intermediate formed during N_2 reduction (i.e. E_2) has been proposed (Burgess *et al.* 1981):

$$E + N_2 + 2e + 2H^+ \overset{D_2}{\rightarrow} EN_2H_2 \rightarrow E + N_2 + 2HD;$$

net reaction:

$$2e + 2H^+ + D_2 \rightarrow 2HD.$$

In all models, one electron is utilized for each HD produced but at this time experimental difficulties prevent the determination of how many slow steps occur before HD formation can occur.

1.5 *Nif* PRODUCTS OF *K. pneumoniae*

The structure and organization of the *nif*-gene cluster of *K. pneumoniae* is described in detail in Chapter 4; here the biochemical functions and properties of those *nif* products which have been characterized will be considered.

The initial assignment of many of the *nif* products of *K. pneumoniae* was based on comparison of the properties of extracts of wild-type and point or Mu insertion Nif⁻ mutants (Roberts *et al.* 1978; Roberts and Brill 1980; Houmard *et al.* 1980; Merrick *et al.* 1978).

Subsequently, cloning techniques were used to link specific EcoRI fragments of *nif* to constitutive promoters of genes for antibiotic resistance in order to identify polypeptide products synthesized in *E. coli* minicells (Pühler and Klipp 1981). Saturation mutagenesis of cloned fragments *in vivo* with the transposon Tn5 has allowed the specific assignment of *nif* proteins to *nif* genes. In addition to allowing the assignment of products to some genes which were intractable in studies with mutants of *K. pneumoniae*, this approach identified two new *nif*-specific polypeptides which were assigned to *nifX* and *nifY*, genes of unknown function. In most cases data obtained were in good agreement with data previously existing.

The products of the structural genes for Kp1 (*nifD* and *nifK*) and Kp2 (*nifH*) were assigned by comparison of the migration on two-dimensional gels of purified nitrogenase components with proteins present in extracts of Nif⁻ mutants. These products require post-translational modification to form a functional nitrogenase *in vivo* (Eady *et al.* 1978). Nif gene products involved in this processing were identified by *in vitro* activity-complementation of mutant extracts with purified Kp1, Kp2, or FeMoco. Such studies indicated that *nifM* and *nifS* products were involved in the processing of Kp2 and that *nifB*, *nifE*, and *nifN* were involved in the synthesis or insertion of FeMoco into Kp1. Kp1 polypeptides are not necessary for FeMoco synthesis since extracts of mutants lacking both subunits of Kp1 have FeMoco activity (Ugalde *et al.* 1984). In such mutants FeMoco is protein bound, presumably accumulating in a carrier protein as a consequence of a blockage in the normal biosynthetic route since in wild-type organisms Kp1 is the major molybdo-protein. *nifF* and *nifJ* products are involved in the transfer of reducing equivalents to nitrogenase *in vivo* since extracts are unable to couple nitrogenase to pyruvate, formate, or malate, but extracts are active with $S_2O_4^{2-}$ as an electron donor (Hill and Kavanagh 1980; Nieva Gomez *et al.* 1980). *nifV* product modified FeMoco so as to alter the substrate specificity of Kp1 from an acetylene reductase to a nitrogenase (Hawkes and Smith 1983). As described in Chapter 4, the *nifA* and *nifL* products function to activate and repress other transcripts of the *nif* regulon in response to O_2 and NH_4^+. The relative molecular masses, isoelectric points, and functions of known *nif* products are summarized in Table 1.5.

Only five of the *nif*-gene products have been purified with retention of function: Kp1 (*nifD*, *nifK*), Kp2 (*nifH*); *nifJ* (Bogusz *et al.* 1981; Shah *et al.* 1983); and *nifF* (Nieva Gomez *et al.* 1980). Kp1 has been purified from *nifB*⁻ (Hawkes and Smith 1983) and *nifV*⁻ mutants (McLean and Dixon 1981), and Av1 from a mutant UW45 of *Azotobacter vinelandii* which is deficient in FeMoco synthesis (Burgess *et al.* 1981).

1.5.1 *nifKDH*

The properties of these products are described in detail elsewhere in this chapter: studies with strains with mutations in these genes have

TABLE 1.5
nif gene products of **K. pneumoniae**

Nif gene	Assigned polypeptide	Relative molecular mass of native protein	Prosthetic group	pI	Function	Comment
Q	not assigned	—	—	—	Utilization of molybdenum at low concentrations	
B	51 K	—	—	—	FeMoco activity	—
A	57–60 K	—	—	7.05	Activates expression of *nif* transcripts except *nifL*	—
L	45–50 K	—	—	5.1	Inhibits expression of *nif* transcripts except *nifL*	Two spots assigned
F	10–22 K	22 000	FMN	5.0–4.5	Electron transfer	
M	27–28 K	—	—	7.0	Kp2 activity	
V	38–42 K	—	—	5.9	Kp1 substrate activity	
S	45 K	—	—	5	Kp2 activity	
U	22–28 K	—	—	5	Unknown	
X	18 K	—	—	—	Unknown	
N	40–50 K	—	—	6.7	FeMoco activity	
E	40–46 K	—	—	6.8	FeMoco activity	
Y	21 K	—	—	—	Unknown	
K	60 K	β 220 000	2Mo 32Fe 28S	5.7	Nitrogenase component Kp1	
D	50–60 K	α $\alpha_2\beta_2$ 68 000		6.1, 6.0	Nitrogenase component Kp2	*nifD* resolves into two spots on IEF
H	35 K		4Fe4S	4.9		Two spots resolved on SDS 35 000 and 39 000 apparent mol. wt.
J	120 K	245 000	30Fe22S	6.0	Electron transfer	

See text and Lowe *et al.* (1983) for references.

established that these products stabilize each other and may have a role in stabilizing other *nif*-gene products *in vivo*. The *nifJ* product is required for maximum Kp1 activity (Bogusz *et al.* 1981; Hill and Kavanagh 1980).

1.5.2 *nifK*

NifK⁻ strains have negligible C_2H_2 reduction activity *in vivo*; extracts are inactive unless complemented with Kp1, when partial levels of wild-type activity are restored. Such mutants show an altered mobility of the 60K-subunit of Kp1, and many show decreased steady-state levels of other *nif* products including *nifD*, suggesting that this product is necessary for the stability or maximal expression of other transcripts.

1.5.3 *nifD*

Extracts of *NifD⁻* mutants show low activity unless Kp1 is added and an altered mobility of the 56 K-subunit of Kp1. In Mu insertion, but not point mutants of *nifD*, the level and activity of Kp2 is decreased indicating that the *nifD* product is involved in maintaining maximum levels of Kp2. Two proteins of molecular weight 56 K but with pI of 6.0 and 6.1 have altered mobility in point or insertion mutants of *nifD*. In this context, it is interesting to note that purified Av1 has been shown to resolve into three spots on two-dimensional-electrophoresis corresponding to an $\alpha\alpha'\beta_2$ structure (Harker and Wullstein 1981).

1.5.4 *nifH*

Extracts of *nifH⁻* strains show low activity which is restored by the addition of Kp2. Mutants show an altered mobility of two spots 35 K and 39 K which have a pI of 4.9; all strains except *nifM⁻* mutants have a comparable ratio of these proteins. A large number of point mutants have low steady-state levels of other *nif* products.

The effects of mutations in *nifDH* and *K* in decreasing the steady-state levels of the products of these genes and also of the *nifJ*, *N*, and *E* products indicate a possible regulatory role in the synthesis or stability of other *nif* products. There is evidence that molybdenum and the products of the *nif*HDKY operon are necessary for maximum rates of expression from the *nifH* promoter (Dixon *et al.* 1980). A role in stabilization is indicated since many mutants show normal rates of synthesis in pulse-labelled samples but low steady-state levels of product.

1.5.5 *nifB*

Extracts of *NifB⁻* mutants have low activity unless complemented with Kp1 or FeMoco.

1.5.6 *nifN* and *nifE*

Extracts have a similar activity–complementation pattern to *NifB⁻* mutants. Point and insertion mutants in *nifN* or in *nifE* showed decreased rates of synthesis of two proteins of 46 K and 50 K which were unstable in strains with mutations in either gene. On the basis of short-term labelling experiments, these proteins were assigned as 40 K to *nifE* and 50 K to *nifN*, an assignment which was confirmed by data using the *E. coli* minicell system. These data are consistent with both the N and E products being required for each other's stability in *K. pneumoniae*.

In addition to impaired FeMoco formation, *NifE⁻*, but not *NifB⁻* or *NifN⁻*, strains have lower levels of active Kp2.

1.5.7 *nifV*

NifV⁻ strains grow poorly on nitrogen-free media but show wild-type levels of acetylene reduction. A polypeptide of 38 K was assigned to *nifV*, since it was not present in Mu insertion mutants and showed altered migration on two-dimensional-gels in some *NifV⁻* strains. In contrast with extracts of wild-type *K. pneumoniae*, N_2 did not inhibit H_2 evolution by nitrogenase and 3 per cent CO produced a 50 per cent inhibition of H_2 evolution (McLean and Dixon 1981). When purified from extracts of a *NifV⁻* strain, Kp2 showed normal kinetic properties but Kp1 retained the *nifV⁻* phenotype.

When extracted from Kp1 isolated from a *nifV* mutant, FeMoco confers the *NifV⁻* phenotype on reconstituted Kp1 from a *NifB⁻* strain (Hawkes *et al.* 1983). This observation indicates that the role of the *nifV* product is to modify FeMoco in some way so as to change the substrate specificity of Kp1.

1.5.8 *nifS*

Extracts of *nifS⁻* strains have low nitrogenase activity which is partially restored by the addition of Kp2. Initially a polypeptide of 18 K was tentatively assigned to the *nifS* product, later this was amended (Roberts and Brill 1980) to 42 K which agrees well with the value of 45 K obtained in the minicell system (Pühler and Klipp 1981).

1.5.9 *nifM*

Extracts of *NifM* mutants have low activity which is stimulated by the addition of Kp2. Two-dimensional electrophoresis shows that normal levels of Kp1 are present but that Kp2, particularly the 39 K polypeptide, are at low levels. The *nifM* product has been assigned to 27 K polypeptide which is absent in Mu insertions and has altered electrophoretic mobility in some mutants.

1.5.10 *nifF*

The observation that *NifF⁻* mutants had low activity *in vivo* but that extracts had wild-type levels of activity implicated the *nifF* product in the donation of electrons to nitrogenase. Insertion mutants lacked two polypeptides of 22 K and 26 K which reappeared in Nif⁺ revertants. The protein of relative molecular mass 22 000 has been purified to homogeneity (Nieva-Gomez *et al.* 1980) and shown to be a flavoprotein containing flavine mononucleotide (FMN). When added to extracts of *NifF⁻* mutants it restores formate- or malate-dependent nitrogenase activity, which formed the basis for the assay of this protein during purification. The significance of the 26 K protein is not clear, since synthesis of this protein is not observed until after nitrogenase activity is derepressed. The relative molecular mass of the *nifF* product has also been reported as 10 K (Pühler and Klipp 1981; Houmard *et al.* 1980). The reason for this discrepancy is not clear.

1.5.11 *nifJ*

The phenotype of *NifJ⁻* mutants is similar to *NifF⁻* mutants in that both have low nitrogenanse activity *in vivo* but high activity *in vitro* when $S_2O_4^{2-}$ is used as an electron donor indicating a role in electron transfer to nitrogenase *in vivo*. A 120 K polypeptide has been assigned to *nifJ* since this protein has altered mobility in several charge mutations, polar mutations also result in the lack of an 80 K polypeptide of unknown function.

Extracts of *NifJ⁻* mutants are also deficient in active Kp1 although normal levels of synthesis are observed. Since activity is enhanced by added FeMoco, synthesis or insertion of FeMoco is impaired in the absence of the *nifJ* product.

Using the ability of *nifJ* product to reconstitute pyruvate-supported nitrogenase activity to extracts of *nifJ⁻* mutants as an assay, the *nifJ* product has been purified to homogeneity (Bogusz *et al.* 1981). Activity is O_2-labile and unstable ($t_{1/2} \simeq 3$ h), but the addition of 10 per cent glycol to buffers during purification stabilizes the activity. The native protein is a dimer of relative molecular mass 245 000, contains 30 Fe and 24 acid labile S per mol but less than 0.1 mol of molybdenum per mol.

A *nifJ* product with rather different properties has been purified to homogeneity using a pyruvate-thiamine pyrophosphate,-CoA, flavodoxin-linked nitrogenase assay (Shah *et al.* 1983). This protein, which has the same relative molecular mass and subunit structure contains only 8 Fe and 6 S^{2-} per molecule, has been shown to be a CoA-dependent pyruvate flavodoxin oxidoreductase. The enzyme is specific for flavodoxin, since it will not utilize ferredoxin or viologen dyes. *Azotobacter* flavodoxin has only one-third the activity of *Klebsiella* flavodoxin.

1.5.12 *nifA*

NifA⁻ mutants do not reduce acetylene *in vivo* or *in vitro* and lack all identifiable *nif* products in extracts with the exception of *nifA* and *nifL* products. The properties of point mutants in *nifA* and Mu insertion in *nifA* have allowed the assignment of a peptide of 57–66 K to *nifA*. The *nifA* product is essential for transcription from all *nif* promoters except *nifL*.

1.5.13 *nifL*

Many *NifL⁻* strains are strongly polar on *nifA* and therefore share the same phenotype. Two peptides of 52 K have been assigned to *nifL*. Some revertants of Mu insertions strains lacked both spots and escaped O_2-repression and repression by low levels of NH_3.

1.5.14 *nifQ*

NifQ strains are defective in nitrogen fixation due to an elevated requirement for molybdenum (Imperial *et al.* 1984). Since molybdenum uptake is not impaired, the *nifQ* product is implicated in molybdenum processing. Optimum growth of *nifQ* mutants requires 100 μM molybdenum, in contrast to the wild-type which shows a maximal response at 10 nM. No polypeptide has been assigned to the *nifQ* product, and extracts contain normal levels of other *nif* products.

1.5.15 Stability of *nif* products

Pulse-labelling experiments (Roberts and Brill 1980) have revealed that under derepressing conditions the polypeptides of *nifL* (52 K), *nifF* (22 K), *nifM*, and *nifV* are less stable in the 18 h following synthesis than are other *nif* products. The *nifF* product was most stable in *nifJ⁻* strains, a finding which was utilized in the purification of the *nifF* product. The *nifN* and *nifE* products were required for each other's stability as were the products of the structural genes of nitrogenase.

The addition of repressive concentrations of NH_4^+ has little effect over 3 h on the stability of *nif* products already synthesized. This is in agreement with chemostat studies which indicate that nitrogenase activity decays with dilution following the addition of NH_4^+.

Aeration for 3 h resulted in the degradation of *nif*-specific polypeptides, including *nifF* product which is stable to O_2 during purification. The structural polypeptides are more stable in the presence of chloramphenicol (Eady *et al.* 1978), suggesting that oxidative degradation may involve a newly synthesized protease.

1.6 PROTECTION FROM INACTIVATION BY OXYGEN

Both the MoFe protein and the Fe protein are rapidly and irreversibly inactivated by O_2; typical half-lives of activity for the purified proteins in air are 10 min and 45 s respectively. In crude extracts of *A. vinelandii* and *A. chroococcum*, nitrogenase is more stable to O_2 than are the recombined purified proteins. O_2-tolerance is not due to association with cytoplasmic membranes but to tight (1:1:1) complex formation with an FeS protein (Haaker and Veeger 1977; Robson 1979). The protective protein is a 2Fe–2S protein of relative molecular mass 23 000 (AvP) when isolated from *A. vinelandii* and 14 000 when isolated from *A. chroococcum*. Mg^{2+} is essential for complex formation and protection from O_2. In *A. vinelandii* (Veeger *et al.* 1980) gel filtration experiments are consistent with the complex being a stoichiometric association of $Av1_{ox}Av2_{ox}AvP_{ox}$ of 1–2×10^6 K. Dissociation of the complex is induced by $S_2O_4^{2-}$ or ADP. Stopped-flow studies have shown that although Av2 will reduce AvP ($K = 6.5 \times 10^5$ M^{-1} s^{-1}) there is no evidence for complex formation between these two proteins alone.

In vivo protection from oxygen inactivation is provided by high respiration rates (*Azotobacter*) compartmentalization of nitrogenase into specialized cells in heterocystous Cyanobacteria, O_2-buffering in the case of symbiotic systems, and avoidance by facultative diazotrophs (see Robson and Postgate 1980).

Studies on azotobacters have shown that when cultures are subjected to O_2-stress such that respiration is unable to maintain low DOT, nitrogenase activity is 'switched off' and the enzyme is conformationally protected but inactive. The discovery of protective proteins discussed above rationalizes this. The lack of activity could also be due in part to auto-oxidation of flavodoxin hydroquinone: oxygen represses nitrogenase synthesis in *K. pneumoniae*, *A. chroococcum*, and free-living *R. japonicum* (see Chapter 8).

1.7 IRON AND MOLYBDENUM INVOLVEMENT IN DIAZOTROPHY

1.7.1 Regulation of *nif* expression by molybdenum

Since both iron and molybdenum form part of the redox centres of nitrogenase, they are essential for nitrogen fixation. Very little is known about the peripheral biochemistry and physiology of iron in N_2-fixation, probably because it is also essential for many other metabolic processes during diazotrophic growth. In contrast, molybdenum has a specific involvement in N_2 fixation and consequently molybdenum uptake; storage and processing has been studied in some detail as described below.

Apart from its role as a component of nitrogenase, molybdenum has been shown to be involved in the regulation of the synthesis of nitrogenase in *A. vinelandii*, *C. pasteurianum*, *R. capsulata*, and *K.*

pneumoniae but not *Plectonema boryanum* (see Eady *et al.* 1982; Hallenbeck *et al.* 1982). In *K. pneumoniae* the presence of molybdenum is required for maximum expression of the structural genes although this has been questioned (Ugalde *et al.* 1984).

It has recently been suggested that *A. vinelandii* has an alternative molybdenum-independent N_2-fixing system (Bishop *et al.* 1980, 1982; Riddle *et al.* 1982; Page and Collinson 1982). The basis for this proposal is the diazotrophic growth of WO_4^{2-}-resistant mutants and the growth of pseudorevertants of conventional Nif$^-$ mutants on molybdenum-deficient media. In both cases conventional nitrogenase components cannot be detected in extracts by two-dimensional-electrophoresis but different NH_4^+-repressible polypeptides are observed. Similar changes are observed when wild-type *A. vinelandii* is grown on molybdenum-deficient media.

These interesting observations have to be viewed against the general perspective that (i) *A. vinelandii* has an exceptionally effective molybdenum uptake system which allows N_2-dependent growth on purified molybdenum-deficient media (Eady and Robson 1984); (ii) in molybdenum-sufficient media nitrogenase levels are apparently in excess of the N_2 fixation rate required to support growth (see Cardenas and Mortenson 1975); (iii) molybdenum-limitation results in the induction of the synthesis of membrane proteins (Page and von Tigerstrom 1982). In summary, an efficient molybdenum-uptake system may allow very low levels of conventional nitrogenase to support N_2-dependent growth, the altered pattern of protein synthesis being a fortuitous effect of molybdenum-limitation. Although these factors could account for the behaviour of wild-type *A. vinelandii*, it is more difficult to rationalize the molybdenum inhibition of N_2-dependent growth of some of the mutants strains. Recent work has shown Av2-like activity to be present in extracts of some *Nif$^-$* strains of *A. vinelandii* under conditions of molybdenum deprivation (Premakumar *et al.* 1984). This protein is of a similar size to conventional Av2, has a slightly more basic pI, and is repressed by 100 nM molybdenum. It may be that *A. vinelandii*, like *A. chroococcum*, has multiple copies of *nifH* (Robson *et al.* 1984), and that such genes are under regulatory control by the availability of molybdenum.

1.7.2 Molybdenum uptake

The uptake of MoO_4^{2-} has been studied in most detail in N_2-grown *C. pasteurianum* (Elliott and Mortenson 1975, 1976, 1977). $^{99}MoO_4^{2-}$ uptake is energy-dependent with a K_m of 48 μM, and is competitively inhibited by SO_4^{2-} (K_i = 30 μM) and WO_4^{2-} (K_i = 24 μM). VO_3^{2-} has no effect on MoO_4^{2-} uptake. Cold-chase experiments with a hundred-fold excess of unlabelled molybdenum showed that very little of the

molybdenum transported remained free, presumably due to incorporation into the molybdenum storage protein described below. This
behaviour contrasts with WO_4^{2-} and SO_4^{2-} which are exchanged, indicating that in this organism they are unable to substitute for MoO_4^{2-}.
These data suggest that in *C. pasteurianum* WO_4^{2-} antagonism of
MoO_4^{2-} occurs at the level of transport. The molybdenum-uptake
system is repressed by NH_4^+ and by high concentrations of MoO_4^{2-},
which at 1 mM represses uptake activity by 80 per cent although such
organisms contain the molybdenum-storage protein.

When *C. pasteurianum* is grown in molybdenum-deficient medium
supplemented with different levels of molybdenum, no correlation of
growth rate and nitrogenase activity is observed. Maximum growth
rates were obtained at molybdenum levels which did not allow maximum nitrogenase activity, suggesting that in molybdenum-sufficient
cells higher levels of nitrogenase were present than were required for
the rate of nitrogen-fixation for growth.

MoO_4^{2-} uptake by *A. vinelandii* and *K. pneumoniae* has been studied
during derepression of Mo-starved NH_4^+-grown cultures (Pienkos and
Brill 1981). When measured at 3 h, derepression of *A. vinelandii*
nitrogenase activity was maximal at 0.5 μM Mo but accumulation of
molybdenum continued up to 5 μM Mo. Nitrogenase activity, but not
molybdenum uptake, was repressed by NH_4^+. In contrast, both
nitrogenase activity and Mo uptake were maximal at 1 μM MoO_4^{2-} in
K. pneumoniae and uptake was inhibited by NH_4^+.

Molybdenum-starved, N_2-grown *A. vinelandii* shows a synergistic
growth response to added molybdenum and iron (Eady and Robson
1984). In this context, it is interesting to note that the extracellular
pigments N,N'-bis (2,3 dihydroxybenzoyl)-L-lysine and a peptide
containing a fluorescent chromophore, which are produced during N_2
growth on iron-deficient media, also chelate molybdenum. It is not
clear at present whether such compounds are siderophores which are
involved in molybdenum-uptake.

1.7.3 Molybdenum storage

Molybdenum-storage proteins have been identified in *C. pasteurianum*
(Elliott and Mortenson 1977) and *A. vinelandii* (Pienkos and Brill 1981).
Fractionation of crude extracts by DEAE-cellulose chromatography
showed for both organisms, that when grown on low molybdenum
(0.1 μM) Av1 or Cp1 were the only or major molybdenum proteins
detected. Extracts of organisms grown on high molybdenum concentrations showed an additional Mo protein which eluted at lower ionic
strength. These data contrast with that for *K. pneumoniae* where only
Kp1 was detected in organisms grown on high molybdenum.

The molybdenum storage proteins have not been subject to detailed study. The protein from *C. pasteurianum* has a relative molecular mass of 67 000 and contains six or more Mo atoms/mol. The protein from *A. vinelandii* has two subunit types of 21 and 24 K and has a proposed $\alpha_2\beta_2$ structure of relative molecular mass 90 000, and contains 14–15 atoms of Mo/mol. This protein, which is unstable, does not have FeMoco activity, but appears to provide molybdenum for growth on transfer to molybdenum-deficient media.

1.7.4 Molybdenum processing

Molybdenum is generally supplied to organisms as MoO_4^{2-}, and little is known about the steps involved in the transformation of Mo=O to the Mo—S ligation present in FeMoco. In *K. pneumoniae* four *nif*-specific genes (*nifB, E, V,* and *N*) are involved in the synthesis of FeMoco. The similarity of FeMoco when isolated from MoFe proteins of different diazotrophs strongly suggests that proteins of analogous functions to *nifB, E,* and *N* are present in these organisms. However, the lack of nitrogenase activity of *narD* mutants of *E. coli* carrying Klebsiella *nif* on the plasmid pRD1 indicates that nitrate reductase and nitrogenase share some Mo processing functions involving the *narD* product (Kennedy and Postgate 1977).

Insertion of FeMoco into purified-apoproteins of the mutants *A. vinelandii* UW45 and *NifB⁻ K. pneumoniae* (UNF5058) is spontaneous and the consequent restoration of activity forms the basis for the assay of these proteins during purification. This finding together with the properties of the purified apoproteins is consistent with the mutant apoproteins containing only the 'P' cluster redox centres.

When *K. pneumoniae* (Kahn *et al.* 1982) and *Plectonema boryanum* (Nagatani and Haselkorn 1978) are derepressed under conditions of molybdenum deprivation, nitrogenase activity can be reconstituted *in vivo* by the addition of MoO_4^{2-} in the absence of protein synthesis. Similar observations have been reported for *A. vinelandii* provided that WO_4^{2-} is present (Pienkos *et al.* 1981). Extracts of WO_4^{2-}-derepressed *A. vinelandii* can be partially reactivated by MoO_4^{2-} in an ATP-dependent reaction. The factor responsible for reactivation is unstable and membrane-bound, but this system provides some promise as an area for future work on molybdenum processing.

REFERENCES

ALEF, K. and ZUMFT, W. G. (1981). *Z. Naturf.* **36**, 784–9.
ANTONIO, M. R., TEO, B. K., ORME-JOHNSON, W. H., NELSON, M. J., GROH, S. E., LINDAHL, P. A., KAVZLARICH, S. M., and AVERILL, B. A. (1982). *J. Am. chem. Soc.* **104**, 4703–5.

BERNDT, H., LOWE, D. J., and YATES, M. G. (1978). *Eur. J. Biochem.* **86**, 133–42.

BISHOP, P. E., JARLENSKI, D. M. L., and HETHERINGTON, D. R. (1980) *Proc. natn. Acad. Sci. U.S.A.* **77**, 7342–6.

—— (1982). *J. Bact.* **150**, 1244–51.

BOGUSZ, D., HOUMARD, J., and AUBERT, J.-P. (1981). *Eur. J. Biochem.* **120**, 421–6.

BRAAKSMA, A., HAAKER, H., GRANDE, H. J., and VEEGER, C. (1982). *Eur. J. Biochem.* **121**, 483–91.

— and VEEGER, C. (1938). *Eur. J. Biochem.* **133**, 71–6.

BURGESS, B. K., JACOBS, D. B., and STIEFEL, E. I. (1980). *Biochem. biophys. Acta,* **614**, 196–209.

— WHERLAND, S., NEWTON, W. E., and STIEFEL, E. I. (1981). *Biochem.* **20**, 5140–6.

BURRIS, R. H. and ORME-JOHNSON, W. H. (1976). In *Proc. 1st Int. Symp. Nitrogen Fixation* (ed. W. E. Newton and C. J. Nyman), Vol. 1, pp. 208–33 Washington State University Press, Pullman.

— ARP, D. J., HAGEMAN, R. V., HOUCHINS, J. P., SWEET, W. J., and TSO, M.-Y. (1981). In *Current perspectives in nitrogen fixation* (ed. A. H. Gibson and W. E. Newton), pp. 56–66. Australian Academy of Science, Canberra.

CARDENAS, J. and MORTENSON, L. E. (1975). *J. Bact.* **123**, 978–84.

CARITHERS, R. P., YOCH, D. C., and ARNON, D. I. (1979). *J. Bact.* **137**, 779–89.

CARTER, C. W., KRAUT, J., FREER, S. T., and ALDEN, R. A. (1974). *J. biol. Chem.* **249**, 6339–46.

CHATT, J. (1980). In *Nitrogen fixation* (ed. W. D. P. Stewart and J. R. Gallon), pp. 1–11. Academic Press, London.

CORDEWENER, J., HAAKER, H., and VEEGER, C. (1983). *Eur. J. Biochem.* **132**, 47–54.

CRAMER, S. P., GILLUM, W. O., HODGSON, K. O., MORTENSON, L. E., STIEFEL, E. I., CHISNELL, J. R., BRILL, W. J., and SHAH, V. K. (1978). *J. Am. chem. Soc.* **100**, 3814–19.

DAVIS, L. C., HENZL, M. T., BURRIS, R. H., and ORME-JOHNSON, W. H. (1979). *Biochem.* **18**, 4860–9.

DILWORTH, M. J. and THORNELEY, R. N. F. (1981). *Biochem J.* **193**, 971–83.

DIXON, R. A., EADY, R. R., ESPIN, G., HILL, S., IACCARINO, M., KAHN, D., and MERRICK, M. (1980). *Nature, Lond.* **286**, 128–32.

DOWLING, T. E., PRESTON, G. G., and LUDDEN, P. W. (1982). *J. biol. Chem.* **257**, 13987–92.

EADY, R. R. (1980a). *Meth. Enzymol.* **69**, 751–76.

— (1980b). In *Methods for evaluating biological nitrogen fixation* (ed. F. J. Bergersen), pp. 213–64. Wiley–Interscience, New York.

— (1981). In *Current perspectives in nitrogen fixation* (ed. A. H. Gibson and W. E. Newton) pp. 171–81. Australian Academy of Sciences, Canberra.

— SMITH, B. E., COOK, K. A., and POSTGATE, J. R. (1972). *Biochem. J.* **128**, 655–75.

— LOWE, D. J., and THORNELEY, R. N. F. (1978a). *FEBS Lett.* **95**, 211–13.

44 *Nitrogen fixation*

— Issack, R., Kennedy, C., Postgate, J. R., and Ratcliffe, H. (1978*b*). *J. gen. Microbiol.* **104**, 277–85.

—— and Robson, R. L. (1984). *Biochem. J.* **224**, 853–62.

— and Smith, B. E. (1979). In *Dinitrogen fixation* (ed. R. W. F. Hardy, F. Bottomley, and R. C. Burns), pp. 399–490. John Wiley, New York.

— Imam, S., Lowe, D. J., Miller, R. W., Smith, B. E., and Thorneley, R. N. F. (1980). In *Nitrogen fixation* (ed. W. D. P. Stewart and J. R. Gallon), pp. 19–35. Academic Press, New York.

— Kahn, D., and Buchanan-Wollaston, V. (1982). *Israel J. Bot.* **31**, 45–60.

Elliott, B. B. and Mortenson, L. E. (1975). *J. Bact.* **124**, 1295–301.

—— (1976). *J. Bact.* **127**, 770–9.

—— (1977). In *Recent developments in nitrogen fixation* (ed. W. E. Newton, J. R. Postgate, and C. Rodriguez-Barrueco), pp. 205–17. Academic Press, New York.

Emerich, D. W. and Burris, R. H. (1978). *Biochim. biophys. Acta* **536**, 172–83.

—— (1979). *Meth. Enzymol.* **53**, 314–29.

— Hageman, R. V., and Burris, R. H. (1981). *Adv. Enzymol.* **52**, 1–22. John Wiley, New York.

Falk, G., Johansson, B. C., and Nordlund, S. (1982). *Arch. Microbiol.* **132**, 251–3.

Fuhrmann, M. and Hennecke, H. (1984). *J. Bact.* **158**, 1005–11.

Gillum, W. O., Mortenson, L. E., Chen, J. S., and Holm, R. H. (1977). *J. Am. chem. Soc.* **99**, 584–95.

Gotto, J. W. and Yoch, D. C. (1982*a*). *J. biol. Chem.* **257**, 2868–73.

—— (1982*b*). *J. Bact.* **152**, 714–21.

Guth, J. H. and Burris, R. H. (1983). *Biochem.* **22**, 5111–22.

Haaker, H. and Veeger, C. (1977). *Eur. J. Biochem.* **77**, 1–10.

— Laane, C., Hellingwerf, K., Houwer, B., Konings, W. N., and Veeger, C. (1982). *Eur. J. Biochem.* **127**, 639–45.

Hageman, R. V. and Burris, R. H. (1978). *Biochem.* **75**, 2699–702.

—— (1979). *J. biol. Chem.* **254**, 11189–92.

—— Orme-Johnson, W. H., and Burris, R. H. (1980). *Biochem.* **19**, 2333–42.

Hallenbeck, P. C., Kostel, P. J., and Benemann, J. R. (1979). *Eur. J. Biochem.* **98**, 275–84.

— Meyer, C. M., and Vignais, P. M. (1982). *J. Bact.* **149**, 708–17.

Hardy, R. W. F. (1979). In *Dinitrogen fixation* (ed. R. W. F. Hardy, F. Bottomley, and R. C. Burns), pp. 515–68. John Wiley, New York.

Harker, A. R. and Wullstein, L. H. (1981). *J. biol. Chem.* **256**, 11983–6.

Hase, T., Nakano, T., Matsubara, H., and Zumft, W. G. (1981). *J. Biochem.* **90**, 295–8.

— Wakabayashi, S., Nakano, T., Zumft, W. G., and Matsubara, H. (1984). *FEBS Lett.* **166**, 39–43.

Hausinger, R. P. and Howard, J. B. (1982). *J. biol. Chem.* **257**, 2483–90.

—— (1983). *J. biol. Chem.* **258**, 13486–92.

Hawkes, T. R. and Smith, B. E. (1983). *Biochem. J.* **209**, 43–50.

— Lowe, D. J., and Smith, B. E. (1983). *Biochem. J.* **211**, 495–7.

HILL, S. and KAVANAGH, E. (1980). *J. Bact.* **141**, 470–5.

HILLMER, P. and FAHLBUSCH, K. (1979). *Arch. Microbiol.* **122**, 213–18.

HOFFMAN, B. M., VENTNERS, R. A., ROBERTS, J. E., NELSON, M., and ORME-JOHNSON, W. H. (1982). *J. Am. chem. Soc.* **104**, 4711–12.

HOUMARD, J., BOGUSZ, D., BIGAULT, R., and ELMERICH, C. (1980). *Biochimie* **62**, 267–75.

HUYNH, B. H., HENZL, M. T., CHRISTNER, J. A., ZIMMERMANN, R., ORME-JOHNSON, W. H., and MÜNCK, E. (1980). *Biochim. biophys. Acta* **623**, 124–38.

IMAM, S. and EADY, R. R. (1980). *FEBS Lett* **110**, 35–8.

IMPERIAL, J., UGALDE, R. A., SHAH, V. K., and BRILL, W. J. (1984). *J. bact.* **158**, 187–94.

JOHANSSON, B. C. and GEST, H. (1977). *Eur. J. Biochem.* **81**, 365–71.

JOHNSON, M. K., THOMSON, A. J., ROBINSON, A. E., and SMITH, B. E. (1981). *Biochim. biophys. Acta* **671**, 61–70.

JONES, B. L. and MONTY, K. J. (1979). *J. Bact.* **139**, 1007–13.

JONES, R. (1984). *D. Phil. thesis*, University of Sussex.

KAHN, D., HAWKINS, M., and EADY, R. R. (1982). *J. gen. Microbiol.* **128**, 779–87.

KENNEDY, C. K. and POSTGATE, J. R. (1977). *J. gen. Microbiol.* **98**, 551–7.

— EADY, R. R., KONDOROSI, E., and KLAVANS REKOSH, D. (1976). *Biochem. J.* **155**, 383–9.

KIMBER, S. J., BISHOP, E. O., and SMITH, B. E. (1982). *Biochim. biophys. Acta* **705**, 385–95.

LI, J., BURGESS, B. K., and CORBIN, J. L. (1982). *Biochem.* **21**, 4393–402.

— and BURRIS, R. H. (1983). *Biochem.* **22**, 4472–80.

LIKHTENSTEIN, G. I. (1980). In *Nitrogen fixation* (ed. W. E. Newton and W. H. Orme-Johnson), pp. 195–209. University Park Press, Baltimore.

LJONES, T. and BURRIS, R. H. (1972). *Biochim. biophys. Acta* **275**, 93–101.

—— (1978a). *Biochem.* **17**, 1866–72.

—— (1978b). *Biochem. biophys. Res. Commun.* **80**, 22–5.

LOWE, D. J. (1978). *Biochem. J.* **175**, 955–7.

— EADY, R. R., and THORNELEY, R. N.F. (1978). *Biochem. J.* **173**, 277–90.

— SMITH, B. E., and EADY, R. R. (1980). In *Recent advances in biological nitrogen fixation* (ed. N. S. Subba Rao), pp. 34–87. Arnold, London.

— and THORNELEY, R. N. F. (1984a). *Biochem. J.* **224**, 877–86.

—— (1984b). *Biochem. J.* **224**, 895–901.

— and SMITH, B. E. (1983). In *Metalloproteins*, Vol. 1 (ed. P. M. Harrison). Macmillan, London.

LUDDEN, P. W. (1981). *Biochem. J.* **197**, 503–5.

— BURRIS, R. H. (1976). *Science* **194**, 424–6.

—— (1978). *Biochem. J.* **175**, 251–9.

—— (1979). *Proc. natn. Acad. Sci. U.S.A.* **76**, 6201–5.

— PRESTON, G. G., and DOWLING, T. E. (1982a). *Biochem. J.* **203**, 663–8.

— HAGEMAN, R. V., ORME-JOHNSON, W. H., and BURRIS, R. H. (1982b). *Biochim. biophys. Acta* **700**, 213–16.

LUNDELL, D. J. and HOWARD, J. B. (1978). *J. biol. Chem.* **253**, 3422–6.

—— (1981). *J. biol. Chem.* **256**, 6385–91.

MAZUR, B. J., RICE, D., and HASELKORN, R. (1980). *Proc. natn. acad. Sci. U.S.A.* **77**, 186–90.
— and CHUI, C.-F. (1982). *Proc. natn. Acad. Sci. U.S.A.* **79**, 6782–6.
McKENNA, C. E., McKENNA, M.-C., and HIGA, M. T. (1976). *J. Am. chem. Soc.* **98**, 4657–9.
— HUANG, C. W. (1979). *Nature, Lond.* **280**, 609–10.
— JONES, J. B., ERAN, H., and HUANG, C. W. (1979). *Nature, Lond.* **280**, 611–12.
— ERAN, H., NAKAJIMA, T., and OSUMI, A. (1981). In *Current perspectives in nitrogen fixation* (ed. A. H. Gibson and W. E. Newton), p. 358. Australian Academy of Science, Canberra.
McLEAN, P. A. and DIXON, R. A. (1981). *Nature, Lond.* **292**, 655–6.
— SMITH, B. E., and DIXON, R. A. (1983). *Biochem. J.* **211**, 589–97.
MERRICK, M., FILSER, M., KENNEDY, C., and DIXON, R. A. (1978). *Molec. & gen. Genet.* **165**, 103–11.
—— DIXON, R. A., ELMERICH, C., SIBOLD, L., and HOUMARD, J. (1980). *J. gen. Microbiol.* **117**, 509–20.
MEVARECH, M., RICE, D., and HASELKORN, R. (1980). *Proc. natn. Acad. Sci. U.S.A.* **77**, 6476–80.
MICHALSKI, W. P., VIGNAIS, P. M., and NICHOLAS, D. J. D. (1982). *Proc. 2nd Eur. Bioenergetics Conf.* (Lyon), pp. 295–6.
MILLER, R. W., ROBSON, R. L., YATES, M. G., and EADY, R. R. (1980). *Can. J. Biochem.* **58**, 542–7.
MORTENSON, L. E. (1978). *Curr. Topics cell. Regul.* **13**, 179–232.
— and THORNELEY, R. N. F. (1979). *A. Rev. Biochem.* **48**, 387–418.
— and UPCHURCH, R. G. (1981). In *Current perspectives in nitrogen fixation* (ed. A. H. Gibson and W. E. Newton), pp. 75–8. Australian Academy of Science, Canberra.
MÜNCK, E., RHODES, H., ORME-JOHNSON, W. H., DAVIS, L. C., BRILL, W. J., and SHAH, V. K. (1975). *Biochim. biophys. Acta,* **400**, 32–53.
NAGATANI, H. H. and HASELKORN, R. (1978). *J. Bact.* **134**, 597–605.
NETTESHEIM, D. G., JOHNSON, W. V., and FEINBERG, B. A. (1980). *Biochim. biophys. Acta,* **593**, 371–83.
NIEVA-GOMEZ, D., ROBERTS, G. P., KLEVICKIS, S., and BRILL, W. J. (1980). *Proc. natn. Acad. Sci. U.S.A.* **77**, 2555–8.
NORDLUND, S., ERIKSSON, U., and BALTSCHEFFSKY, H. (1977). *Biochim. biophys. Acta* **462**, 187–95.
— and LUDDEN, P. W. (1983). *Biochem. J.* **209**, 881–4.
NUTI, M. P., LEPIDI, A. A., PRAKASH, R. K., SCHILPEROORT, R. A., and CANNON, F. C. (1979). *Nature, Lond.* **282**, 533–5.
O'DONNELL, M. J. and SMITH, B. E. (1978). *Biochem. J.* **173**, 831–9.
ORME-JOHNSON, W. H. and DAVIS, L. C. (1977). In *Iron-sulfur proteins* (ed. W. Lovenberg) Vol. 3, pp. 15–60. Academic Press, New York.
— DAVIS, L. C., HENZL, M. T., AVERILL, B. A., ORME-JOHNSON, N. R., MÜNCK, E., and ZIMMERMANN, R. (1977). In *Recent developments in nitrogen fixation* (ed. W. E. Newton, J. R. Postgate, and C. Rodriguez-Barrueco), pp. 131–78. Academic Press, London.
— LINDAHL, P., MEADE, J., WARREN, W., GROH, S., ORME-JOHNSON, N. R.,

MÜNCK, E., HUYNH, B. H., EMPTAGE, M., RAWLINGS, J., SMITH, J., ROBERTS, J., HOFFMANN, B., and MIMS, W. B. (1981). In *Current perspectives in nitrogen fixation* (ed. A. H. Gibson and W. E. Newton), pp. 79–83. Australian Academy of Science, Canberra.

PAGE, W. J. and COLLINSON, S. K. (1982). *Can. J. Microbiol.* **28**, 1173–80.

— and VON TIGERSTOM, M. (1982). *J. Bact.* **151**, 237–42.

PIENKOS, P. T., SHAH, V. K., and BRILL, W. J. (1977). *Proc. natn. Acad. Sci. U.S.A.* **74**, 5468–71.

— and BRILL, W. J. (1981). *J. Bact.* **145**, 743–51.

— KLEVICKIS, S., and BRILL, W. J. (1981). *J. Bact.* **145**, 248–56.

PREMAKUMAR, R., LEMOS, E. M., and BISHOP, P. E. (1984). *Biochem. biophys. Acta* **797**, 64–70.

PRESTON, G. G. and LUDDEN, P. W. (1982). *Biochem. J.* **205**, 489–94.

PÜHLER, A. and KLIPP, W. (1981). In *Biology of inorganic nitrogen and sulphur* (ed. H. Bothe and A. Trebst), pp. 276–86. Springer-Verlag, Berlin.

RAWLINGS, J., SHAH, V. K., CHISNELL, J. R., BRILL, W. J., ZIMMERMANN, R., MÜNCK, E., and ORME-JOHNSON, W. H. (1978). *J. biol. Chem.* **253**, 1001–4.

REES, D. C. and HOWARD, J. B. (1983). *J. biol. Chem.* **258**, 12733–4.

RICHARDS, R. L. (1980). In *Trends in the chemistry of nitrogen fixation* (ed. J. Chatt, L. M. da Câmara Pina, and R. L. Richards), pp. 199–214. Academic Press, London.

RIDDLE, G. D., SIMONSON, J. G., HALES, B. J., and BRAYMER, H. D. (1982). *J. Bact.* **152**, 72–80.

RIVERA-ORTIZ, J. M. and BURRIS, R. H. (1975). *J. Bact.* **123**, 537–45.

ROBERTS, G. P., MacNEIL, T., MacNEIL, D., and BRILL, W. J. (1978). *J. Bact.* **136**, 267–79.

— and BRILL, W. J. (1980). *J. Bact.* **144**, 210–16.

ROBSON, R. L. (1979). *Biochem. J.* **181**, 569–75.

— (1984). *FEBS Lett.* **173**, 394–8.

— JONES, R., KENNEDY, C. K., DRUMMOND, M., RAMOS, J., WOODLEY, P. R., WHEELER, C., CHESSHYRE, J., and POSTGATE, J. R. (1984). In *Advances in nitrogen fixation research* (ed. C. Veeger and W. E. Newton), pp. 543–651. Nijhoff/Junk, The Hague.

— and POSTGATE, J. R. (1980). *A. Rev. Microbiol.* **34**, 183–207.

RUVKUN, G. B. and AUSUBEL, F. M. (1980). *Proc. natn. Acad. Sci. U.S.A.* **77**, 191–5.

SCOTT, K. F., ROLFE, B. G., and SHINE, J. (1981). *J. molec. appl. Genet.* **1**, 71–81.

——— (1983a). *DNA* **2**, 141–8.

——— (1983b). *DNA* **2**, 149–55.

SHAH, V. K. and BRILL, W. J. (1977). *Proc. natn. Acad. Sci. U.S.A.* **74**, 3249–53.

—— (1981). *Proc. natn. Acad. Sci. U.S.A.* **78**, 3438–40.

— CHISNELL, J. R., and BRILL, W. J. (1978). *Biochem. biophys. Res. Commun.* **81**, 232–6.

— STACEY, G., and BRILL, W. J. (1983). *J. biol. Chem.* **278**, 12064–8.

SIMPSON, F. B. and BURRIS, R. H. (1984). *Science* **224**, 1095–7.

48   Nitrogen fixation

SMITH, B. E. (1977). *J. Less Common Metals* **54**, 465–75.

— (1980). In *Molybdenum chemistry of biological significance* (ed. W. E. Newton and S. Otsuka), pp. 170–90. Plenum Press, New York.

— (1983). In *Nitrogen fixation: the chemical, biochemical, genetic interfaces* (ed. A. Müller and W. E. Newton). Plenum Press, New York.

— LOWE, D. J., and BRAY, R. C. (1973). *Biochem. J.* **135**, 331–41.

— and LANG, G. (1974). *Biochem. J.* **137**, 169–80.

— THORNELEY, R. N. F., YATES, M. G., EADY, R. R., and POSTGATE, J. R. (1976). *Proc. 1st Int. Symp. Nitrogen Fixation* (ed. W. E. Newton and C. J. Nyman), pp. 150–76. Washington State University Press, Pullman.

—— EADY, R. R., and MORTENSON, L. E. (1976). *Biochem. J.* **157**, 439–47.

— O'DONNELL, M. J., LANG, G., and SPARTALIAN, K. (1980). *Biochem. J.* **191**, 449–55.

— LOWE, D. J., CHEN, C. X., O'DONNELL, M. J., and HAWKES, T. R. (1983). *Biochem. J.* **209**, 207–13.

SMITH, J., EMPTAGE, M. H., and ORME-JOHNSON, W. H. (1982). *J. biol. chem.* **257**, 2310–13.

STEPHENS, P. J., McKENNA, C. E., SMITH, B. E., NGUYEN, H. T., McKENNA, M.-C., THOMSON, A. J., DEVLIN, F., and JONES, J. B. (1979). *Proc. natn. Acad. Sci. U.S.A.* **76**, 2585–9.

—— McKENNA, M.-C., NGUYEN, H. T., MORGAN, T. V., and DEVLIN, F. (1981). In *Current perspectives in nitrogen fixation*. (ed. A. H. Gibson and W. E. Newton), p. 357. Australian Academy of Sciences, Canberra.

SUNDARESAN, V. and AUSUBEL, F. M. (1981). *J. biol. Chem.* **256**, 2808–12.

SWEET, W. J. and BURRIS, R. H. (1981). *J. Bact.* **145**, 824–31.

—— (1982). *Biochim. biophys. Acta.* **680**, 17–21.

TANAKA, M., HANIU, M., and YASNUOBU, K. T. (1977). *J. biol. Chem.* **252**, 7093–100.

THORNELEY, R. N. F., YATES, M. G., and LOWE, D. J. (1976). *Biochem. J.* **155**, 137–44.

— and EADY, R. R. (1974). *Biochem. J.* **133**, 405–8.

—— (1978). *Biochem. J.* **167**, 457–61.

—— LOWE, D. J. (1978). *Nature, Lond.* **272**, 557–8.

— and LOWE, D. J. (1982). *Israel J. Bot.* **32**, 61–71.

—— (1981). In *Current perspectives in nitrogen fixation* (ed. A. H. Gibson and W. E. Newton), pp. 360–1. Australian Academy of Science, Canberra.

—— (1983). *Biochem. J.* **215**, 393–403.

—— (1984*a*). *Biochem. J.* **224**, 887–94.

—— (1984*b*). *Biochem. J.* **224**, 887–94.

TÖRÖK, I. and KONDOROSI, A. (1981). *Nucl. Acids Res.* **9**, 5711–23.

TRIPLETT, E. W., WALL, J. D. and LUDDEN, P. W. (1982). *J. Bact.* **152**, 786–91.

TSO, M. Y. and BURRIS, R. H. (1973). *Biochim. biophys. Acta* **309**, 263–70.

UGALDE, R. A., IMPERIAL, J., SHAH, V. K., and BRILL, W. J. (1984). *J. Bact.* **159**, 888–93.

VEEGER, C., LAANE, C., SCHERINGS, G., MATZ, L., HAAKER, H., and VON ZEELAND-WOLBUS, L. (1980). In *Nitrogen fixation* (ed. W. E. Newton and W. H. Orme-Johnson), pp. 11–38. University Park Press, Baltimore.

VOORDOUW, G., HAAKER, H., VAN BREEMAN, J. F. L., VAN BRUGGEN, E. F. J., and EADY, R. R. (1983). *Eur. J. Biochem.* **136**, 397–401.

WALKER, G. A. and MORTENSON, L. E. (1973). *Biochem. biophys. Res. Commun.* **53**, 904–9.

WALL, J. D. and GEST, H. (1979). *J. Bact.* **137**, 1459–63.

WANG, Z-C. and WATT, G. D. (1984). *Proc. natn. Acad. Sci. U.S.A.* **81**, 376–9.

WATT, G. D., BULEN, W. A., BURNS, A., and HADFIELD, K. L. (1975). *Biochem.* **14**, 4266–72.

— BURNS, A., LOUGH, S., and TENNET, D. L. (1981). *Biochem.* **19**, 4926–32.

WEINMAN, J. J., FELLOWS, F. F., GRESSHOFF, P. M., SHINE, J., and SCOTT, K. T. (1984). *Nucl. Acids Res.* **12**, 8320–344.

WHERLAND, S., BURGESS, B. K., STIEFEL, E. I., and NEWTON, W. E. (1981). *Biochem.* **20**, 5132–40.

WHITING, M. J. and DILWORTH, M. J. (1974). *Biochim. biophys. Acta.* **371**, 337–51.

YAMANE, I., WEININGER, M. S., MORTENSON, L. E., and ROSSMANN, M. G. (1982). *J. biol. Chem.* **257**, 1221–3.

YATES, M. G. and PLANQUE, K. (1975). *Eur. J. Biochem.* **60**, 467–76.

YOCH, D. C. (1979). *J. Bact.* **140**, 987–95.

— (1980). *Biochem. J.* **187**, 273–6.

— CANTU, M. (1980). *J. Bact.* **142**, 899–907.

YUN, A. C. and SZALAY, A. A. (1984). *Proc. natn. Acad. Sci. U.S.A.* **81**, 7358–62.

ZIMMERMANN, R., MÜNCK, E., BRILL, W. J., SHAH, V. K., HENZL, M. T., RAWLINGS, J., and ORME-JOHNSON, W. H. (1978). *Biochem. biophys. Acta* **623**, 124–38.

ZUMFT, W. G. (1981). In *The biology of inorganic N and S* (ed. H. Bothe and A. Trebst), pp. 116–32. Springer-Verlag, Berlin.

— MORTENSON, L. E., and PALMER, G. (1974). *Eur. J. Biochem.* **46**, 525–35.

— NORDLUND, S. (1981). *FEBS Lett.* **127**, 79–82.

— ALEF, K., and MUMMLER, S. (1981). In *Current perspectives in nitrogen fixation* (ed. A. H. Gibson and W. E. Newton), pp. 190–3. Australian Academy of Sciences, Canberra.

2 Biochemical physiology of *Rhizobium* dinitrogen fixation

Robert A. Ludwig and Gert E. de Vries

2.1 INTRODUCTION

The constraints that biological dinitrogen fixation imposes on living systems dictate that, in many instances, it occurs in specialized cells or cell-states that are optimized for conditions favouring fixation. This has indeed proven to be a successful evolutionary strategy, as witnessed by symbiotic dinitrogen fixation between bacteria of the genus *Rhizobium* and leguminous plants, which is by far the greatest net contributor of fixed nitrogen to the biosphere. In this chapter we shall present a review of the biochemical physiology of symbiotic dinitrogen fixation in so far as it applies to *Rhizobium*–legume symbioses. This will serve as an introduction to detailed discussions of the molecular biology of the various specific *Rhizobium*–legume systems discussed in subsequent chapters. A second, well-characterized symbiosis involves the cyano-bacterium *Anabaena* and the water fern *Azolla*, which is discussed specifically in Chapter 7.

Symbiotic dinitrogen fixation between rhizobia and leguminous plants is at once the most prolific and ubiquitous form of this important bio-ecological process. The best estimates suggest that approximately 20 per cent of all biological dinitrogen fixation is conducted by such symbioses and that the net fixed-nitrogen contribution to the biosphere is a much larger fraction since the counteractive processes of nitrification and denitrification are minimized (Burns and Hardy 1975).

In subsequent chapters, many examples of the broad range of *Rhizobium*–legume symbioses will be explored. Such symbioses are often highly specific for both partners, although there exist examples of rhizobia that form symbiotic relations with diverse host plants and vice versa. Symbiotic dinitrogen fixation in all cases occurs in analogous but not necessarily homologous symbiotic organs—root nodules.

The genus *Rhizobium* comprises gram-negative, obligate aerobic bacteria widely distributed in the biosphere. Rhizobia have been further 'classified' into fast-growing and slow-growing strains, for obvious reasons, principally on the bases of metabolic and physiological studies (Vincent 1974). Species have been defined according to ability to nodulate a given legume, and such designations have proven useful. However, as will be presented in subsequent chapters, recent studies demonstrate that nodulation specificity resides mostly in genetic instructions encoded on plasmid DNAs that are generally transmissible among *Rhizobiaceae*. Hence the 'species' may be taken to indicate the constellation of plasmid DNAs residing in a given *Rhizobium* at a given time. The distinction between fast-growing and slow-growing rhizobia transcends plasmid DNA compositions however and the profound differences in fundamental metabolism between these groups suggests that, although both groups contrive to nodulate host legumes, they should otherwise be considered separate genera.

A root nodule is initiated by a specific interaction between a single *Rhizobium* and its host legume. The nature of these recognition events, their sites, and specificities have been extensively studied and reviewed and will not be discussed in detail here (Verma and Long 1983). Several of the following chapters will address the attempts to exploit molecular biology techniques to address questions of specificity. Recognition certainly involves a diverse set of morphological structures on the part of both partners. Infection of the root proceeds with an elaboration and invagination of plant cell-wall material to form an 'infection thread' in which infecting rhizobia multiply enormously.

The process of root-nodule development represents either of two archetypes. In one nodule-type, root cortical tissues proliferate synchronously and more or less with spherical symmetry. This synchronous development extends to subsequent invasion by infection threads full of rhizobia and thereafter maturation of nodule tissue with a coordinated onset of dinitrogen fixation. *Glycine max* (soyabean) and most tropical legumes are representative of plants forming such nodules. In the second nodule-type, meristematic tissues direct the elongation of proliferating cortical cells, and development of such nodules is progressive. Invasion by infection threads full of rhizobia follows such meristematic growth resulting in distinct zones of a given nodule at varying stages of development and senescence. Most temperate legumes produce nodules of this progressive type. Slow-growing rhizobia tend to infect legumes that produce synchronous nodules; fast-growing rhizobia tend to infect legumes that produce progressive nodules. Subsequently, rhizobia escape infection threads and invade cortical cells. At this stage, rhizobia undertake a remarkable differentiation into 'bacteroid' organelles which are specialized for and in which dinitrogen fixation occurs. The biochemical physiology of this process will be discussed in detail, in Section 2.2.

In addition to the capability to fix dinitrogen as symbiotic 'bacteroid' organelles, many slow-growing rhizobia also conduct this process as free-living bacteria, ostensibly to promote their own survival or growth in nitrogen-depleted environments. Recently, a *Rhizobium* able to stem-modulate *Sesbania rostrata*, strain *R. sp.* ORS571, has also been shown to exhibit dinitrogen fixation-dependent growth. We shall discuss and contrast symbiotic and free-living dinitrogen fixation in detail.

2.1 *RHIZOBIUM* FREE-LIVING N_2 FIXATION

2.1.1 Metabolic requirements

Under micro-aerobic conditions, certain strains of slow-growing rhizobia are capable of free-living dinitrogen fixation (Kurz and LaRue 1975; McComb *et al.* 1975; Pagan *et al.* 1975). Recently, one such strain 9, *R. sp.* ORS571, has also been shown to be capable of (Dreyfus and Dommergues 1981; Elmerich *et al.* 1982). However, of N_2-dependent growth other slow-growing strains have defied attempts to demonstrate dinitrogen fixation (Nif)-dependent growth, and therefore an important question has been: why do orthodox slow-growing rhizobia fix dinitrogen in the free-living state?

The metabolic requirements for conducting dinitrogen fixation in rhizobia are similar to those for diazotrophic bacteria (Chapter 1). Since rhizobia are obligate aerobes they must couple substrate oxidation with electron transport to electron accepturs: (1) O_2, so that ATP synthesis can occur in large quantities necessary for dinitrogen fixation; and (2) N_2 as substrate for dinitrogen fixation. However, the O_2 partial pressure must not exceed that resulting in an O_2 flux at the cell surface over and above that able to be utilized, since intracellular O_2 can quickly react with reduced electron carrier proteins and nitrogenase components leading to inactivation of nitrogenase. Both dinitrogen-fixing 'bacteroids' and free-living rhizobia synthesize cytochrome P450-type mono-oxygenases (Appleby 1969; Daniel and Appleby 1972) and one function of such cytochrome P450 may be to scavenge intracellular O_2.

2.1.2 Nitrogen metabolism

Studies with isolated bacteroids have showed that, when presented with $^{15}N_2$ as substrate for dinitrogen fixation, all heavy-labelled nitrogen was exported as ammonium and none was assimilated into either amino acid or protein (Brown and Dilworth 1975). Likewise, when free-living slow-growing rhizobia fix dinitrogen in liquid culture, essentially all fixed-nitrogen is exported as ammonium ion (Ludwig 1980*a*). Biochemical physiology studies support this conclusion. *Rhizobium sp.* 'cowpea' strain 32H1 has a single, poorly-functioning pathway for ammonium assimilation (Ludwig 1978). This contrasts with ammonium assimilation in enteric bacteria, which proceeds by either of two pathways (Tyler 1978): (1) via the action of glutamate dehydrogenase (GDH) in excess ammonium conditions; or (2) via the concerted actions of glutamine synthetase (GS) and glutamine-oxoglutarate amidotransferase (GOGAT). In the GDH biosynthetic reaction, NAPDH normally acts as reductant. Glutamate dehydrogenase has an extremely high K_m for ammonium ($\sim$ 15 mM) and therefore cannot efficiently assimilate ammonium at levels below millimolar. Under these limiting ammonium conditions the GS, GOGAT pathway is employed. Depending on the organism, GOGAT may either utilize NADPH or NADH as reductant. Since GS is the committed nitrogen-donor for a wide spectrum of amino acid, purine, and pyrimidine biosynthetic pathways, its catalytic activity is subject to extensive allosteric end-product inhibition. The catalytic activity of GS is also modulated by a reversible adenylylation of a specific tyrosine residue on each dodecameric subunit. The unadenylylated form of GS is biosynthetically active, the adenylylated form is inactive, and further-more the biosynthetic activity is inversely proportional to the average fraction of adenylylated subunits per holoenzyme (Ginsburg and Stadt-man 1973).

This scheme contrasts with that of the orthodox slow-growing *Rhizo-bium* strain 32H1 where only a catabolic, NAD$^+$-dependent GDH ac-tivity has been observed (Ludwig 1978). Therefore, such slow-growing rhizobia have only a single enzymatic pathway for ammonium assimilation, but interestingly they possess two distinct GS activities (Darrow and Knotts 1977). GSI seems to be a classical prokaryotic-type GS: the holoenzyme is a dodecamer of identical subunits arrayed in a hexameric bilayer, the subunit molecular weight is usually around 50 Kdal, and each subunit can be reversibly adenylylated. GSII seems to be a classical eukaryotic-type GS: the holoenzyme is an octomer of identical subunits that reversibly dissociates into tetramers, the subunit molecular weight is usually around 37 Kdal, and it displays no revers-ible adenylylation. GSII is widely distributed among *Rhizobiaceae*.

Why do only bacteria of this family contain two distinct GS enzymes? This question remains an intriguing one. Circumstantial evidence sug-gests that GSII may participate preferentially in *de novo* purine biosyn-thesis as amide-nitrogen donor, possibly as part of a complex because it is an unstable protein. Remarkably, recent evidence shows that anti-bodies raised against purified *R. leguminosarum* GSII cross-react in immuno-diffusion experiments with one or more GS activities specifically induced in *Pisum sativum* (pea) nodules. GSII is undetec-table, either antigenically or biochemically, in bacteroids isolated from such nodules, however! Conversely, antiserum raised against nodule

GS protein(s) cross-reacts with bacterial GSII (de Vries *et al.* 1983). This strongly suggests that such induced nodular GS is localized in the pea cytoplasm, and raises three fascinating hypotheses: (1) that homologous genes exist in bacterial and plant repertoires; (2) that GSII is encoded by only the bacterium but exported to the plant cytosol; and (3) that gene transfer between bacterium and plant occurs. This third hypothesis, however, is complicated by the fact that GSII can at least be considered a prokaryotic-type gene since it functions normally in free-living rhizobia. Furthermore, a cDNA clone for *Phaseolus vulgaris* GS (Cullimore and Miflin 1983) shows no detectable homology with the *R. phaseoli* genome.

Relatively little is known about the regulation of nitrogen-metabolism and N_2-fixation in rhizobia, in contrast with enteric bacteria and the related soil organism *Klebsiella pneumoniae* (Tyler 1978). In these organisms a principal regulatory protein has been identified that is responsible for inducing genes encoding a broad spectrum of nitrogen-yielding catabolic enzyme systems, including the *nifAL* operon of *K. pneumoniae*, Chapter 4. This protein, referred to as both *GlnG* (Pahel and Tyler 1979) in *E. coli* and *GlnR* or *NtrC* (Kustu *et al.* 1979) in *S. typhimurium* has also been identified as an important regulator of dinitrogen fixation in *K. pneumoniae* (Ausubel *et al.* 1979). Recently it has been demonstrated that the *nifA* protein of *K. pneumoniae* fucntions as an inducer of *nif* operons in a way analogous to *GlnG/NtrC* protein. Indeed, the *nifA* gene is homologous with that of *GlnG/NtrC* (Drummond *et al.* 1983; Ow and Ausubel 1983). The reader is referred to Chapter 4, for a detailed discussion of the regulation of dinitrogen fixation in *K. pneumoniae*.

A '*Gln*-system' also regulates dinitrogen fixation in rhizobia. In *Rhizobium*, certain *Gln*-mutations also confer a *nif*-phenotype in culture and *in planta* (Ludwig and Signer 1977; Kondorosi *et al.* 1977). GSII is not involved in *nif* regulation since revertants of such *Gln*-auxotrophs that re-acquire this enzyme still fail to induce nitrogenase activity under all conditions, whereas revertants that produce a constitutively un-adenylylated GSI are nitrogenase overproducers. This implicates some function(s) of the GSI-system in regulation (Ludwig 1980*a*,*b*). However GSI itself does not regulate *nif* gene expression. The GSI gene of *R. meliloti* 104A14 has been cloned, and derivative strains carrying Tn5 insertions in this gene remain Nif$^+$ (Somerville and Kahn 1983). Tn5 insertions in *R. meliloti* and *R. leguminosarum* have been identified that have a pleiotrophic-negative effect on all symbiotic *nif* functions in root nodules produced by these organisms, and this suggests that such insertions might inactivate a *Rhizobium nifA* gene, Chapters 9 and 10. Since a cloned *R. meliloti nifHDK* operon can be induced by *K. pneumoniae nifA* in *E. coli* (Sundaresan *et al.* 1983) further analogy between regulatory schemes in diverse dinitrogen fixing organisms seems likely. However, the dinitrogen fixation regulatory schemes of diazotrophic bacteria and rhizobia cannot be identical, since fixed-nitrogen is assimilated in diazotrophs but exported as ammonium in rhizobia. The exception to this may be *R. sp.* ORS571, but it remains possible that this organism may sequentially: (1) fix N_2 and export ammonium and (2) reimport and assimilate ammonium. Thus ORS571 may oscillate between both modes, each mode representing a distinct cell-state.

2.1.3 Energetics of dinitrogen fixation

As discussed above, rhizobia must conduct active oxidative phosphorylation at low oxygen tensions in order to support the enormous ATP requirements of dinitrogen fixation. This dilemma poses a paradox for the organisms, and goes some way to explain why diazotrophic growth is problematical. Clearly, dinitrogen fixation occurs both in the free-living and bacteroid states at sub-micromolar dissolved oxygen tensions. In the root nodule, oxygen tension is buffered and probably transported to bacteroids by leghaemoglobin (Section 2.2.2). Electron transport to oxygen is facilitated in such environments by the biosynthesis of new set(s) of cytochrome oxidase(s) with extremely low $K_m(O_2) = 10^{-8}$ to 10^{-9} M (Bergersen and Turner 1980). The P/O ratios, i.e. the stoichiometries of (ATP produced):(electrons transported), under these conditions remain unclear. However, there is no *a priori* reason to expect diminished efficiences.

Otherwise, dinitrogen fixation is conducted in both free-living and symbiotic conditions in a conventional fashion via the nitrogenase complex (Mortenson and Thorneley 1979). As discussed in Chapter 1, this soluble enzyme complex consists of two dissociable components, I and II. Component I comprises a tetramer of two different subunits and an acid extrudable co-factor (FeMoco), the iron-molybdenum–sulphur cofactor, which is, in fact, the active site for dinitrogen reduction. Component II is an iron-sulphur protein comprising a dimer of identical subunits. The mechanism of dinitrogen reduction can be summarized as follows: component II binds two equivalents of ATP and is reduced in one electron equivalents by an as yet incompletely characterized electron transport system, although elements of this system, such as a soluble ferredoxin (Carter *et al.* 1980), have been identified in rhizobia. Reduced-component II < 2 ATP > concomitantly transfers one electron to partially-reduced component I and hydrolyses 2 ATP to 2 ADP,P_i. The free energy of ATP hydrolysis is also thought to be transferred to component I as 'conformational energy' which can somehow be utilized by component I to lower the Arrhenius activation energy of the high stable dinitrogen triple bond for reductive protonation. Oxidized component II is released from fully-reduced component I and this cycle is then repeated a total of six times in order to transfer six electron equivalents to dinitrogen and ultimately release the product ammonium. Thus an ideal stoichiometry of 2 ATPs hydrolysed/electron transferred or 12 ATPs hydrolysed/dinitrogen reduced to ammonium is obtained. This excess ATP is utilized both to lower the Arrhenius activation energy of the dinitrogen–FeMoco complex and to displace the equilibirium of the overall reaction towards ammonium as follows:

$$N_2 + 3\ H_2 = 2\ NH_3 \qquad \Delta G_0' = -29\ kJ\ mol^{-1}$$
$$2\ NH_3 + 2\ H^+ = 2\ NH_4^+ \qquad \Delta G_0' = -50\ kJ\ mol^{-1}$$
$$6\ H^+ + 6\ e^- = 3\ H_2 \qquad \Delta G_0' = +242\ kJ\ mol^{-1}$$

$$N_2 + 8\ H^+ + 6e^- = 2\ NH_4^+ \qquad \Delta G_0' = +163\ kJ\ mol^{-1}$$
$$12\ ATP + 12\ H_2O = 12\ ADP,P_i \qquad \Delta G_0' = -366\ kJ\ mol^{-1}$$

Therefore, the overall biological dinitrogen fixation reaction is, under standard conditions, highly exergonic ($\Delta G_0' = -203\ kJ\ mol^{-1}$) *in vivo*, and ammonium production is limited instead by the extremely slow turnover number of the nitrogenase complex. An important unaddressed question therefore is: what happens to all this excess free energy generated on the 'product' or 'backside' of the reaction by ATP hydrolysis? There may exist some chemical energy conservation of free energy *in vivo* by an as yet undiscovered mechanism. If the reaction is (instead of soluble) membrane-bound, some energy of ATP hydrolysis might be utilized to further raise the membrane potential and facilitate other processes (see chapter 3).

2.1.4 Hydrogenase

Even so the hypothetical limit of 12 ATPs hydrolysed per dinitrogen reduced probably vastly under-estimates the real *in vivo* stoichiometry. Under conditions of dinitrogen sub-saturation, electron transport to protons can result in the production of hydrogen at the FeMoco. However, many rhizobia possess an uptake hydrogenase system that can couple re-oxidation of hydrogen with electron transport to oxygen by oxidative phosphorylation and thus recover squandered ATP in a theoretical mol:mol yield . The *Rhizobium* genus has been extensively surveyed for uptake hydrogenase activities which may contribute to the overall efficiency of symbiotic and presumably also free-living dinitrogen fixation (Schubert and Evans 1976). Uptake hydrogenase systems in rhizobia are complex, because mutants of *R. japonicum* have been isolated that allow methylene blue- but not oxygen-dependent hydrogenase activity (Maier 1981). Biochemical complementation has also been achieved *in vitro* between extracts of different mutant cells

(Maier and Mutaftschiev 1982). Recently, recombinant DNA clones for *R. japonicum* uptake hydrogenase genes have also been identified (Cantrell *et al.* 1983) and this should greatly facilitate functional analysis of rhizobial uptake hydrogenases.

2.1.5 Syntrophic growth of slow-growing rhizobia under dinitrogen-fixing conditions

Do rhizobia qualify as diazotrophs? This is both an interesting biological and semantic question. Slow-growing rhizobia do not simultaneously both grow and fix dinitrogen; dinitrogen-fixing rhizobia instead reside in a cell-state distinct from that of vegetative growth. The distinctive biochemical physiology of such discrete cell-states is not yet known in any detail because physical separations of both (at least two) cell-types present in all dinitrogen-fixing populations has not been effected. Dinitrogen-fixing cells are non-growing and ammonium-exporting. By inference, vegetatively-growing cells are ammonium-importing and assimilating, but not fixing. This allows syntrophic, i.e. by a mechanism of cross-feeding, growth of rhizobia under dinitrogen-fixing conditions. This has been conclusively demonstrated by co-culturing *nif*-prototrophic and *nif*-pleiotrophic-negative strains of *Rhizobium sp.* 32H1. Under *nif*-dependent growth conditions, the viable titre of *nif*⁺ cells decreases whereas *nif*− cells 'grow'. Thus it is the *nif*− cells that grow under *nif*-dependent growth conditions, having been cross-fed by *nif*⁺ cells. (Ludwig 1984). Further experiments will be necessary to understand fully the mechanism of *nif*-dependent syntrophic growth in slow-growing rhizobia.

The anomalous *R. sp.* ORS571 grows in liquid culture under *nif*-dependent growth conditions (Dreyfus and Dommergues 1981; Elmerich *et al.* 1982). Is this strain therefore an aerobic diazotroph like *Azotobacter vinelandii*? Alternatively, does *R. sp.* ORS571 fix dinitrogen syntrophically, but rather oscillate between distinct cell-states of vegetative growth and dinitrogen-fixing non-growth? If free-living rhizobia grow syntrophically on dinitrogen in some soil colonial organization, then it would be those cells superfically located that would preferentially grow, being presented with highest levels of nutrients and oxygen. Some interior colonial stratum would thus be optimized for dinitrogen fixation, i.e. would experience both N- and O_2-limitation. These cells would switch cell-state to allow fixation, would export ammonium, and would cross-feed superficial cells growing vegetatively. This mechanism therefore presumes that such exported ammonium *is* preferentially utilized by sibling rhizobia, since only then would it confer the necessary selective advantage on syntrophism. Since soil microflora can certainly be presumed to be organized in complex, heterogeneous populations, the very existence of syntrophism suggests that such 'metabolic cooperation' be specialized and extensive in these populations and may be responsible for proliferation of the community on a complicated set of substrates.

2.2 BIOCHEMICAL PHYSIOLOGY OF ROOT NODULES

2.2.1 General physiology

After rhizobia are released in root nodule cells from their infection threads by a process resembling pinocytosis (Goodchild and Bergersen 1966; Kijne 1975), the bacterial cytoplasm is separated from the plant-cell cytoplasm by a series of three membranes: (1) the bacterial cytoplasmic membrane; (2) the bacterial outer membrane; and (3) a vacuolar-like peribacteroid membrane, originally derived from the plant-cell plasmalemma by endocytosis. Within this membrane, slow-growing rhizobia will generally still divide several times resulting in as many as sixteen individual cells within the peribacteroid enclosure. At this stage, fast-growing rhizobia do not show further cell division, but rapidly differentiate into dinitrogen-fixing bacteriods that may deform morphologically during this process (Basset *et al.* 1977; van Brussel *et al.* 1977; Tu 1977). Within the plant-cell cytoplasm, large numbers of ribosomes, vesicles, membrane fragments, Golgi-bodies, and endoplasmic reticulum are observed by electron microscopy (Kijne and Planque 1977), indicating a major re-differentiation on the part of the plant cell as well.

Respiration by the endogenous rhizobia and plant-cell mitochondria produces a rapid lowering of oxygen tension within the nodular structure which correlates with a variety of phenomena. The synthesis of leghaemoglobin in the nodule tissue, induction of alternative electron transport proteins, accumulation of poly-3-hydroxybutyric acid in bacteroids, and fermentative pathways in the nodule cells will be discussed.

2.2.2 Leghaemoglobin

Globins are induced in the nodule cells and synthesized on cytoplasmic polysomes (Verma *et al.* 1974; Verma and Bal 1976) possibly in response to the low oxygen tension created in the plant cells. These apoproteins form leghaemoglobins when combined with haem-prosthetic groups, which are synthesized and excreted by rhizobia (Cutting and Schulman 1972; Godfrey *et al.* 1975; Avissar and Nadler 1978). *Ex planta* haem excretion was exhibited by *Rhizobium japonicum* when cultured under low oxygen tensions (Avissar and Nadler 1978). *Rhizobium* mutants, defective in haem synthesis, induce ineffective nodules devoid of detectable leghaemoglobin (Noel *et al.* 1982*a*; Leong *et al.* 1982). Globin synthesis was not necessarily impaired (Noel *et al.*

1982*a*), indicating that availability *per se* of the haem-prosthetic group does not induce globin biosynthesis. In *Pisum sativum* root nodules, immunoprecipitable leghaemoglobin could be demonstrated before nitrogenase component II was detectable in bacteroids (Bisseling *et al.* 1979). Although this does not necessarily demonstrate that nitrogenase biosynthesis and activity is prevented in the absense of leghaemoglobin in the nodule cells, it demonstrates that conditions leading to leg-haemoglobin synthesis occur very early in nodule development, and before actual dinitrogen fixation commences. Earlier studies (Nash and Schulman 1977) indicate that the amount of leghaemoglobin present within a *Glycine max* (soyabean) nodule seems more closely related to the volume of the nodule than its nitrogenase activity, which suggests that synthesis of leghaemoglobin is related to conditions in nodular internal milieux rather than directed as a response to specific needs by endogenous rhizobia. While the function of leghaemoglobin in legume nodules as an oxygen carrier to dinitrogen-fixing bacteroids is well established, there has been a lengthy dispute over the presence or absence of leghaemoglobin within peribacteroid envelopes. Electron microscopy, coupled with autoradiography (Dilworth and Kidby 1967), histochemical (Bergersen and Goodchild 1973) or immunochemical (Verma and Bal 1976) staining procedures, and interpretation of leghaemoglobin partitioning during bacteroid isolation (Robertson *et al.* 1978), has given rise to contradictory interpretations (Dilworth 1980). Recent reports on the careful isolation of bacteroids, still largely surrounded by their peribacteroid envelopes, showed copurification of bacteroids and part of the nodular leghaemoglobin (Bergersen and Appleby 1981; Zhiznevskaya *et al.* 1981) indicating the presence of leghaemoglobin in the immediate bacteroid surroundings. However, since peribacteroid membranes repeatedly bud off and refuse, this result may not be entirely unexpected if, alternatively, leghaemoglobin is cytologically non-compartmentalized (Verma and Long 1983). Future *in vitro* experiments should functionally distinguish proposed localization versus dispersal of leghaemoglobin.

2.3.3 *Rhizobium* cytochromic oxidases

Dramatic increases in respiratory and nitrogenase activities are observed in *R. japonicum* bacteroid suspensions supplemented with leghaemo-globin preparations (Bergersen *et al.* 1973). A similar effect is observed when very dilute bacteroid suspensions are employed to reduce the in-fluence of the oxygen-buffering capacity of leghaemoglobin (Wit-tenberg *et al.* 1974). The existence of a basal level of respiration insufficient to support dinitrogen fixation led to the hypothesis and demonstration (Bergersen and Turner 1975) of two distinct bacteroid cytochrome oxidase systems. An oxidase with low oxygen-affinity, possibly a flavometalloprotein, functions above $1\mu M$ dissolved oxygen and is largely ineffectual in ATP production. This oxidase may offer respiratory protection for nitrogenase (Appleby 1977). A high oxygen affinity cytochrome oxidase system functions efficiently at 1–100 nM dissolved oxygen concentrations, buffered by leghaemoglobin, to faciliate oxidative phosphorylation activity (Appleby *et al.* 1975). Although these strains of *R. japonicum* show the induction of specific terminal oxidase systems with concomitant loss of the conventional oxidases aa3 and o, this seems atypical in *Rhizobium*. *R. leguminosarum* bacteroids and bacteroids from non-leguminous *Parasponia* nodules (Appleby *et al.* 1981), and a recently described strain of *R. japonicum* (Keister *et al.* 1983) reportedly induce high-affinity oxidase systems but also retain oxidases aa3 and o. The same appears to be true for *Rhizobium* sp. 32H1 when fixing nitrogen under free-living conditions (Appleby *et al.* 1981).

The physical involvement of deoxyleghaemoglobin, if present in the bacteroid vicinity, at the site of electron transfer to oxygen, remains to be established. Experiments employing oxygen-binding proteins with very low oxygen-dissociation constants that prevent sufficient supply of dissolved free oxygen to *R. japonicum* bacteroids nevertheless showed appreciable nitrogenase activities (Wittenberg *et al.* 1974). This indicates that oxygenated proteins may directly participate in electron transfer or that oxygen-binding proteins, thus leghaemoglobin, facilitate oxygen diffusion (through a thin unstirred layer of solution surrounding the respiring cells) to terminal oxidases with very high afinities for oxygen. Since multiple leghaemoglobins (Dilworth and Appleby 1979), with presumably distinct binding constants for oxygen, and multiple bacteroid cytochrome oxidases (Bergersen and Turner 1980), also with distinct binding constants for oxygen, can exist in nodules, it is attractive to hypothesize that oxygen transport to bacteroids might be catalysed. The presence of restrained, diffusible molecules in the peribacteroid space would fulfil this function. Inefficient electron transfer to oxygen in dinitrogen-fixing rhizobia under free-living conditions may account for the low (or absent) rates of dinitrogen fixation when compared with bacteroids. However, the existence of effective *Rhizobium* symbioses without functional leghaemoglobin seems contradictory (Trinick 1973).

2.2.4 Poly-3-hydroxybutyric acid

During the period of leghaemoglobin induction and the concomitant changes in cytochrome patterns in the bacteroids, unusually large amounts of poly-3-hydroxybutyric acid (PHB) are accumulated within the bacteroid compartment (Mosse 1964; Romanov *et al.* 1975). In short, PHB synthesis represents an attempt to balance an excess of oxidizable-carbon sources under oxygen limitation. PHB may subsequently be reoxidized under: (1) C-limitation or (2) oxygen excess. The amount of PHB gradually declines from 14 per cent to 3 per cent of dry matter in the dinitrogen fixing stage of *R. lupini* bacteroids while ineffective strains accumulate PHB up to 35 per cent of dry matter (Fedulova *et al.* 1980). The specific activities of enzymes involved in PHB synthesis (acetyl CoA acetyltransferase and acetoacetyl CoA hydrolase) and its catabolism (PHB depolymerase and 3-keto acids-CoA transferase) increase during this period, while high activities of 3-hydroxybutyrate dehydrogenase are present in *R. lupini* and *R. japonicum* bacteroids throughout nodule development (Wong 1971; Fedulova *et al.* 1980). Induction of isocitrate lyase activity occurs in *R. japonicum* bacteroids when PHB utilization is stimulated by interruption of photosynthate flow to the nodules; however, PHB is not able to support high activities of dinitrogen fixation (Wong 1971). Vast quantities of PHB accumulate during diazotrophic growth of *Alcaligenes latus* under low oxygen pressure (Malik *et al.* 1981) and in *Azotobacter beijerinckii* when cultured in a chemostat and growth is limited by the availability of oxygen (Senior *et al.* 1972). Considering the available data, one may confidently theorize that *Rhizobium* bacteroids accumulate PHB as a result of the oxygen stress that they experience within the nodular structure when photosynthate is in excess.

2.2.5 Fermentative pathways in nodule cells

The beneficial effects of enhanced concentrations of carbon dioxide in the rooting environment of leguminous plants on dinitrogen fixation has long been known (Mulder and van Veen 1960). Relatively high levels of carbonic anhydrase occur in root nodules but not in uninfected root tissue for several *Rhizobium*–legume symbioses (Atkins 1974). This enzyme greatly increases the interconversion of carbon dioxide and bicarbonate ions (Enns 1967). The prevention of carbaminohaemoglobin formation and/or a catalysing action on leghaemoglobin oxygen loading and unloading under conditions of, respectively, dehydration of bicarbonate or hydration of carbon dioxide, may be facilitated by carbonic and hydrase activity. No information on these reactions *in planta* has yet been forthcoming. An important function of carbonic anhydrase may be to rapidly distribute carbon dioxide as bicarbonate anion from the respiring bacteroids or the exterior of the nodule throughout the plant cytoplasm so as to supply sufficient substrate levels of bicarbonate anion to the nodular phosphoenolpyruvate carboxylase (PEPC). The concentration of bicarbonate required for half saturation of PEPC activity from *Glycine max* nodules was reported to be 0.41 mM (Petersen and Evans 1978).

A dual role for PEPC activity in root nodules has been proposed:

1. end-product oxaloacetate may serve as C-skeleton for amino-acid synthesis, e.g. asparagine in lupin nodules (Laing *et al.* 1979) and *Vicia* nodules (Lawrie and Wheeler 1975), and derived aspartate in soyabean nodules may facilitate ureide synthesis (Reynolds *et al.* 1982*c*). Strong correlation between dinitrogen fixation activity and PEPC activity is observed in lupin nodules (Christeller *et al.* 1977). ATP, citrate, and ammonium inhibit substrate level phosphorylation from phosphoenolpyruvate by pyruvate kinase, which facilitates oxaloacetate production and amino-acid synthesis under these conditions (Petersen and Evans 1978).

2. PEPC activity in root nodules allows accumulation of malate, which could serve as energy source for actively respiring and dinitrogen-fixing bacteroids. In *Pisum sativum* enhanced induction of PEPC and malate dehydrogenase activities occurs in root tissue when oxygen supply is limited (de Vries *et al.* 1980*a*). More extreme oxygen limitation in root systems leads to the induction of alcohol dehydrogenase. Elevated levels of PEPC, malate dehydrogenase, and alcohol dehydrogenase are the status quo in pea-root nodules. Indeed, millimolar concentrations of malate have been found in soyabean nodules (Strumpf and Burris 1981) and pea nodules (de Vries *et al.* 1980*a*). Malate may be actively exported towards the numerous nodule cell-membrane vesicles, that contain bacteroids, by analogy to storage of accumulated malate in plant-cell vacuoles to prevent or regulate acidification of the cytoplasm (Davies 1973; Crawford 1978). In support of this theory, intact plant cell vacuoles from *Bryophyllum diagremontiansus* (a crassulacean acid metabolism (CAM) plant) can be isolated and a malate permease is demonstrable in these membranes (Buser-Suter *et al.* 1982). Alcohols and aldehydes, generated under the microaerobic conditions in the plant cells, may constitute alternative energy sources during dinitrogen fixation in bacteroids (Petersen and LaRue 1981). Utilization of these substrates will simultaneously detoxify the nodular tissue from these endproducts of fermentative plant cell metabolism.

Other energy sources for nitrogen fixing bacteroids

Oxidation of hydrogen may also represent a significant energy source for bacteroids via uptake hydrogenase systems (Albrecht *et al.* 1979; Elmerich *et al.* 1979; Ruiz-Argueso *et al.* 1979; Nelson and Salminen 1982). Nodules devoid of this activity squander photosynthate to varying extents by allowing gaseous hydrogen to escape (Schubert and Evans 1976) and may in addition suffer inhibition of nitrogenase activity when intracellular hydrogen levels build up (Dixon *et al.* 1981). A variety of other putative energy sources have been found in root-nodule tissue in abundance. Sucrose, glucose, and fructose, the primary products of the photosynthate flow, are easily detected, while presence of the more unusual cyclic sugar pinitol (Streeter and Bosler 1976) and the organic acid malonate (Stumpf and Burris 1981) have also been found in soyabean nodules. However, these compounds were not specifically produced in root nodules (Smith and Phillips 1982; Stumpf and Burris 1981).

Are explicit energy sources for dinitrogen fixation optimized for each specific symbiosis? If so, does this specialization determine in large part the efficiency and effectiveness of a given symbiosis? (Keyser *et al.* 1982) Studies have been conducted on carbon-source utilization by isolated bacteroids. Invariably, organic acids like succinate and malate stimulate nitrogenase activity most effectively. Transport studies on isolated bacteroids demonstrate the presence of dicarboxylate transport system in *R. leguminosarum* bacteroids (Glenn *et al.* 1980; de Vries *et al.* 1980*a,b*), as also found in the free-living state (Finnan *et al.* 1981). Glucose transport is absent in *R. leguminosarum* bacteroids (Hudman and Glenn 1980; de Vries *et al.* 1982) and may result from catabolite repression by organic acids on glucose (de Vries *et al.* 1982) and mannose (Arias *et al.* 1982) utilization. However, glucose utilization has been reported for *R. phaseoli* (Trinchant *et al.* 1981) and *R. lupini* (Romanov *et al.* 1980) bacteroids. *R. trifolii* mutants, unable to utilize glucose or fructose for growth, show normal dinitrogen-fixing abilities in nodules (Ronson and Primrose 1979). In contrast, a variety of mutant strains of rhizobia with defects in organic acid metabolism have been described which lead to an ineffective phenotype in root nodules: 2-ketoglutarate dehydrogenase (Duncan and Fraenkel 1979), dicarboxylate transport (Ronson *et al.* 1981), succinate dehydrogenase (de Vries 1980*b*; Gardiol *et al.* 1982) and a possible mutant in 2-ketoglutarate transport (Wong 1982). Studies on metabolic differences between fast- and slow-growing rhizobia have been restricted to carbohydrate metabolism, and such differences may not lead to principal differences in energy metabolism in the bacteroid state. Typical differences in carbohydrate catabolism of slow-growing rhizobia are the absence of the decarboxylating NADP-dependent 6-phosphogluconate dehydrogenase (Martinez-Drets and Arias 1972; Mulongoy and Elkan 1977), and the general inability to catabolize dissacharides like sucrose (Martinez-Drets *et al.* 1974; Glenn and Dilworth 1981).

2.2.6 Ammonium assimilation and nitrogen storage

Dinitrogen fixation by bacteroids leads to the production of ammonium that is excreted into the plant cytosol. Two schemes of incorporation and transport occur in leguminous plants, which subdivide them into: (1) asparagine; and (2) ureide producing plants (Reynolds *et al.* 1982*a*).

In nodules of both types, glutamine is the first heavy-labelled amino acid detectable when $^{15}N_2$ is incorporated. Subsequently, glutamate, aspartate (and asparagine in the case of asparagine-producing plants) are also labelled (Kennedy 1966a,b). Plant glutamine synthetase, in conjunction with glutamate synthase and aminotransferases, catalyse these conversions, while asparagine synthetase operates in lupin nodules that mainly export asparagine from nodules to the xylem (Scott *et al.* 1976). Glutamate dehydrogenase has been implicated to play a role in ammonium assimilation in lupin nodules (Stone *et al.* 1979), but because of its apparent low activities in other symbioses, its low affinity for ammonium and its mitochondrial location in *Phaseolus* nodules (Awonaike *et al.* 1981), its role in ammonium assimilation is suspect.

Incorporation of ammonium in ureide-producing plants is more complex. If grown under dinitrogen-fixing conditions, the majority of translocated nitrogen in such plants takes the form of the ureides allantoin and allantoic acid. However, when the plants are fed with a source of fixed nitrogen, ureide levels in the xylem sap are low and asparagine is the main exported compound (Pate *et al.* 1980). Allantoin and allantoic acid are the ultimate oxidation products from *de novo* purine biosynthesis in nodules. The complete purine biosynthetic pathway (Thomas and Schrader 1981; Reynolds *et al.* 1982c) has not been elucidated in these systems, although the induction of some enzymes involved in purine biosynthesis (e.g. 5-phosphoribosyl-1-pyrophosphate synthetase and 5-phosphoribosyl amidotransferase) occur during nodule development in ureide-producing legumes (Schubert 1981; Reynolds *et al.* 1982b). Nitrogen for purine biosynthesis is supplied by glutamine and aspartate; data on the incorporation of labelled glycine (Atkins *et al.* 1980) and carbon dioxide (Boland and Schubert 1981) are consistent with the proposed pathway. Although the final product of purine biosynthesis is unknown, subsequent oxidation involves xanthine dehydrogenase (Triplett *et al.* 1980), uricase (Rainbird and Atkins 1981), and allantoinase, in classical fashion. Purine oxidases are induced in ureide-producing legumes concomitantly with the appearance of nitrogenase (Reynolds *et al.* 1982a), while no significant activities are present in nodules of asparagine-producing plants. Because these oxidations generate free-radicals, such reactions and peroxisomal-mediated detoxifications probably occur in ancillary uninfected nodule cells.

Effect of combined nitrogen on dinitrogen fixation

Induction of root nodules on leguminous plants is prevented by defined levels of fixed nitrogen, for reasons yet unclear. When nodulated root systems are exposed to these levels of fixed nitrogen, dinitrogen fixation is abolished but not necessarily immediately. In soyabean nodules, an

80 per cent decline in nitrogenase activity was observed over a 4 to 7 day period (Streeter 1981). A corresponding decrease in nodular protein synthesis is observed, but no specific nitrogenase degradation is seen (Noel et al. 1982*b*). Rates of *de novo* synthesis of nitrogenase components I and II in pea nodules are unaffected when plants are treated for four days with 10 mM ammonium nitrate (Bisseling *et al.* 1978), and isolated bacteroids from treated and untreated nodules possess similar nitrogenase activities (Houwaard 1979*a*). Decline of *in vivo* rates of dinitrogen fixation are ascribed to the decrease of photosynthate flow to the nodules, an interpretation substantiated by similar observed nitrogenase activities in intact but excised treated and untreated nodules incubated in the presence of a utilizable carbon source (Houwaard 1979*b*; Noel *et al.* 1982*b*). The presence of a carbon source during nodulation reportedly counteracts inhibition of nodule formation by 15 mM nitrate in *Lens esculenta* and reduces nitrate utilization up to 83 per cent when compared to control plants (Wong 1980). However, nitrate added to nodulated soyabean plants decreases glucose but increases sucrose levels in the nodules (Streeter 1981). These results of nitrate addition are inconsistent with the hypothesized decrease of photosynthate flow. Further studies on the influence of lowered oxygen concentrations in root nodules and the subsequent changes in carbohydrate metabolism should lead to a more defined understanding of the influence of combined nitrogen on the growth inhibition of invading rhizobia and on the development of dinitrogen-fixing nodules.

REFERENCES

ALBRECHT, S. L., MAIER, R. J., HANUS, F. J., RUSSEL, S. A., EMERICH, D. W., and EVANS, H. J. (1979). *Science* 203, 1255.

APPLEBY, C. A. (1977) In *Proc. 11th Mtg. Fed. Eup. Biol. Sci.*, Copenhagen, (ed. H. Degn, D. Lloyd, and G. C. Hill) 49, 11. Pergamon Press, New York.

— TURNER, G. L., and MACNICOL, P. K. (1975). *Biochim. biophys. Acta* 387, 461.

— BERGERSEN, F. J., CHING, T. M., GIBSON, A. H., GRESSHOF, P. M., and TRINICK, M. J. (1981). *Proc. 4th Int. Symp. Nitrogen Fixation*, Canberra, p. 369.

ARIAS, A., GARDIOL, A., and MARTINEZ-DRETS, G. (1982). *J. Bact.* 151, 1069.

ATKINS, C. A. (1974). *Phytochem.* 13, 93.

— RAINBIRD, R. M., and PATE, J. S. (1980). *Z. PflPhysiol.* 97, 249.

AUSUBEL, F. M., BIRD, S. C., DURBIN, K., JANSSEN, K., MARGOLSKEE, R., and PESKIN, A. (1979). *J. Bact.* 140, 597.

AVISSAR, Y. T. and NADLER, K. D. (1978) *J. Bact.* 135, 782.

AWONAIKE, K. O., LEA, P. J., and MIFLIN, B. J. (1981). *Pl. Sci. Lett.* 23, 189.

BASSET, B., GOODMAN, R. N., and NOVACKY, A. (1977). *Can. J. Microbiol.* **23**, 573.

BERGERSEN, F. J. and GOODCHILD, D. J. (1973). *Aust. J. biol. Sci.* **26**, 741.

— APPLEBY, C. A. (1981). *Proc. 4th Int. Symp. Nitrogen Fixation*, Canberra, p. 366.

— and TURNER, G. L. (1975). *J. gen. Microbiol.* **91**, 345.

—— (1980). *J. gen. Microbiol.* **118**, 235.

—— APPLEBY, C. A. (1973). *Biochim. biohys. Acta* **292**, 271.

BISSELING, T., VAN DEN BOS, and VAN KAMMEN, A. (1978). *Biochim. biophys. Acta* **539**, 1.

—— WESTSRATE, M. W., HAKKAART, M. J. J., and VAN KAMMEN,(1979). *Biochim. biophys. Acta* **562**, 515.

BOLAND, M. J. and SCHUBERT, K. R. (1981). *Arch. Biochem. Biophys.* **213**, 486.

BROWN, C. M. and DILWORTH, M. J. (1975). *J. gen. Microbiol.* **86**, 38.

BRUSSEL, A. A. N. VAN, PLANQUE, K., and QUISPEL, A. (1977). *J. gen. Microbiol.* **101**, 51.

BURNS, R. C. and HARDY, R. W. F. (1975). *Nitrogen fixation in bacteria and higher plants.* Springer-Verlag, New York.

BUSER-SUTER, C., WIEMKEN, A., and MATILE, P. (1982). *Pl. Physiol.* **69**, 456.

CANTRELL, M. A., HAUGLAND, and EVANS, H. J. (1983). *Proc. natn. Acad. Sci. U.S.A.* **80**, 181.

CARTER, K. R., RAWLINGS, J., ORME-JOHNSON, W. H., BECKER, R., and EVANS, H. J. (1980). *J. Biol. Chem.* **255**, 4213.

CHRISTELLER, J. T., LAING, W. A., and SUTTON, W. D. (1977). *Pl. Physiol.* **60**, 47.

CRAWFORD, R. M. M. (1978). In *Plant life in anaerobic environments* (ed. D. D. Hook, R. M. M. Crawford), p. 126. Ann Arbor, Michigan.

CULLIMORE, J. V. and MIFLIN, B. J. (1983). *FEBS Lett.* **158**, 107.

CUTTING, J. A. and SCHULMAN, H. M. (1972). *Biochim. biophys. Acta* **261**, 321.

DARROW, R. A., KNOTTS, R. R. (1977). *Biochem. biophys. Res. Commun.* **78**, 554.

DAVIES, D. D. (1973). *Soc. exp. Biol. Symp.* **27**, 513.

DILWORTH, M. J. (1980). In *Nitrogen fixation*, Vol. 2, p. 4. (ed. W. E. Newton and W. H. Orme-Johnston) University Park Press, Baltimore.

— and APPLEBY, C. A. (1979). In *A treatise on dinitrogen fixation* (ed. R. W. F. Hardy, F. Bottomley, and R. C. Burns), p. 691. John Wiley, New York.

— and KIDBY, D. K. (1967). *Exp. cell. Res.* **49**, 148.

DIXON, R. O. D., BLUNDEN, E. A. G., and SEARL, J. W. (1981). *Pl. Sci. Lett.* **23**, 109.

DREYFUS, B. L. and DOMMERGUES, Y. R. (1981). *FEMS Microbiol. Lett.* **10**, 313.

DRUMMOND, M., CLEMENTS, J., MERRICK, M., and DIXON, R. A. (1983). *Nature, Lond.* **301**, 302.

DUNCAN, M. J. and FRAENKEL, D. G. (1979). *J. Bact.* **37**, 415.

ELMERICH, C., DREYFUS, B L., REYSSET, G., and AUBERT, J.-P. (1982). *Eur. Molec. Biol. Orgn. J.* **1**, 499.

ELMERICH, D. W., RUIZ-ARGUESO, T., CHING, T. M., and EVANS, H. J. (1979). *J. Bact.* **137**, 153.

ENNS T. (1967). *Science* **155**, 44.

FEDULOVA, N. G., CHERMENSKAYA, I. E., ROMONOV, V. I., KRETOVICH, V. L. (1980). *Fiziologiya Rast. (Moscow)* **27**, 544.

FINNAN, T. M., WOOD, J. M., and JORDAN, D. G. (1981). *J. Bact.* **148**, 193.

GARCIA, E., BANCROFT, S., RHEE, S. G., and KUSTU, S. (1977). *Proc. natn. Acad. Sci. U.S.A.* **74**, 1662.

GARDIOL, A., ARIAS, A., CERVENANSKY, C., and MARTINEZ-DRETS, G. (1982). *J. Bact.* **151**, 1621.

GINSBERG, A. and STADTMAN, E. R. (1973). In *The enzymes of glutamine metabolism*, p. 9–43. Academic Press, New York.

GLENN, A. R., POOLE, P. S., and HUDMAN, F. (1980) *J. gen. Microbiol.* **119**, 267.

— DILWORTH, M. J. (1981). *Arch. Microbiol.* **129**, 233.

GODFREY, C. A., COVENTRY, D. R., and DILWORTH, M. J. (1975). In *Nitrogen fixation by free living micro-organisms* (ed. W. D. P. Steward), p. 311. Cambridge University Press.

GOODCHILD, D. J. and BERGERSEN, F. J. (1966) *J. Bact.* **92**, 204.

HOUWAARD, F. (1979*a*). Thesis, Landbouwhogeschool, Wageningen, Holland.

— (1979*b*). *Appl. & environ. Microbiol.* **37**, 73.

HUDMAN, J. F. and GLENN, A. R. (1980). *Arch. Microbiol.* **128**, 72.

KEISTER, D. L., MARSH, S. S., and EL MOKADEM, M. T. (1983). *Pl. Physiol.* **71**, 194.

KENNEDY, I. R. (1966*a*). *Biochim. biophys. Acta* **130**, 285.

— (1966*b*). *Biochim. biophys. Acta* **130**, 295.

KEYSER, H. H., VAN BERKUM, P., and WEBER, D. F. (1982). *Pl. Physiol.* **70**, 1626.

KIJNE, J. W. (1975). *Physiol. Pl. Pathol.* **5**, 75.

— and PANQUE, K. (1972). *Physiol. Pl. Pathol.* **14**, 339.

KONDOROSI, A., SVAB, Z., KISS, G. B., and DIXON, R. A. (1977). *Molec. & gen. Genet.* **151**, 221.

KURZ, W. G. W. and LA RUE, T. A. (1975). *Nature, Lond.* **256**, 407.

KUSTU, S., McFARLAND, N., HUI, S., ESMON, B., and FERRO-LUZZI AMES G. (1979). *Proc. natn. Acad. Sci. U.S.A.* **79**, 4576.

LAING, W. A., CHRISTELLER, J. T., and SUTTON, W. D. (1979). *Pl. Physiol.* **63**, 450.

LAWRIE, A. C. and WHEELER, C. T. (1975). *New Phytol.* **74**, 437.

LEONG, S. A., DITTA, G. S., and HELINSKI, D. R. (1982). *J. biol. Chem.* **257**, 8724.

LUDWIG, R.A. and SIGNER, E. R. (1977). *Nature, Lond.* **267**, 245.

— (1978) *J. Bact.* **135**, 114.

— (1980*a*). *Proc. natn. Acad. Sci. U.S.A.* **77**, 5817.

— (1980*b*). *J. Bact.* **141**, 1209.

— (1984). *Proc. natn. Acad. Sci. U.S.A.* **81**, 1566.

MAIER, R. J., (1981). *J. Bact.* **145**, 533.

— and MUTAFTSCHIEV, S. (1982). *J. biol. Chem.* **257**, 2092.

MALIK, K. A., JUNG, C., CLAUS, D., and SCHLEGEL, H. G. (1981). *Arch. Microbiol.* **129**, 254.

MARTINEZ-DRETS, G. and ARIAS, A. (1972). *J. Bact.* **109**, 467.

—— ROVIRA DE CUTINELLA, M. (1974). *Can. J. Microbiol.* **20**, 605.

McCOMB, J. A., ELLIOTT, and DILWORTH, M. J. (1975). *Nature, Lond.* **256**, 245.

MORTENSON, L. E. and THORNELEY, R. N. F. (1979). *A. Rev. Biochem.* **48**, 387.

MOSSE, B. (1964). *J. gen. Microbiol.* **36**, 49.

MULDER, E. G. and VAN VEEN, W. L. (1960). *Plant & Soil* **13**, 265.

MULONGOY, K. and ELKAN, G. H. (1977). *Can. J. Microbiol.* **23**, 1293.

NASH, D. T. and SCHULMAN, H. M. (1977). *Can. J. Bot.* **54**, 2790.

NELSON, L. and SALMINEN, S. O. (1982). *J. Bact.* **151**, 989.

NOEL, K. D., STACEY, G., TANDON, S. R., SILVER, L. E., and BRILL, W. J. (1982*a*). *J. Bact.* **152**, 485.

NOEL, K. D., CARNEOL, M., and BRILL, W. J. (1982*b*). *Pl. Physiol.* **70**, 1236.

OW, D. and AUSUBEL, F. M. (1983). *Nature, Lond.* **301**, 307.

PAGAN, J. D., CHILD, J. J., SCOWCROFT, W., and GIBSON, A. H. (1975). *Nature, Lond.* **256**, 406.

PAHEL, G. and TYLER, B. M. (1979). *Proc. natn. Acad. Sci. U.S.A.* **74**, 4544.

PATE, J. S., ATKINS, C. A., WHITE, S. T., RAINBIRD, R. M., and WOO, K. C. (1980). *Pl. Physiol.* **65**, 961.

PETERSON, J. B. and EVANS, H. J. (1978). *Pl. Physiol.* **61**, 909.

— LARUE, T. A. (1981). Pl. Physiol. **68**, 489.

RAINBIRD, R. M. and ATKINS, C. A. (1981). *Biochim. biophys. Acta* **659**, 132.

REYNOLDS, P. H. S., BLEVINS, D. G., BOLAND, M. J., SCHUBERT, K. R., and RANDALL, D. D. (1982*a*). *Physiologiya Pl.* **55**, 255.

— BOLAND, M. J., BLEVINS, D. G., SCHUBERT, K. R., and RANDALL, D. D. (1982*b*). *Pl. Physiol.* **69**, 1334.

——— RANDALL, D. D., and SCHUBERT, K. R. (1982*c*). *Trends biochem. Sci.* **7**, 366.

RIVERA-ORTIZ, J. M. and BURRIS, R. H. (1975). *J. Bact.* **123**, 537.

ROBERTSON, J. G., WARBURTON, M. P., LYTTLETON, P., FORDYCE, A. M., and BULLIVANT, S. (1978). *J. Cell Sci.* **30**, 151.

ROMANOV, V. I., IVANOF, B. F., FEDULOVA, N. G., RAIKHINSHTEIN, M. V., CHERMENSKAYA, I. E., ZEMLYANUKHIN, A. A., and KRETOVICH, V. L. (1980). *Biochemistry (Moscow)* **45**, 1628.

— YUSHKOVA, L. A., and KRETOVICH, V. L. (1975). *Mikrobiologiya* **44**, 820.

RONSON, C. W. and PRIMROSE, S. B. (1979). *J. gen. Microbiol.* **112**, 77.

— LYTTLETON, P., and ROBERTSON, J. G. (1981). *Proc. natn. Acad. Sci. U.S.A.* **78**, 4284.

RUIZ-ARGUESO, T., ELMERICH, D. W., and EVANS, H. J. (1979). *Biochem. biophys. Res. Commun.* **86**, 259.

SCHUBERT, K. R. and EVANS, H. J. (1976). *Proc. natn. Acad. Sci. U.S.A.* **73**, 1207.

SCHUBERT, K. R. (1981). *Pl. Physiol.* **68**, 1115.

SCOTT, D. B., FARNDEN, K. J. F., and ROBERTSON, J. G. (1976). *Nature, Lond.* **263**, 703.

SENIOR, P. J., BEECH, G. A., RITCHIE, G. A. F., and DAWES, E. A. (1972). *Biochem. J.* **128**, 1193.

SMITH, A. E. and PHILLIPS, D. V. (1982). *Pl. Physiol.* **54**, 31.

SOMERVILLE, J. E. and KAHN, M. J. (1983). *J. Bact.* **156**, 168.

STONE, S. R., COPLAND, L., and KENNEDY, I. R. (1979). *Phytochem.* **18**, 1273.

STREETER, J. G. and BOSLER, M. E. (1976). *Pl. Sci. Lett.* **7**, 321.

— (1981). *Pl. Physiol.* **68**. 840.

STUMPF, D. K. and BURRIS, R. H. (1981). *Pl.* Physiol. **68**, 989.

SUNDARESAN, V., JONES, J., OW,D., and AUSUBEL, F. M. (1983). *Proc. natn. Acad. Sci. U.S.A.* **80**, 4030.

THOMAS, R. J. and SCHRADER, L. E. (1981). *Phytochem.* **20**, 361.

TRINCHANT, J. C., BIROT, A. M., and RIGAUD, J. (1981). *J. gen. Microbiol.* **125**, 159.

TRINICK, M. J. (1973). *Nature, Lond.* **244**, 459.

TRIPLETT, E. W., BLEVINS, D. G., and RANDALL, D. D. (1980). *Pl. Physiol.* **65**, 1203.

TU, J. C. (1977). *Can. J. Bot.* **55**, 35.

TYLER, B. (1978). *A. Rev. Biochem.* **47**, 1127.

VERMA, D. P. S. and LONG, S. R. (1983). *Intl. Rev. Cytol.*, **suppl. 14**, 211.

— NASH, D. T., and SCHULMAN, H. M. (1974). *Nature, Lond.* **251**, 74.

— and BAL, A. K. (1976) *Proc. natn. Acad. Sci. U.S.A.* **73**, 3843

VINCENT, J. M. (1974). In *The biology of nitrogen fixation* (ed. A. Quispel), p. 265. North Holland, Amsterdam.

VRIES, G. E. de, IN 'T VELD, P., and KIJNE, J. W. (1980*a*) *Pl. Sci. Lett.* **20**, 115.

— (1980*b*). Thesis, R.U. Leiden, the Netherlands.

— VAN BRUSSEL, A. A. N., and QUISPEL, A. (1982). *J. Bact.* **149**, 872.

— OOSTERWIJK, E., and KIJNE, J. W. (1983). *Pl. Sci. Lett.*, **32**, 333.

WITTENBERG, J. B., BERGERSEN, F. J., APPLEBY, C. A., and TURNER, G. L. (1974). *J. biol. Chem.* **249**, 4057.

WONG, P. P. (1971). *Pl. Physiol.* **47**, 750.

— (1980). *Pl. Physiol.* **66**, 78.

— (1982). *Pl. Physiol., Suppl.* **69**, p. 36.

ZHIZNEVSKAYA, G. YA., LIVANOVA, G. I., and ANDREEVA, I. N. (1981). *Proc. 4th Int. Symp. Nitrogen Fixation*, p. 367.

3 Electron transport to nitrogenase in diazotrophs

H. Haaker

3.1 INTRODUCTION

Studies with nitrogenase, isolated from a variety of micro-organisms, showed that biological dinitrogen fixation requires an anaerobic environment, MgATP and a strong reductant. Despite these demands, fixation is observed in a variety of micro-organisms, including obligate aerobes.

Aerobic diazotrophs have to protect the site of the nitrogenase system against oxygen. Several mechanisms are known to be inolved in oxygen protection—respiratory protection, slime production, a thick cell wall and oxygen-binding proteins like leghaemoglobin. All have been reviewed earlier (Yates 1977; Bergerson 1984) and will not be discussed here.

The integration of nitrogenase in cellular metabolism does not seem to involve new or unique principles in biochemistry. This is certainly true for the obligatory anaerobic diazotrophs. In these fermentative organisms, the dual requirement of ATP generation and electron donation can readily be met. Reduced ferredoxin is the physiological electron donor for nitrogenase, and the pathways for ferredoxin reduction and ATP generation are well understood and are by no means unique in dinitrogen fixation (Thauer et al. 1977).

In other diazotrophs the generation of reducing power for dinitrogen fixation is still unsolved. Diazotrophs include the obligatory aerobic *Azotobacter, Rhizobium* and *Rhizobium* bacteroids, as well as the cyanobacteria and the purple photosynthetic bacteria (Haaker et al. 1980). Little information is available on the absolute value of the redox potential necessary for nitrogenase activity. It has been demonstrated that nitrogenase complexes isolated from *A. vinelandii* and *Chromatium vinosum* show full activity with electrochemically-reduced viologens and dithionite as electron donors at redox potentials below -500 mV and no significant activity at redox potentials higher than -430 mV

(Evans and Albrecht 1974; Watt and Bulen 1976). With photochemically-reduced *Azotobacter* flavodoxin as electron donor, nitrogenase isolated from *A. vinelandii* shows full activity at redox potentials below -500 mV and no activity above -460 mV (Scherings *et al.* 1977; Braaksma *et al.* 1982).

When photochemically-reduced viologens, *Megasphaera elsdenii* flavodoxin or the redox couple dithionite/sulphite are used as reductant, *A. vinelandii* nitrogenase shows maximum activity up to a redox potential of -440 mV. At higher potentials the nitrogenase activity declines and no significant activity was detectable at potentials above -350 mV (Braaksma *et al.* 1982). Furthermore, the redox potentials of the Fe protein of *Clostridium pasteurianum* and *A. vinelandii* were measured (Zumft *et al.* 1974; Braaksma *et al.* (1982). It was observed that the Fe protein of *C. pasteurianum* has a midpoint potential of -294 mV. In the presence of MgATP the potential shifted to -400 mV. For *A. vinelandii* these values are -393 mV and -435 mV respectively (Braaksma *et al.* 1982). From these data it is clear that the actual redox potential at which nitrogenase is fully active may well depend upon the bacteria from which the nitrogenase is isolated.

Other aspects which are important in the pathway for electron transport to nitrogenase, are the type of electron carriers and the electron carrier-reducing systems present in the diazotrophs studied. Low-potential electron carriers have been isolated from all kinds of diazotrophs, but the existence of electron carrier-reducing systems are less well documented (Yoch 1979). Even when a possible ferredoxin-reducing enzyme system has been demonstrated to be present in a bacterium, the physiological direction of the reaction is not always clear (Haaker *et al.* 1980; Bothe 1980).

In this chapter the different low-potential electron carriers isolated from diazotrophs will be described and their function in cellular metabolism discussed. This will also include the ferredoxin and flavodoxin reductases and their physiological relevance. For several bacteria a relationship exists between the proton motive force and whole-cell nitrogenase activity. The relevance of this relationship with electron transport to nitrogenase will also be discussed along with different models that describe how electrons for nitrogenase can be generated.

3.2 ELECTRON CARRIERS PRESENT IN DIAZOTROPHS

3.2.1 Ferredoxin

Ferredoxins are a group of iron- and sulphur-containing proteins. The distinguishing feature of iron–sulphur proteins is the structure of the cluster which contains iron bound to the peptide chain through cysteinyl ligands. In nearly all cases ferredoxins function as electron carriers. The ferredoxins can be classified by their type of Fe–S cluster. These clusters are referred to as 1 Fe, 2Fe–2S, 3Fe–3S and 4Fe–4S centres. Ferredoxins containing two Fe–S clusters per molecule and also two different clusters, found within one molecule, have been isolated. *C. pasteurianum* ferredoxin (Palmer *et al.* 1966) contains two 4Fe–4S centres, while *A. vinelandii* ferredoxin contains a 3Fe–3S and a 4Fe–4S centre (Emptage *et al.* 1980). Recently Morgan *et al.* (1984) showed that the protein could be transformed into two 4Fe–4S centres containing protein.

Classification of ferredoxins by their type(s) of Fe–S cluster(s) is arbitary, since the structure and redox behaviour of the cluster is not determined by the amount of iron and sulphide in the cluster itself. The protein has an important effect on the structure of the cluster and therefore on its redox potential. For instance, the so-called P centres of component I of nitrogenase can be made to yield 4Fe–4S cores, although from Mössbauer, electron proton resonance (EPR), and magnetic circular dichroism (MCD) spectroscopy it is clear that the structure of these clusters differs significantly from the 4Fe–4S clusters found in bacterial 4Fe–4S centres containing ferredoxins (Huynh *et al.* 1980; Johnson *et al.* 1982; Orme-Johnson and Orme-Johnson 1982).

The effect of the protein on the properties of the Fe–S cluster can also be seen from the differences in redox potential of the Fe–S clusters. The two 4Fe–4S centres in *C. pasteurianum* ferredoxin have a midpoint potential at pH 7.0 of -390 mV (Palmer *et al.* 1966), but the 4Fe–4S centre in the ferredoxin from *C. vinosum* has a midpoint potential at pH 7.0 of $+350$ mV (Bartsch 1963); the two different 4Fe–4S centres also differ in the charge state. For the 4Fe–4S centres from *C. pasteurianum* the core charge state changes from $2+$ to $1+$ upon reduction, while that of the 4Fe–4S centre in *C. vinosum* changes from $3+$ to $2+$. The latter couple was first detected in ferredoxins that in their 'as isolated' state could be oxidized but not reduced. They were designated as 'high potential' proteins (HIPIP). Features of the proteins determining stability of the different oxidation states of the Fe–S cluster are the most important factors in the value of the redox potential of the Fe–S cluster in the protein. Factors that are involved in the determination of the properties of the Fe–S cluster include:

> The coordination geometry. The polypeptide provides a stereo-specific binding cavity and accommodates the intrinsically asymmetric 4Fe–4S clusters.
>
> NH—S hydrogen bonding. H-bonding from amide NH to both inorganic S and S-cysteine ligands stabilizes the negative charge

and favours two of the three possible redox states of the 4Fe-4S centres.

Environment of hydrophobic and aromatic amino acid residues and the amount of solvent exposure of the cluster. Exclusion of the cluster from water by the polypeptide raises the redox potential.

1Fe *ferredoxins (rubredoxins)*

The simplest group of small non-haem iron proteins found in diazotrophs are the rubredoxins. The class of iron–sulphur proteins contain a single iron atom bound to the peptide through four cysteinyl–sulphur bands (Lovenberg 1977). Rubredoxins operate at redox potentials around 0 mV, a value far too positive to be involved in electron transport to nitrogenase. The physiological role of the rubredoxins in cellular metabolism of these organisms is still not understood.

2Fe-2S *ferredoxins*

The second class of iron-sulphur proteins that are found in diazotrophs are the 2Fe-2S centre containing ferredoxins. Although it was thought that the 2Fe-2S centre was restricted to ferredoxins isolated from plant chloroplasts and the related Cyanobacteria (Tagawa and Arnon 1962), they are also found in diazotrophs. The 2Fe-2S ferredoxin from the Cyanobacteria strongly resembles that of plants. They have a molecular mass of about 12 kDa and contain one 2Fe-2S centre which operates between the core charge states $2+$ and $1+$ with a midpoint potential at pH 7.0 around -400 mV (Yasunobu and Tanaku 1973; Hall *et al.* 1975; Yoch and Carithers 1979). As in plant chloroplasts, the 2Fe-2S ferredoxin from cyanobacteria is involved in conventional electron transport. Ferredoxin is reduced by photosystem I and can reduce the pyridine nucleotides, but can also be used to reduce nitrogenase component II. In some cyanobacteria, the presence of pyruvate : ferredoxin oxidoreductase has been demonstrated. It is therefore possible that in the dark ferredoxin can also be reduced by pyruvate oxidation (Neuer and Bothe 1982). It is likely that the 2Fe-2S ferredoxins in the cyanobacteria are involved in electron transport to nitrogenase.

Other 2Fe-2S centres containing proteins are isolated from diazotrophs. From *Clostridion pasteurianum* a paramagnetic protein with a molecular mass of 24 kdal was isolated. This protein contained a 2Fe-2S with a redox potential of -300 mV (Cardenas *et al.* 1976); the biological function of the protein remains unknown. From *Azotobacter*, several 2Fe-2S centres containing proteins have been isolated (Shethna *et al.* 1968; Robson 1979). The 2Fe-2S protein II from *A. vinelandii* and the 2Fe-2S protein from *A. chroococcum* play a role in the protection of nitrogenase against dioxygen inactivation (Haaker and Veeger 1977; Robson 1979; Veeger *et al.* 1980). The native protective proteins from

A. vinelandii and *A. chroococcum* have different molecular masses, 26 kdal and 14 kdal respectively (Shethna *et al.* 1968; Scherings *et al.* 1977; Robson 1979), but recently it was found (G. Scherings *et al.* 1983) that the protective protein from *A. vinelandii* shows one band of an apparent-molecular mass of 14 kdal in the SDS-polyacrylamide gel electrophoresis. The protein contains 1.9 atom iron and 1.9 atom sulphide per 14 kdal. This suggests that the native protective protein from *A. vinelandii* is a dimer with two identical subunits each containing a 2Fe-2S cluster.

4Fe-4S *ferredoxins*

Ferredoxins containing 4Fe-4S centres can be divided into electron carriers with a single 4Fe-4S centre, and those with two 4Fe-4S centres that function at the same redox potential. Ferredoxins, with one 4Fe-4S centre and one 3Fe-3S centre both present in one molecule and the two clusters operating at different redox potentials, can also be classified within the 4Fe-4S group.

In principle, the 4Fe-4S clusters are thought to have three possible oxidation states (Carter *et al.* 1972; Holm and Ibers 1977). These are: paired-spin (EPR silent); oxidized ($S = \frac{1}{2}$, $g = 2.01$); and reduced ($S = \frac{1}{2}$, $g = 1.94$). The protein selects for two of the three possible electronic states of the cluster. Consequently, the cluster has a redox potential of $+0.35$ V (HIPIP) or -0.42 V (low potential). High-potential proteins select for the spin-paired state/oxidized ($g = 2.01$) state (HP$_{red}$/HP$_{ox}$). Low-potential proteins select for the spin-paired state/reduced ($g = 1.94$) state (Fd$_{ox}$/Fd$_{red}$).

The redox properties of the protein-bound 4Fe-4S centres can be correlated with cluster orientation, suggesting that the protein matrix selects two of the three oxidation states. In the HIPIP proteins, two portions of the molecule form an interface at the cluster-binding site such that the 4Fe-4S cluster is embedded in hydrophobic amino acid residues (Carter *et al.* 1974). The 4Fe-4S centre(s) in the low-potential ferredoxins are relatively more exposed to the solvent than the 4Fe-4S high-potential centre of the HIPIP proteins (Peisach *et al.* 1977). It is known that the potential of synthetic [Fe$_4$S$_4$ (SPh)$_4$]$^{2- /3-}$ is shifted to more negative values when exposed to more aqeous media (Hill *et al.* 1977). Besides the coordination geometry of the Fe-S centre, NH—S hydrogen bonds may be an important factor in determining redox properties of the Fe-S centre (Carter 1977; Adman 1979). In the low potential 4Fe-4S ferredoxin from *P. aerogenas*, the number of NH—S hydrogen bonds stabilize a more reduced cluster (Carter 1977; Adman 1979).

The physiological role of the HIPIP proteins in the cell is still unclear, but the function of one or two 4Fe-4S centres containing ferredoxins is known. These ferredoxins are primarily found in anaerobic and photosynthetic bacteria. In these cells they catalyse several electron transfer reactions such as electron transport between pyruvate and H^+, between pyruvate and the pyridine nucleotides or between pyruvate and component II of nitrogenase. These ferredoxins, which function between the 2^+ and 1^+ oxidation core charge states, have midpoint potentials around -420 mV.

The first example of a different class of electron carriers was *Azotobacter* ferredoxin I (Yoch and Arnon 1972; Sweeney *et al.* 1975; Emptage *et al.* 1980). It was suggested that this ferredoxin contains two 4Fe–4S clusters operating at two different redox potentials. Only recently (Emptage *et al.* 1980; Stout *et al.* 1980), it was discovered that *A. vinelandii* ferredoxin I contains a 3Fe–3S cluster. Recently Morgan *et al.* (1984) showed that the 3Fe–3S cluster is a decomposition product of a 4Fe–4S cluster. It is suggested that probably all ferredoxins with two different redox potentials and a content of about eight iron atoms and a similar number of sulphide atoms belong to this group of electron carriers. The midpoint potential of the 3Fe–3S cluster operates around -420 mV, and the 4Fe–4S centre belongs to the HIPIP type of clusters with a midpoint potential around $+300$ mV. 3Fe–3S centres raise questions about their role in biochemical reactions. Data for the *D. gigas* ferredoxin clearly demonstrate that ferredoxins from this organism can be isolated in different oligomeric forms with the same basic polypeptide (molecular mass 6 kdal. Ferredoxin I and ferredoxin II are two of the oligomeric forms in which the ferredoxins from *D. gigas* can be isolated (Bruschi *et al.* 1976). The active centre in isolated *D. gigas* ferredoxin II was fully characterized by EPR and Mössbauer spectroscopy. In the protein, the presence of a 3Fe–3S cluster was firmly established (Cammack *et al.* 1978; Huynh *et al.* 1980). Yet, after reconstitution of the *D. gigas* ferredoxin II apoprotein with iron and sulphide under reducing conditions, the protein contains only 4Fe–4S centres (LeGall *et al.* 1982). Ferredoxin I of *D. gigas* also contains a 3Fe–3S cluster as well as a 4Fe–4S centre (Le Gall *et al.* 1982). These results indicate that the same polypeptide chain can accommodate two different types of Fe–S clusters. The physiological relevance of the 7Fe–7S containing ferredoxins with two different Fe–S centres in one molecule and operating at quite different redox potentials is not yet clear.

3.2.2 Flavodoxins

Flavodoxins are flavoproteins which function as mediators in electron transfer between prosthetic groups of other proteins, as do ferredoxins. In this respect, they are involved in electron transport to nitrogenase in several diazotrophs. For a general review about flavodoxins see Mayhew and Ludwig (1975), and for a specific review about flavodoxins isolated from diazotrophs, see Yates (1977). Flavodoxins contain flavine mononucleotide (FMN) as the prosthetic group. Molecular masses of flavodoxins vary from 14.5 kdal to 23 kdal, and they can be divided into two groups—one group with a molecular mass around 15 kdal, the other around 22 kdal (Mayhew and Ludwig 1975; Yates 1977). Flavodoxins can accept two electrons per flavin molecule, in contrast to ferredoxins which can only accept one electron per redox centre. In all flavodoxins studied, three different redox states of the flavin moiety are possible. The flavin molecule can be reduced completely, thus containing two electrons, or can be reduced by one electron to the semiquinone state. Depending upon the protein backbone, the different redox states of the flavin molecule are stabilized differently. This is the reason why flavodoxins can catalyse quite different reactions. The redox potential of the redox couple flavodoxin semiquinone/flavodoxin hydroquinone (E_1) is always more negative than the redox couple flavodoxin quinone/flavodoxin semiquinone (E_2). Below pH = 7 the redox couple E_1 is pH-dependent, but not above pH = 7 (Mayhew and Ludwig 1975). Flavodoxins also react with each other in the so-called 'comproportionation reaction'; flavodoxin hydroquinone reacts with flavodoxin quinone and forms flavodoxin semiquinone. The rate of this reaction is strongly dependent on pH and ionic strength and this may probably reflect the large net negative charge on the protein. In the reverse reaction, the 'disproportionation reaction' rate is smaller for flavodoxins since the equilibrium favours the semiquinone. For electron donation to nitrogenase, the redox couple flavodoxin semiquinone/flavodoxin hydroquinone is important. This is probably the redox couple at which flavodoxins shuttle as they replace ferredoxins; the other redox couple of the flavodoxins may have a quite different function in the cell. Examples of the involvement of the later redox couple in cellular metabolism are scarce, however.

In *E. coli,* the redox couple E_1 is thought to regulate the activity of pyruvate-formate lyase, and the redox couple E_2 can activate methionine synthetase (Fujii *et al.* 1977; Blaschkowski *et al.* 1982).

In most cases, the involvement of flavodoxins in electron transfer has been demonstrated in crude extracts or with partly purified enzymes. Several reactions have been used to test the ability of flavodoxins to function as electron carriers. Stimulation of NADPH production by washed chloroplasts (Smillie and Entsch 1971), stimulation of the phosphoroclastic reaction (Mayhew and Massey 1969), stimulation of the production of hydrogen with dithionite and hydrogenase (Knight and · Hardy 1966), coupling between NADPH oxidation and cytochrome-c-reduction in the presence of purified ferredoxin : NADP$^+$ oxidoreductase (Shin 1971) have all been tested. Unfortunately, these measurements have the limitation that the measuring system does not discriminate between ferredoxin or flavodoxin. In my opinion it is meaningless to use the reactivity of an electron carrier in one of the above-mentioned reactions as an indication for its function in an electron transport chain to nitrogenase. In principle, all properly reduced low-potential electron carriers donate electrons to nitrogenase. In general, ferredoxins and flavodoxins do not react specifically with redox enzymes. This is probably the reason why the exact physiological function of several ferredoxins and flavodoxins is not yet known.

As in other bacteria, the synthesis of flavodoxin in some diazotrophs is only observed when the bacteria are grown on a medium with a low iron concentration. The best known example is *C. pasteurianum* (Knight and Hardy 1966), but it has been demonstrated in cyanobacteria (Bothe 1977), and in *Desulvofibrio* (Dubourdien and Le Gall 1970). It is generally accepted that in these organisms flavodoxin replaces ferredoxin. Flavodoxin functions as electron acceptor of pyruvate : ferredoxin oxidoreductase and NADH : ferredoxin oxidoreductase. Favodoxin, reduced by these enzymes, can efficiently donate electrons to nitrogenase, hydrogenase, ferredoxin : NADP$^+$ oxidoreductase, CO_2 reductase, and sulphite reductase (Thauer *et al.* 1977). In other diazotrophs like *K. pneumoniae* (Yoch 1974; Nieva-Gomez *et al.* 1980) and *Azotobacter* (Van Lin and Bothe 1972) the synthesis of flavodoxin, however, is not suppressed by the presence of high concentrations of iron, while from other diazotrophs, *M. flavum* and from *R. japonicum* bacteroids (Carter *et al.* 1980), no flavodoxin and only ferredoxins can be isolated (Bothe and Yates 1976).

The function of flavodoxin as a specific electron carrier to nitrogenase has only been proven for *K. pneumonia*. Genetically, it has been demonstrated that the flavodoxin gene is a part of the *nif* cluster and that the *nifF* product, flavodoxin, functions as a specific electron carrier between pyruvate : flavodoxin oxidoreductase and nitrogenase (Nieva-Gomez *et al.* 1980; Bogusz *et al.* 1981). It is also demonstrated that the pyruvate : flavodoxin oxidoreductase is *nif*-gene J (Bogusz *et al.* 1981). A specific role of flavodoxin in electron transport to nitrogenase in *Azotobacter* has not yet been proven, but when flavodoxin is reduced by illuminated chloroplasts (Yoch *et al.* 1969), by dithionite (Yates 1972), or photochemically (Scherings *et al.* 1977), it acts as a very efficient electron donor to nitrogenase. The main problem is that the flavodoxin-reducing system in *Azotobacter* has not been identified. Flavodoxin-reducing enzymes have been demonstrated in *Azotobacter* (Haaker and Veeger 1977; Veeger *et al.* 1980), but the activity of these enzymes in cell-free extracts is low. Furthermore, NADPH or NADH are the electron donors and *Azotobacter* flavodoxin is not reduced below the semiquinone state. It is therefore not surprising that no significant nitrogenase activity was observed when NAD(P)H oxidation was coupled to electron transport to nitrogenase. This problem will be discussed later (Sections 3.4 and 3.5).

3.3 ELECTRON CARRIERS REDUCING ENZYMES

3.3.1 Pyruvate : ferredoxin oxidoreductase

In obligate anaerobic diazotrophs, the strong reducing power of pyruvate oxidation is used for the reduction of ferredoxins or flavodoxins. The reducing power of ferredoxins is used for reactions requiring stronger reductants than the reduced pyridine nucleotides. These reactions are nitrogen fixation, synthesis of 2-oxo-acids, reduction of CO_2 to formate or the production of dihydrogen. Reduced ferredoxin can be used to generated NADPH for biosynthetic purposes. Pyruvate : ferredoxin oxidoreductases isolated from different organisms have molecular masses of 200 to 300 kdal. The enzyme complex contains thiamine diphosphate and a Fe–S cluster (Kerscher and Oesterhelt 1982).

The presence of pyruvate : ferredoxin oxidoreductase has been demonstrated in *Clostridium* sp, *Desulvofibrio* sp, *K. pneumoniae, R. rubrum* as well as in cyanobacteria (Yoch 1979*b*; Ludden and Burris 1981; Neuer and Bothe 1982). It is generally accepted that in *Clostridium*, ferredoxin is mainly reduced by pyruvate : ferredoxin oxidoreductase and that reduced ferredoxin is the physiological electron donor for nitrogenase. In *K. pneumoniae*, the generation of reducing equivalents for dinitrogen fixation was only proven recently. *K. pneumoniae* is a coliform bacterium and metabolism is closely related to *E. coli*. The main flow of carbon in the anaerobic metabolism of glucose is via CoA-acetylating pyruvate formate lyase. Besides this enzyme, another pyruvate oxidoreductase was found in *E. coli* (Vetter and Knappe 1971; Blaschkowski *et al.* 1982). The concentration of this enzyme, identified as CoA-acetylating pyruvate : flavodoxin (ferredoxin) oxidoreductase, is smaller by a factor of 100 to 1000 than that of pyruvate formate-lyase. The main function of this pyruvate : flavodoxin oxidoreductase is the activation of pyruvate formate-lyase.

Yoch (1974) showed that pyruvate, formate, malate and NADPH oxidation can be coupled to the generation of reducing equivalents for nitrogen fixation in *K. pneumoniae*; the predominant electron carrier is flavodoxin, although a ferredoxin is possibly also involved.

Which enzyme system is used to reduce the electron carriers in *K. pneumoniae* has not been unequivocally proven. Yet, since it has been demonstrated in crude extracts that nitrogenase activity with different electron donors was possible, it is reasonable to accept the presence of the appropriate dehydrogenases. More important, the activity was considerably less than with dithionite as electron donor and, in addition, the nitrogenase activity with the physiological electron donors was variable (Yoch 1974). Recently, Nieva-Gomez *et al.* (1980), Hill and Kavanagh (1980), and Bogusz *et al.* (1981) have shown with *nif* mutants that pyruvate: flavodoxin (CoA-dependent) oxidoreductase (*nifJ*) and flavodoxin (*nifF*) are obligatory for electron transport to nitrogenase in intact *K. pneumoniae* cells. This indicated that the other proposed electron donors for nitrogenase in *Klebsiella* (formate, malate, and NADPH) cannot donate electrons *in vivo* to nitrogenase. In crude extracts, the concentration of flavodoxin limits the pyruvate or formate-driven nitrogenase activity (Hill and Kavanagh 1980). Bogusz *et al.* (1981) showed that *nifJ* is coding for pyruvate : flavodoxin (CoA-acylating) oxidoreductase. Surprisingly, extracts of *nifJ* mutants do not show any activity with formate, while it was demonstrated that formate can effectively be used as electron donor for H_2 production (Hill and Kavanagh 1980), probably via formate dehydrogenase, ferredoxin, and hydrogenase. The reason might be different growth conditions of the cells which determine whether formate dehydrogenase is present.

A role for pyruvate : ferredoxin oxidoreductase in electron transport to nitrogenase in cyanobacteria is not yet clear. In cyanobacteria the enzyme was first demonstrated by Leach and Carr (1971) and characterized by Bothe *et al.* (1974). Since the active site of nitrogenase in the filamentous cyanobacteria are the heterocysts, the presence of pyruvate : ferredoxin oxidoreductase in the heterocysts has to be demonstrated. Recently, Neuer and Bothe (1982) demonstrated the presence of pyruvate : ferredoxin oxidoreductase in extracts of heterocysts; they also showed pyruvate-dependent nitrogenase activity in these preparations. A problem which always arises with preparation of isolated heterocysts is that it is difficult to isolate large numbers of intactheterocysts. Heterocyst preparations are a mixture of closed heterocysts and permeable heterocysts. Closed heterocysts are impermeable to small molecules like ATP and dithionite, and nitrogenase activity is only observed with H_2 or pyruvate (electron donors taken up by intact heterocysts), and light (ATP generation by cyclic phosphorylation in closed heterocysts).

The permeable heterocysts show only nitrogenase activity with dithionite and an ATP regenerating system. If one compares the dithionite activity with the pyruvate plus CoA dependent activity, it is clear that the pyruvate-driven nitrogenase activity is less than 10 per cent of the dithionite activity. Also, the fact that ferredoxin or methylviologen did not stimulate the nitrogenase activity, indicates that the activity of the pyruvate : ferredoxin oxidoreductase in isolated broken heterocysts is not enough to contribute significantly to electron donation to nitrogenase.

For the same reason, a main role of pyruvate : ferredoxin oxido-reductase in the electron transport chain to nitrogenase in the photo-synthetic bacterium *R. rubrum* is questionable. With high amounts of added ferredoxin, around 10 per cent of the dithionite activity was measured with pyruvate as electron donor (Ludden and Burris 1981), while in *C. pasteurianum* or in *K. pneumoniae* with added ferredoxin, equal or higher activities with pyruvate were obtained than with dithionite (Hill and Kavanagh 1980). This indicates that it is still not certain that the pyruvate : ferredoxin oxidoreductase pathway is responsible for full nitrogenase activity in whole cells of photosynthetic bacteria.

3.3.2 Ferredoxin-NAD⁺ oxidoreductase

In fermentation balance studies of *C. kluyveri* (Thauer *et al.* 1968) and *C. pasteurianum* (Daesch and Mortenson 1967) it was observed that *Clostridium* produces large quantities of hydrogen during growth. It was also clear that pyruvate via pyruvate : ferredoxin oxidoreductase can-not be the only electron donor for the hydrogen produced. It was pro-posed that NADH might be a physiological ferredoxin reductant via NADH : ferredoxin oxidoreductase. The enzyme responsible for the transfer of electrons from NADH to ferredoxin is NADH : ferredoxin oxidoreductase. *In vitro* this enzyme catalyses the reaction in both direc-tions (Jungermann *et al.* 1973). Indeed, with an efficient NADH regenerating system (galactose + galactose dehydrogenase), it was possible to demonstrate electron transport from NADH to nitrogenase (Jungermann *et al.* 1974). As suggested by Jungermann *et al.* (1973, 1974, 1976), the *in vivo* intracellular conditions may be quite different from the optimal conditions *in vitro*. The intracellular NADH/NAD⁺ ratio is about 0.3 and it is doubtful whether the redox potential of the redox couple NADH/NAD⁺ under these conditions, at a pH between 6 and 7, is low enough to transfer electrons efficiently to nitrogenase (Haaker *et al.* 1980).

Perhaps another possibility should be considered, however. In all experiments with electron transfer from NADH to ferredoxin, acetyl-CoA (Ac∿CoA) must be present. It is not clear whether Ac∿CoA is only required for the activation of NADH-ferredoxin : oxidoreductase or whether Ac∿CoA is hydrolysed during the reaction and its energy used to reduce ferredoxin with NADH. The possibility of Ac∿CoA hydrolysis coupled to the production of H_2 from NADH cannot be ex-cluded by fermentation balance studies (Jungermann *et al.* 1976). Although the enzyme has not been purified completely, there are in-dications that two protein components are necessary for activity. The second protein component might be the regulating or energy-transducing unit of the enzyme system (Jungermann *et al.* 1976).

This protein might be absent in fermentative organisms that do not have branched fermentations. Furthermore, it is important to remember that the kinetic properties of the enzyme indicate that it is only active in the direction of electron transport from NADH to ferredoxin when there is enough Ac∿CoA (Jungermann *et al.* 1976). If Ac∿CoA hydrolysis and electron transfer from NADH to ferredoxin are coupled, the thermodynamic problem that would otherwise arise with uphill electron transport is solved, and NADH can be considered as a physiological electron donor for nitrogen fixation in *Clostridium*.

3.3.3 Ferredoxin-NADP⁺ oxidoreductase

In several diazotrophs the presence of a ferredoxin : NADP⁺ oxidoreductase has been demonstrated by the isolation of the enzyme (Yoch 1973), or indirectly by showing an NADPH-dependent diaphorase activity, NADPH-dependent H_2 evolution, or NADPH-dependent acetylene reduction by crude extracts (Bothe 1970; Yoch 1974; Tel-Or and Stewart 1976; Lockau *et al.* 1978; Apte *et al.* 1978). In spinach chloroplasts, the enzyme catalyses electron transfer from ferredoxin to NADP⁺. In *C. pasteurianum* the direction of the enzyme reaction is thought to be the same. In other bacteria, it is suggested that the direction of the reaction is uphill in terms of redox potential. The oxidation of organic substrates via NADPH and ferredoxin-NADP⁺ oxidoreductase is coupled to electron transport to nitrogenase (Yoch 1973, 1979*b;* Lockau *et al.* 1978; Apte *et al.* 1978; Neuer and Bothe 1982). Electron transport to nitrogenase via ferredoxin : NADP⁺oxidoreductase has been especially studied in cyanobacteria. In the light, ATP requiring nitrogenase activity is generated by cyclic phosphorylation and ferredoxin can be reduced by photosystem I. In the dark, oxidative phosphorylation provides ATP for nitrogenase operation and electrons are donated by NADPH. This hypothesis has been tested with heterocysts, the site of nitrogen fixation in filamentous cyanobacteria. Isolated broken heterocysts contain active NADPH-dependent dehydrogenases.

Electrons can be donated to nitrogenase from glucose 6-phosphate, 6-phosphogluconate, and isocitrate, via NADP⁺ electrons in isolated broken heterocysts (Lockau *et al.* 1978; Apte *et al.* 1978). Perhaps the electron flow is directed from the organic substrate to NADP⁺, then from NADPH to ferredoxin against the redox potential by a high ratio of NADPH/NADP⁺. Reduced ferredoxin is the electron donor from nitrogenase component II. To test this hypothesis, Apte *et al.* (1978) studied the kinetic properties of enzymes involved in this part of the electron transport chain. Ferredoxin : NADP⁺ oxidoreductase was found to be inhibited by NADP⁺ and glucose 6-phosphate dehydrogenase was strongly inhibited by NADPH. Apte *et al.* showed that at specific NADPH/NADP$^+$ ratios both enzymes can transfer electrons from glucose 6-phosphate via NADP$^+$ and ferredoxin to 2,6-dichlorophenol indophenol or to cytochrome c, themselves artificial electron acceptors of ferredoxin : NADP$^+$ oxidoreductase. They did not show, however, that at these specific NADPH/NADP$^+$ ratios, uphill electron transport via ferredoxin to nitrogenase was possible.

At a redox potential of -320 mV at pH 7.0 with [NADPH] = [NADP$^+$] we were not able to show any significant nitrogenase activity in *Azotobacter* extracts, whether supplemented with articifical electron carriers like methyl viologen or benzyl viologen or not (H. Haaker, unpublished results). In cynobacteria, electron transport from NADPH to nitrogenase is possible via ferredoxin : NADP$^+$ oxido-reductase, but how the cell overcomes the thermodynamic problem to transport electron from NADPH uphill to ferredoxin has not been solved (see also Section 3.5).

3.3.4 NAD(P)H : flavodoxin oxidoreductase

In all cells where flavodoxin is induced when the cells are grown under iron limited conditions in which flavodoxin replaces ferrodoxin, flavodoxin is reduced and oxidized efficiently by the same enzymes that react with ferredoxin. In *Azotobacter,* the presence of a flavodoxin has been demonstrated, and by using artificial reductants (illuminated chloroplasts, or dithionite) it was also demonstrated that flavodoxin could donate electrons to nitrogenase (Benemann *et al.* 1969; Yates 1972). The presence of NAD(P)H flavodoxin oxidoreductase activity in a cell-free extract of *A. vinelandii* was demonstrated by Haaker and Veeger (1977), and a NAD(P)H : flavodoxin oxidoreductase was isolated from *A. vinelandii* (Veeger *et al.* 1980). NAD(P)H reduce flavodoxin quinone by two electrons to flavodoxin hydroquinone in the presence of the isolated enzyme. When the flavodoxin hydroquinone concentration increases, flavodoxin hydroquinone reacts with flavodoxin quinone to give two flavodoxin semiquinone molecules. The formation of flavodoxin hydroquinone by a one-electron reduction step from flavodoxin semiquinone ($E'_0 = -495$ mV) is not possible with reduced pyridine nucleotides ($E'_0 = -320$ mV). The two-electron reduction of flavodoxin quinone to flavodoxin hydroquinone ($E'_0 = -270$ mV) is both thermodynamically possible and occurs (Veeger *et al.* 1980). But the NAD(P)H : flavodoxin oxidoreductase ac-tivity in cell-free extracts, as well as the specific activity of the isolated enzyme, was quite low. The activity of the enzyme was too low to be of importance for electron transport to nitrogenase, and with NAD(P)H as electron donor not enough flavodoxin hydroquinone was produced to give nitrogenase activity. The role of the enzyme in cellular metabolism of *Azotobacter* might be regulatory as proposed for

NADPH : flavodoxin oxidoreductase from *E. coli* (Blaschkowski *et al.* 1982). With the latter enzyme, pyruvate formate lyase can be activated effectively at physiological NADPH/NADP$^+$ratios, although the activity of NADPH : flavodoxin oxidoreductase is low, and flavodoxin was only partly reduced beyond the semiquinone state. (Blaschkowski *et al.* 1982).

3.4 RELATIONSHIP BETWEEN PROTON MOTIVE FORCE AND NITROGENASE ACTIVITY IN DIAZOTROPHS

In 1974, Haaker *et al.* showed that nitrogenase activity in intact cells of *A. vinelandii* is dependent on a high proton motive force across the cytoplasmic membrane. Since *Azotobacter* is obligatory aerobic, ATP synthesis is dependent upon the proton motive force. Uptake of most substrates used for energy generation is energy-linked, and therefore substrate uptake and whole-cell oxidation are inhibited by proton motive force lowering agents. An exception is acetate uptake; it was found that acetate uptake by *A. vinelandii* is not energy-linked (Visser *et al.* 1973). It is possible, therefore, that the respiration rate and the steady-state levels of adenine nucleotides of cells with oxidize acetate, are much less influenced by the addition of a H$^+$-conducting ionophore than cells oxidizing substrates that have energy-linked uptake patterns.

When acetate-oxidizing *Azotobacter* cells were partly de-energized by the addition of low concentrations of the H$^+$-conducting uncoupler 4,-5,-6,-7,-tetrachloro-2-trifluoromethylbenzimidazol (TTFB), the respiration rate was not inhibited and the intracellular ATP/ADP levels were not affected, but nitrogenase activity was completely inhibited. It was also found that by lowering the oxygen supply to the cells, the nitrogenase activity becomes inhibited before an effect on the intracellular ATP/ADP ratio was observed. A severe oxygen limtation inhibits the nitrogenase activity completely and lowers the ATP/ADP ratio. When the respiation of the cells was partly inhibited by 2-heptyl-4-hydroxyquinoline-N-oxide (HQNO), a respiratory chain inhibitor, nitrogenase activity was again inhibited first. At higher concentrations of HQNO, nitrogenase activity is completely inhibited and the ATP/ADP ratio decreased. Upon the addition of the membrane-bound H$^+$-ATPase inhibitor oligomycine, the ATP/ADP ratio is first increased and, at higher oligomycin concentrations, lowered. These effects on the ATP/ADP ratio did not effect the nitrogenase activity (Veeger *et al.* 1977).

Such experiments indicate that a slight decrease of the proton motive force, irrespective of how it is induced (oxygen limitation, H$^+$-conducting uncoupler, or respiratory chain inhibitor), immediately influences the nitrogenase activity. When more inhibitor was added, the intracellular ATP/ADP ratio was also influenced. The experiments with oligomycin showed that the nitrogenase activity is not so sensitive to changes in the intracellular ATP/ADP ratios. A similar relationship was demonstrated for *R. leguminosarum* bacteroids (Laane *et al.* 1978). From these experiments it was concluded that whole-cell nitrogenase activity is coupled to the proton motive force. The relationship between the components of the proton motive force ($\Delta\mu H^+$), the membrane potential ($\Delta\psi$), and the pH gradient across the cytoplasmic membrane (ΔpH) and whole-cell nitrogenase activity, was studied in *R. leguminosarum* bacteroids. The effects of the addition of K^+-ionophores (valinomycine and nigericin) were studied with *R. leguminosarum* bacteroids; the addition of these ionophores does not influence the intracellular ATP/ADP ratio, and thus the effect of the components of the proton motive force on electron transport to nitrogenase could be studied. This is not possible with bacteria like *Azotobacter* or *Rhodopseudomonas* because valinomycin and nigericin lower the ATP/ADP ratio (Veeger *et al.* 1980).

Low concentrations of valinomycin in the presence of K^+, inhibit nitrogenase activity in *R. leguminosarum* bacteroids (Laane *et al.* 1979). The effect of valinomycin on *R. leguminosarum* bacteroids was explained as an effect on electron transport to nitrogenase. This conclusion could be reached because valinomycin did not inhibit the succinate respiration, did not influence the ATP/ADP ratio, nor did it inhibit nitrogenase activity in a cell-free extract. The effect of valinomycin on *R. leguminosarum* bacteroids was found to be in accordance with the generally accepted mechanism of action of the K^+-ionophore. It was shown that valinomycin lowered the $\Delta\psi$ and increased the ΔpH across the cytoplasmic membrane (Laane *et al.* 1979). Nigericin increased the $\Delta\psi$ and decreased the ΔpH, and it was found that low concentrations of nigericin stimulated the nitrogenase activity. The results indicated that electron transport to nitrogenase is not dependent upon the total proton motive force but on the value of the electrical potential across the cytoplasmic membrane. The same phenomena has been described for other organisms — *Azotobacter* (Laane *et al.* 1980); for the cyanobacteria *A. variabilis* and *Plectonema boryanum* (Hawkesford 1981, 1982); and for the photosynthetic bacterium *Rhodopseudomonas sphaeroides* (Haaker *et al.* 1982). Laane *et al.* (1980) showed with *A. vinelandii* that the NH_4^+ inhibition of whole-cell nitrogenase activity coincides with a decease in membrane potential without an effect on the intracellular ATP/ADP ratio. Veeger *et al.* (1980) and Haaker *et al.* (1982) showed that the lipophilic cation, tetraphenylphosphonium (TPP^+), at higher concentrations than used in flow dialysis experiments, inhibits nitrogenase activity in *A. vinelandii,* in *R. leguminosarum* bacteroids, and in the photosynthetic bacterium *Rps. sphaeroides.* In all cases, it was possible to

inhibit whole-cell nitrogenase activity almost completely without affecting the ATP/ADP ratio. Since higher concentrations of lipophilic
cations depolarize the membrane, the inhibition of nitrogenase activity
was explained as an inhibition of electron transport to nitrogenase.
Hawkesford *et al.* (1981) showed relationships between nitrogenase
activity and the individual components of the proton motive force in
the heterocystous cyanobacterium *A. variabilis*. Nigericin was found to
collapse the ΔpH in favour of the $\Delta\psi$. The increase in $\Delta\psi$ was correlated
with an increase in nitrogenase activity. As in *R. leguminosarum* bacter-
oids, valinomycin and the H^+-conducting uncoupler, carbonylcyanide
m-chlorophenylhydrazone (CCCP), inhibits nitrogenase activity. The
inhibitor of the membrane-bound H^+-ATPase, N,N^+-dicyclo-
hexylcarbodiimide (DCCD), did not inhibit the nitrogenase activity,
just like oligomycin in *A. vinelandii*, although under these conditions the
supply of ATP is reduced. This again suggests that in whole cells it is
not the supply of ATP that limits the activity of nitrogenase, but other
factors. One possibility is electron transport to nitrogenase. Because
the experiments were carried out with intact filaments, it was important to demonstrate that nigericin, valinomycin, and carbonylcyanide
m-chlorophenylhydrazone had the same effects on isolated heterocysts
as on intact filaments. And, indeed, the same effects were found for the
non-heterocystous filamentous cyanobacterium *P. boryanum*. Using
nigericin and valinomycin, there was a linear increase in nitrogenase
activity with increasing magnitude of the membrane potential between
-25 mV and -100 mV (inside negative). Hawkesford *et al.* (1982)
suggested that, beside the membrane potential, the ATP/ADP ratio
regulates the nitrogenase activity. The last conclusion was derived
from an experiment where, at a high nigericin concentration and at pH
6.0, which induces a low intracellular pH, the nitrogenase activity was
inhibited compared to a control experiment at pH 7.5. The intracellular
ATP/ADP concentration at pH 6.0 was lower compared with the control, but the membrane potentials were almost equal. Thus, in addition
to the membrane potential (Hawkesford *et al.* 1982), the ATP/ADP
ratio may regulate the nitrogenase activity *in vivo*. This conclusion may
not be correct, because in the experiment where nigericin is used to inhibit nitrogenase, the intracellular pH is 6.0, whereas in the control incubation the intracellular pH is 7.5. Thus there could be a significant
difference in internal pH between the two experiments, explaining the
observed effect on the nitrogenase activity.

Another organism which has been studied with respect to a membrane potential dependency of the nitrogenase activity is the photo-
synethetic bacterium *Rps. sphaeroides* (Haaker *et al.* 1982). In this
bacterium, the proton motive force and its components were measured
at an external pH of 7.4. It was found that at this pH, there was no

ΔpH gradient, so the total proton motive force is represented by the membrane potential only. Photosynthetic bacteria generate energy with cyclic electron transport. By lowering the light intensity below a critical value, the measured membrane potential decreased, together with the nitrogenase activity. At light intensities where there is a close correlation between the nitrogenase activity and the membrane potential, the ATP/ADP ratio was not affected. When the light intensity declined below a threshold value, cyclic phosphorylation was inhibited as observed by a lower ATP/ADP ratio. The nitrogenase activity in the photosynthetic bacterium *Rps. spaerodies* could also be inhibited with relative high concentrations of the lipophilic cation tetraphenylphosphonium (TPP$^+$) (0.1–0.5 mM). TPP$^+$is normally used at concentrations of 1–10 μM in experiments to measure the membrane potential. At TPP$^+$concentrations up to 0.1 mM, the ATP and ADP concentrations were not affected, but nitrogenase activity was inhibited. It was therefore concluded that a small depolarization of the membrane potential causes an inhibition of electron transport to nitrogenase. In contrast to *A. vinelandii* (Laane *et al.* 1980), NH$_4^+$did not influence the membrane potential in *Rps. sphaeroides,* nor did NH$_4^+$influence the ATP/ADP ratio. Yet NH$_4^+$inhibited whole cell nitrogenase activity immediately and completely (Haaker *et al.* 1982). Since, in photosynthetic bacteria, nitrogenase component II can be covalently modified and is susceptible to activation/inactivation, it was important to check that an inhibition of whole-cell nitrogenase activity is not caused by an inactivation of nitrogenase component II. In all cases of whole-cell inhibition: by light limitation, by the addition of TPP$^+$ or NH$_4^+$, cell-free nitrogenase activity was not inhibited. Cell-free nitrogenase activity was measured in cells made permeable for dithionite and ATP within a few seconds after inhibition of whole-cell activity. The experiments described for *Rps. sphaeroides* indicate also that for photosynthetic bacteria electron transport to nitrogenase in whole cells is dependent upon the membrane potential. The experiments with NH$_4^+$ in *Rps. sphaeroides* further show, that, besides a depolarization of the membrane and covalent modification of nitrogenase component II, other mechanisms exist in photosynthetic bacteria to regulate electron transport to nitrogenase.

3.5 MODELS FOR ELECTRON TRANSPORT TO NITROGENASE

3.5.1 Pyridine nucleotides as electron donor for nitrogenase

In obligate aerobic diazotrophs such as *Azotobacter* and *Rhizobium,* the presence of a phosphoroclastic reaction has not been demonstrated. In *A. vinelandii* an interesting feature of this reaction has been demonstrated, however — the presence of acetate kinase and phospho-transacetylase (Bresters *et al.* 1972), although these enzymes were shown not to be present in *A. chroococcum* (Cambell and Yates 1973). It is suggested by several authors that reduced pyridine nucleotides reduce either ferredoxin or flavodoxin, and may therefore be a physiological electron donor for nitrogenase (Yates and Daniel 1970; Benemann *et al.* 1971, Yoch 1979*b*; Bothe 1980). There are unsolved problems with this suggestion. The first is that it has never been shown that efficient electron transport to nitrogenase is possible with enzymes isolated from *Azotobacter* or *Rhizobium*. The second problem is that thermodynamic reduction of *Azotobacter* flavodoxin semiquinone to its hydroquinone form requires an electron donor with a redox potential of about -500 mV; because the midpoint potential of this redox couple is -495 MV at pH 7.0, and the midpoint potential above pH 7.0 is pH-independent (Mayhew and Ludwig 1975). For the reduction of *Rhizobium* ferredoxin, an electron donor with a potential around -480 mV is required, since its midpoint potential is -484 mV at pH 7.5 (Carter *et al.* 1980). *Azotobacter* ferredoxin I (E^0 at pH 7.0 = -420 mV) can be reduced with an electron donor with a potential of -420 mV. When these electron carriers are reduced to a considerable extent, reduction of component II of the nitrogenase is kinetically and thermodynamically possible (Benemann *et al.* 1969; Yates 1972; Scherings *et al.* 1977; Carter *et al.* 1980).

To overcome this problem several hypotheses have been put forward (see also Fig.3.1(a)). The first hypothesis was proposed by Benemann and Yoch (1972), who suggested that a high ratio of NADPH/NADP$^+$ results in a potential low enough to reduce the electron carriers involved in electron transport to nitrogenase. By using high concentrations of an NADPH regenerating system and spinach chloroplast ferredoxin : NADP$^+$ oxidoreductase, it was possible to show electron transport from glucose-6 phosphate via the pyridine nucleotides, spinach ferredoxin : NADP$^+$ oxidoreductase, and the endogenous electron carriers to *Azotobacter* nitrogenase. This experiment shows that under these conditions the reducing power of glucose-6 phosphate was low enough to act as an electron donor for nitrogenase. Yet the system is non-physiological in several ways. First, and most importantly, the presence of an efficient NADPH : ferredoxin or a flavodoxin oxido-reductase has not been demonstrated in *Azotobacter* and in symbiotic organisms. Secondly, the observed ratio of NADPH/NADP$^+$ in intact cells under nitrogen-fixing conditions is around 1 and not around 20 (Haaker *et al.* 1974; Voordouw *et al.* 1983). One can argue that the pyridine nucleotide pool inside cells is not homogeneous and high rations may actually exist at specific sites. How the high ratios are generated is not clear, however, since for most of the NADP$^+$-

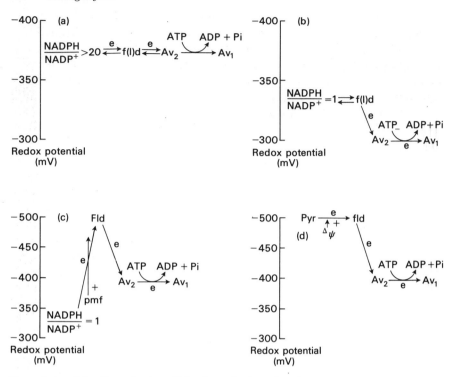

Fig. 3.1. fld = flavodoxin; Fd, ferredoxin; PMF = proton motive force; $\Delta\Psi$ = membrane potential. The location of the different redox couples with respect to the redox potential axis indicates the proposed potentials at which the different redox couples operate *in vivo*. For the explanation of the different models see text on pp. 00–00.

depending dehydrogenases, NADPH is a strong competitive inhibitor. This means that at high ratios of NADPH/NADP$^+$ the activity of the enzyme involved in NADPH-generation is almost completely inhibited. The organic substrates glucose-6 phosphate, malate, or isocitrate, that are supposed to generate the reducing equivalents for nitrogenase, are part of the main flow of carbon through catabolism and the observation is that catabolism, and thus these enzymes, is not inhibited at all during nitrogen fixation.

A variant of this hypothesis was put forward by Gutschick (1981) (see Fig.3.1(b)). He stated that a low redox potential for the reduction of nitrogenase is not necessary. He assumed an intracellular redox potential for NADPH/NADP$^+$ = -350 mV and for the electron carriers involved in electron transport to nitrogenase midpoint potentials around -400 mV. With a one–electron transfer reaction, the 50 mV mismatch means that the ratio-electron carrier red/electron carrier ox

$= 1/7$. Even at this ratio the redox potential of the electron donor is low enough to reduce the Fe protein of *C. pasteurianum* (E^0 at pH 7.5 of Cp_2 $= -290$ mV, Zumft *et al.* 1974). However, in this whole scheme the effect of binding of MgATP to Cp_2 is not included. At cellular ATP concentrations, most of Cp_2 at pH 7.5 is present with MgATP bound, and has a midpoint potential of -400 mV. This means that most of the Fe protein is in the oxidized form at a potential of -350 mV (Zumft *et al.* 1974) and oxidized Fe protein does not simulate nitrogenase activity.

In the case of *A. vinelandii* , Av_2 is oxidized at a potential of -350 mV (Braaksma *et al.* 1982). In principle, the hypothesis of Gutschick is correct for *C. pasteurianum* but one has to assume that there is no redox equilibrium between Cp_2 and Cp *plus* MgATP bound. Theoretically, in the case of *A. vinelandii,* electron transfer to an electron carrier is possible with a lower midpoint potential, because electron transfer from nitrogenase component II to nitrogenase component I is coupled to ATP hydrolysis which is supposed to be irreversible (Mortenson and Thorneley 1979). The effect of low ratios of reduced/oxidized electron carrier has an effect only on the maximal transfer velocity. If the concentration of carriers and enzymes is high enough, reasonable rates can be expected. This means that it must be possible to detect a reasonable nitrogenase activity with pyridine nucleotide as donor in extracts of *Azotobacter* and symbiotic organisms. As mentioned before, this has not been observed.

The involvement of the proton motive force in electron transport to nitrogenase

The experimental evidence for the hypothesis was obtained from experiments with whole cells (see Section 3.4). Until now, there is no support from experiments performed with isolated enzymes; the main experimental problem might be the involvement of the proton motive force in electron transport to nitrogenase. It is technically difficult to generate a high proton motive force in isolated membranes, which is essential since all systems studied show that nitrogenase activity is switched on above a certain threshold membrane potential. A high proton motive force can only be generated by electron transport to oxygen and oxygen is a strong inhibitor of electron transport to and of nitrogenase. Another problem is the uncertainty as to how the proton motive force is involved in electron transport to nitrogenase. Is electron transport to nitrogenase driven by the coupled influx of protons (electron transport to nitrogenase is a real electron transport against the redox potential), or is the proton motive force only involved in the regulation of the activity of the electron transport chain? Both possibilities are discussed below.

In the first hypothesis in electron transport to nitrogenase is driven by the proton motive force (Fig. 3.1(c)). The back flow of protons through an integral membrane NADPH : flavodoxin (ferredoxin) oxidoreductase is essential (Haaker and Veeger 1977; Laane 1980). In this first model, NADPH reduces an integral membrane dehydrogenase. Part of the energy liberated by the back flow of protons is used to lower the potential of a redox centre. The potential of the electrons present on that redox centre is now low enough to reduce flavodoxin or ferredoxin. Reduced *Azotobacter* flavodoxin (E^0 at pH 7.5 = -495 mV) or reduced *Rhizobium* ferredoxin (E^0 at pH 7.5 = -480 mV) are able to reduce nitrogenase component II. The membrane potential may regulate the activity of the integral membrane dehydrogenase but the driving force for electron transport is the total proton motive force (ΔpH and $\Delta\Psi$) (Laane 1980). In this hypothesis, the presence of an integral membrane NADPH : flavodoxin (ferredoxin) oxidoreductase is essential. Isolated membranes of *Azotobacter* and *Rhizobium* contain NADPH dehydrogenase activity, but it has not been demonstrated that in the presence of a proton motive force these NADPH dehydrogenases reduce flavodoxin or ferredoxin. The problems with these assays are outlined above.

Another possibility is shown in Fig. 3.1(d). In this hypothesis, the membrane potential has a regulatory role and is not involved in a reversed electron flow. As a result, of a high membrane potential across the cytoplasmic membrane, an integral membrane enzyme changes its conformation and normally soluble enzymes are bound specifically to this membrane enzyme and form strongly associated enzyme complexes; such an enzyme may be a pyruvate dehydrogenase. Upon being bound to the membrane enzyme, the pyruvate dehydrogenase becomes an active pyruvate : flavodoxin oxidoreductase. The redox potential of pyruvate is low enough to reduce flavodoxin semiquinone or flavodoxin quinone to flavodoxin hydroquinone. To prove this model, a *nif*-specific membrane bound pyruvate dehydrogenase binding protein must be demonstrated in diazotrophs. It is also necessary that the binding of the pyruvate : flavodoxin (ferredoxin) oxidoreductase is dependent upon the membrane potential.

3.6 CONCLUDING REMARKS

Pyruvate is the electron donor for nitrogenase in fermentative growing diazotrophs. In these bacteria, pyruvate reduces ferredoxin or flavodoxin via pyruvate : ferredoxin oxidoreductase. This enzyme is also present in some photosynthetic diazotrophs and in the heterocysts of cyanobacteria, but the activity is too low to account for an effecient electron transport to nitrogenase. In obligate aerobic heterotrophs or photosynthetic bacteria, the proton motive force, and especially a high membrane potential, is obligatory for electron transport to nitrogenase. The next step towards a better understanding of the mechanism of the generation of reducing power dependent upon by the proton motive force should be the development of a cell-free system to study electron transport to nitrogenase.

ACKNOWLEDGEMENTS

The author thanks Miss C. M. Verstege for typing the manuscript and C. Veeger for his critical reading of the manuscript.

The present investigation was supported by the Netherlands Foundation for Chemical Research (S.O.N.) and by financial aid from the Netherlands Organization for the Advancement of Pure Research (Z.W.O.).

REFERENCES

ADMAN, E. T. (1979). *Biochim. biophys. Acta* **549**, 107.

APTE, S. K., ROWELL, P., and STEWART, W. D. P. (1978). *Proc. R. Soc.* B **200,** 1.

BARTSCH, R. G. (1963). In *Bacterial photosynthesis* (eds. H. Gest, A. San Pietra, and L. P. Vernon), p. 315. Antioch. Press, Yellow Springs, Ohio.

BENEMANN, J. R., YOCH, D. C., VALENTINE, R. C., and ARNON, D. I. (1969). *Proc. natn. Acad. Sci. U.S.A.* **64**, 1979.

———— (1971). *Biochim. biophys. Acta* **226**, 205.

—— (1972). *Adv. Microbiol. Physiol.* **8**, 59.

BERGERSEW, F. J. (1984). In *Advances in nitrogen fixation research* (ed. C. Veeger and W. E. Newton), p. 171. Martinus Nijhoff, The Hague.

BLASCHKOWSKI, H. P., NEUER, G., LUDWIG-FEST, M., and KNAPPE, J. (1982). *Eur. J. Biochem.* **123**, 563.

BOGUSZ, D., HOUMARD, J., and AUBERT, J. P. (1981). *Eur. J. Biochem.* **120**, 421.

BOTHE, H. (1970). *Ber. dt. bot. Ges.* **83**, 421.

— FALKENBERG, B. and NOLTEERNSTING, U. (1974). *Arch. Microbiol.* **96**, 291.

— and YATES, M. G. (1976). *Arch. Microbiol.* **107**, 25.

— (1980). In *Nitrogen fixation* (ed. W. D. P. Stewart and J. R. Gallon), p. 249. Academic Press.

BRESTERS, T. W., KRUL, J., SCHEPENS, P. C., and VEEGER (1972). *FEBS Lett.* **22**, 305.

BRUSCHI, M., HATCHIKIAN, C. E., LE GALL, J., MOURA, J. J. G., and XAVIER, A.V. (1976). *Biochim. biophys. Acta* **449**, 275.

CAMBELL, F. C. and YATES, M. G. (1973). *FEBS Letts.* **37**, 203.

CAMMACK, R., RAO, K. K., HALL, D. O., MOURA, J. J. G., XAVIER, A.V., BRUSCHI, M., LE GALL, J., DEVILLE, A., and GAYDA, J. P. (1978). *Biochim. biophys. Acta* **490**, 311.

CARDENAS, J., MORTENSON, L. E., and YOCH, D. C. (1976). *Biochim. biophys. Acta* **434**, 244.

CARTER, C. W., KRAUT, J., FREER, S. T., ALDEN, R. A., SIEKER, L. C., ADMAN, E., and JENSEN, L. H. (1972). *Proc. natn. Acad. Sci. U.S.A.* **69**, 3526.

—— XUONG, Ng. H., ALDEN, R. A., and BARTSCH, R. C. (1974). *J. biol. Chem.* **249**, 4212.

— (1977). In *Iron–sulfur proteins*, Vol. 3 (ed. W. Lovenberg), p. 157. Academic Press, New York.

— RAWLINGS, J., ORME-JOHNSON, W. H., BECKER, R. R., and EVANS, H. J. (1980). *J. biol. Chem.* **255**, 4213.

DAESCH, G. and MORTENSON, L. E. (1967). *J. Bact.* **96**, 346.

EMPTAGE, M. H., KENT, T. A., HUYNH, B. H., RAWLINGS, J., ORME-JOHNSON, W. H., and MÜNCK, E. (1980). *J. biol. Chem.* **255**, 1793.

EVANS, M. C. W., and ALBRECHT, S. L. (1974). *Biochem. biophys. Res. Commun.* **61**, 1187.

FUJII, K., GALIVAN, J. H., and HUENNEKENS, F. M. (1977) *Arch. Biochem. Biophys.* **178**, 662.

GUTSCHICK, V. P. (1982). *Adv. biochem. Engng* **21**, 109.

HAAKER, H., DE KOK, A., and VEEGER, C. (1974). *Biochim. biophys. Acta* **357**, 344.

— and VEEGER, C. (1977). *Eur. J. Biochem.* **77**, 1.

— LAANE, C., and VEEGER, C. (1980). In *Nitrogen fixation* (ed. W. D. P. Stewart and J. R. Gallon), p. 113. Academic Press, London.

—— HELLINGERWERF, K., HOUWER, B., KONINGS, W. N., and VEEGER, C. (1982). *Eur. J. Biochem.* **127**, 639.

HALL, D. O., RAO, K. K., and CAMMACK, R. (1975). *Sci. Prog. (London)* **62**, 285.

HAWKESFORD, M. J., REED, R. H., ROWELL, P., and STEWART, W. D. P. (1981). *Eur. J. Biochem.* **115**, 519.

———— (1982) *Eur. J. Biochem.* **127**, 63.

HILL, C. L., RENAUD, J., HOLM, R. H., and MORTENSON, L. E. (1977). *J. Am. chem. Soc.* **99**, 2549.

— and KAVANAGH, E. P. (1980). *J. Bact.* **141**, 470.

HOLM, R. H. and IBERS, J. A. (1977). In *Iron–sulfer proteins*, Vol. 3 (ed. W. Lovenberg), p. 205. Academic Press, New York.

HUYNH, B. H., HENZL, M. T., CHRISTNER, J. A., ZIMMERMAN, R., ORME-JOHNSON, W. H., and MÜNCK, E. (1980). *Biochim. biophys. Acta* **623**, 124.

— MOURA, J. J. G., MOURA, I., KENT, T. A., LE GALL, J., XAVIER, A. V., and MÜNCK, E. (1980). *J. biol. Chem.* **255**, 3242.

JOHNSON, M. K., ROBINSON, A. E., and THOMSON, A. J. (1982). In *Iron–sulfur proteins*, Vol. 4 (ed. T. G. Spiro) p. 368. Wiley-Interscience, New York.

JUNGERMANN, K., THAUER, R. K., LEIMENSTOLL, G., and DECKER, K. (1973). *Biochim. biophys. Acta* **305**, 268.

— KIRCHNIAWY, H., and KATZ, N. (1974). *FEBS Lett.* **43**, 203.

— KERN, M. RIEBELING, V., and THAUER, R. K. (1976). In *Microbial production and utilization of gases* (ed. H. G. Schlegel, G. Gottschalk, and N. Pfennig), p. 85. E. Goltze, K. G. Göttingen, West Germany.

KERSCHER, L. and OESTERHELDT, D. (1982). *Trends biochem. Sci.* **7**, 371.
KNIGHT, E. and HARDY, R. W. F. (1966). *J. biol. Chem.* **241**, 2752.
LAANE, C., HAAKER, H., and VEEGER, C. (1978). *Eur. J. Biochem.* **87**, 147.
— KRONE, W., KONINGS, W. N., HAAKER, H., and VEEGER (1979). *FEBS Lett.* **103**, 328.
— (1980). Thesis, Pudoc, Wageningen.
— KRONE, W., KONINGS, W. N., HAAKER, H., and VEEGER, C. (1980). *Eur. J. Biochem.* **103**, 39.
LEACH, C. K. and CARR, N. G. (1971). *Biochim. biophys. Acta* **245**, 165.
LE GALL, J., MOURA, J. J. G., PECK, H. D., and XAVIER, A. V. (1982). In *Iron-sulfur proteins* (ed. T. G. Spiro), p. 177. Wiley-Interscience, New York.
LOCKAU, W., PETERSON, R. B., WOLK, C. P., and BURRIS, R. H. (1978). *Biochim. biophys. Acta* **502**, 298.
LOVENBERG, W. (ed.) (1977). *Iron-sulfur proteins*, Vols. 1–3. Academic Press, New York.
LUDDEN, P. W. and BURRIS, R. H. (1981). *Arch. Microbiol.* **130**, 155.
MAYHEW, S. G. and MASSEY, V. (1969). *J. Biol. Chem.* **244**, 794.
— LUDWIG, M. L. (1975). In *The enzymes*, Vol. 12 (ed. P. D. Boyer), p. 75. Academic Press, New York.
MORGAN, T. V., STEPHENS, P. J., BURGESS, B. K., and STOUT, C. D. (1984). *FEBS Lett.* **167**, 137.
NEUER, G. and BOTHE, H. (1982). *Biochim. biophys. Acta* **716**, 358.
NIEVA-GOMEZ, D., ROBERTS, G. P., KLEVICKS, S., and BRILL, W. J. (1980). *Proc. natn. Acad. Sci. U.S.A.* **77**, 2555.
ORME-JOHNSON, W. H. and ORME-JOHNSON, N. R. (1982). In *Iron–sulfur protein* Vol. 4 (ed. T. G. Spiro), p. 68. Wiley-Interscience, New York.
PALMER, G., SANDS, R. H., and MORTENSON, L. E. (1966). *Biochem. biophys. Res. Commun.* **23**, 357.
PEISACH, J., ORME-JOHNSON, N. R., MIMS, W. B., and ORME-JOHNSON, W. H. (1977). *J. biol. Chem.* **252**, 5643.
ROBSON, R. L. (1979). *Biochem. J.* **81**, 569.
SCHERINGS, G., HAAKER, H., and VEEGER, C. (1977). *Eur. J. Biochem.* **77**, 621.
— —WASSINK, H. J., and VEEGER, C. (1983). *Eur. J. Biochem.* **135**, 591.
SHETNA, Y. I., DERVARTANIAN, D. V., and BEINERT, H. (1968). *Biochem. biophys. Res. Commun.* **31**, 862.
SHIN, M. (1971). *Meth. Enzymol.* **23A**, 440.
SMILLIE, R. M. and ENTSCH, B. (1971). *Meth. Enzymol.* **23A**, 504.
STOUT, D. C., GHOSH, D., PATTABHI, V., and ROBBINS, A. (1980). *J. biol. Chem.* **255**, 1979.
SWEENY, W. V., RABINOWITZ, J. C., and YOCH, D. C. (1975). *J. biol. Chem.* **250**, 7842.
TAGAWA, K. and ARNON, D. I. (1962). *Nature, Lond.* **195**, 537.
TEL-OR, E., and STEWART, W. D. P. (1976). *Biochim. biophys. Acta* **423**, 189.
THAUER, R. K., JUNGERMANN, K., HENNINGER, H., WENNING, J., and DECKER, K. (1968). *Eur. J. Biochem.* **4**, 173.
— JUNGERMANN, K., and DECKER, K. (1977). *Bact. Rev.* **41**, 100.
VAN LIN, B. and BOTHE, H. (1972). *Arch. Microbiol.* **82**, 155.

VEEGER, C., HAAKER, H., and SCHERINGS, G. S. (1977). In *Structure and function of energy–transducing membranes* (ed. K. Van Dam, and B. F. Van Gelder), p. 81. Elsevier, Amsterdam.

— LAANE, C., SCHERINGS, G., MATZ, L. L., HAAKER, H., and VAN ZEELAND-WOLBERS, L. (1980). In *Nitrogen fixation*, Vol. 1 (ed. W. E. Newton and W. H. Orme-Johnson), p. 111. University Park Press, Baltimore.

VETTER, H. and KNAPPE, J. (1971). *Hoppe-Seyler's Z. Physiol. Chem.* **352**, 433.

VISSER, A. S. and POSTMA, P. W. (1973). *Biochim. biophys. Acta* **298**, 333.

VOORDOUW, G., VAN DER VIES, S. M., and THEMMEN, A. P. H. (1983). *Eur. J. Biochem.* **131**, 527.

WATT, G. D. and BULEN, W.A. (1976). *Proc. Ist. Int. Symp. Nitrogen Fixation* (ed. W. E. Newton, and C. J. Nyman), p. 248. Washington State University Press.

YASUNOBU, K. T. and TANAKA, M. (1973), In *Iron–sulfur proteins*, Vol. 2 (ed. W. Lovenberg), p. 27. Academic Press, New York.

YATES, M. G. and DANIEL, R. M. (1970). *Biochim. biophys. Acta* **197**, 161.

— (1972). *FEBS Lett.* **27**, 63.

— (1977). In *Recent developments in nitrogen fixation* (ed. W. E. Newton. J. R. Postgate, and C. Rodriguez-Barrueco), p. 219. Academic Press, London.

YOCH, D. C. and ARNON, D. I. (1972). *J. biol. Chem.* **247**, 4515.

— (1973) *J. Bact.* **116**, 384.

— (1974) *J. gen. Microbiol.* **83**, 153.

— and CARITHERS, R. P. (1979a). *Microbiol. Rev.* **43**, 384.

— (1979b). In *A treatise on dinitrogen fixation* (ed. R. W. F. Hardy, F. Bottomley, and R. C. Burns), p. 605. Wiley-Interscience, New York.

ZUMFT, W. G., MORTENSON, L. E., and PALMER, G. (1974). *Eur. J. Biochem.* **46**, 525.

4 Transcriptional analysis of the nitrogen fixation region (*nif*-region) of *Klebsiella pneumoniae* in *Escherichia coli*

W. Klipp and A. Pühler

4.1 INTRODUCTION

Several years ago, we reported the cloning of the whole *K. pneumoniae* *nif*-region using multicopy plasmids of *E. coli* (Pühler *et al.* 1979*a* and 1979*b*). In particular, we constructed plasmid-pWK22o consisting of the plasmid vector pACYC184 (Chang and Cohen 1978) and two *Hind*III fragments of the *Klebsiella* chromosome (Pühler and Klipp 1981). For pWK22o it was shown that *E. coli* cells carrying this plasmid were able to fix atmospheric nitrogen, thus indicating that all essential *nif*-genes are located on the cloned region. Further analysis of the cloned *nif*-region resulted in a gene/protein map (Kennedy *et al.* 1981). We were able to map 15 coding regions and to determine their approximate sizes (Pühler and Klipp 1983*a*). In this article we describe experiments which lead to the identification of four *nif*-specific promoters. These promoters are activated and repressed by the *nifA*- and *nifL*-gene products, respectively. In addition, the transcriptional units belonging to these *nif* specific promoters were determined.

4.2 MAP OF THE *nif* CODING REGIONS of *Klebsiella*

In Fig. 4.1 the map of the *nif* coding region of *Klebsiella* is shown. This map was constructed from a variety of sources (Pühler and Klipp 1983*a*):

First, we cloned various restriction fragments of the *nif* region in different plasmid vectors. Cloning sites which allowed the transcription of the cloned fragments from constitutive plasmid promoters were selected, e.g. the *Eco*RI site in the chloramphenicol resistance gene of plasmid pACYC184 (Chang and Cohen 1978).

The expression of *nif*-genes on the cloned fragments were tested in minicells of *E. coli*. For this purpose minicells harbouring the hybrid plasmid in question were isolated. Such minicells were able to transcribe and translate plasmid-encoded genes. Therefore, in the presence of ^{35}S-methionine, newly-synthesized polypeptides can be identified on SDS polyacrylamide gels by autoradiography. Using minicell analysis, the gene products specified by cloned *nif* fragments were determined.

The exact location of the coding regions on the cloned fragments was determined by transposon mutagenesis. Transposon Tn5 was inserted into the cloned fragment at different sites and the polypeptide pattern of the Tn5-mutagenized plasmids was again determined by the minicell technique. Since a Tn5 insertion in an operon stops transcription and a Tn5 insertion in a coding region terminates translation, the location of coding regions on cloned fragments could be determined.

All these methods were extensively used to map the *nif* coding regions of Klebsiella: 15 *nif* coding regions were identified and could be assigned to *nif*-genes. Two previously undiscovered genes, *nifX* and *nifY*, were found in this way; in contrast, two other genes *nifB* and nifQ could not be identified by the minicell technique. The results are summarized in Fig. 4.1. It should be mentioned that all *nif*-genes, with the exception of *nifF*, are transcribed and translated in the same direction. This direction was determined in the minicell system with the help of Tn5 mutagenesis. For the *nifF*-gene, we initially concluded that its transcriptional direction is same as that of the other *nif*-genes, but we now have evidence that this gene is the only one which is transcribed in the opposite direction (data not shown).

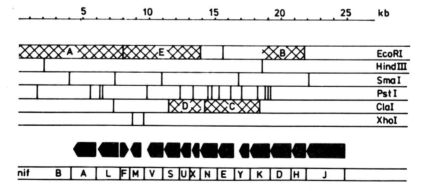

Fig. 4.1. The map of the coding regions of the *Klebsiella nif*-region. In comparison to the restriction map the exact location and direction of translation of 15 individual *nif*-genes are symbolized by big black arrows. The cross-hatched fragments A–E were used in further experiments.

4.3 EXPRESSION OF THE *nifB*-GENE IN MINICELLS OF *E. coli*: IDENTIFICATION OF A *nifA*-DEPENDENT PROMOTER TO THE LEFT OF *nifB*

Until recently, the analysis of the *Eco*RI fragment A (Fig. 4.1) of the *Klebsiella nif*-region in the *E. coli* minicell system only led to the identification of the *nifA* and the *nifL*-gene products (*nifAgp* and *nifLgp*). This is shown in more detail in Fig. 4.2(a): the *Eco*RI fragment A of the *Klebsiella nif*-region was cloned into the plasmid vector pACYC184 in the orientation which puts the *nifAL* operon under the control of the

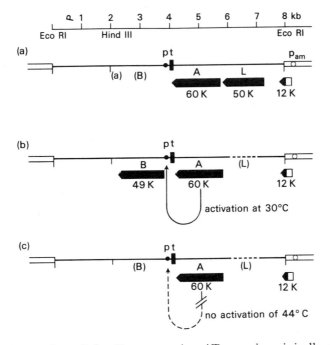

FIG. 4.2. Expression of the *K. pneumoniae nifB*-gene in minicells of *E. coli*. (a) The *Eco*RI fragment A of Fig. 4.1 was cloned into the *Eco*RI site of plasmid pACYC184. One hybrid plasmid with the *nifAL* operon under the control of the constitutive plasmid promoter p_{Cm} was selected and analysed in the *E. coli* minicell system. Polypeptides of molecular weight 12 000, 50 000, and 60 000 were identified and their coding regions were mapped. (b) Following deletion of the *nifL*-gene analysis in the minicell system showed polypeptides of molecular weight 12 000, 49 000, and 60 000. (c) Analysis of the same plasmid in the minicell system at elevated temperature, e.g. at 44 °C instead of at 30 °C, resulted in polypeptides of molecular weight 12 000 and 60 000. The 49 000 molecular weight polypeptide is believed to represent the *nifB* gene product. p_{nifQB} (*p*) represents an internal *nif*-promoter for the *nifQB*-operon and t_{nifAL} (*t*) the transcriptional terminator for the *nifAL*-operon.

chloramphenicol promoter p_{Cm}. In the *E. coli* minicell system, three specific polypeptides with a molecular weight of 12 000, 50 000, and 60 000 respectively were found for this hybrid plasmid. Using transposon mutagenesis, the coding regions of these polypeptides were determined. They are indicated in Fig. 4.2(a): the 12 000 molecular weight polypeptide is a fusion product with the chloramphenicol acetyl transferase (Cat-protein). The polypeptides of 50 000 molecular weight and 60 000 molecular weight are the gene products of *nifL* and *nifA* respectively. On the autoradiogram of Fig. 4.2(a) no polypeptides corresponding to *nifB* and *nifQ* could be identified. We interpret this result by assuming that transcription starting from the p_{Cm}-promoter is terminated beyond *nifA* by a transcriptional terminator t_{nifAL} and that the contiguous *nif* promoter p_{nifQB} is not functional. Therefore, *nifBgp* and *nifQgp* are not synthesized in the *E. coli* minicell system.

The assumption of a terminator beyond *nifAL* and a promoter to the left of *nifQB* on *Eco*RI fragment A could be confirmed by further experiments schematically shown in Fig. 4.2(b) and (c). It is known that *nifA* and *nifL* are regulatory genes (Dixon *et al.* 1981; Kennedy *et al.* 1981). The *nifAgp* is believed to represent an activator necessary for the transcription of *Klebsiella nif* operons. In contrast, *nifLgp* inhibits the action of the activator if conditions for nitrogen fixation, such as low oxygen pressure and absence of ammonia, are not fulfilled. Since the minicell experiments were carried out under normal atmospheric oxygen, it explains why the promoter P_{nifQB} could not be activated in the presence of *nifLgp*. In order to circumvent these difficulties we deleted the *nifL*-gene in the *Eco*RI fragment A. The derived fragment with, the *nifL* deletion, now allows the activation of the p_{nifQB} promoter: in the minicell system a new polypeptide of molecular weight 49 000 was identified (Fig. 4.2(b)). The coding region of this polypeptide maps next to *nifA*. We therefore assume that this polypeptide represents *nifBgp*. In these experiments, no polypeptide corresponding to the *nifQgp* could be identified by the minicell technique.

The model of gene regulation underlying this experiment is the following: the *nifA*-gene is transcribed from the constitutive p_{Cm} promoter and the *nifAgp* subsequently activates the *nif* promoter p_{nifQB} resulting in the expression of *nifB*. This model can be easily tested. It is known that the *nifAgp* is temperature sensitive (Zhu and Brill 1981). Therefore at elevated temperature, e.g. at 44 °C instead of a 30 °C, *nifAgp* should be unable to activate p_{nifQB}. In Fig. 4.2(c) the experimental results are shown. At 44 °C the *nifBgp* of 49 000 molecular weight is not found on the autoradiogram of the SDS polyacrylamide gel, thus supporting the suggested model of regulation.

4.4　ANALYSIS OF THE *nif*B TRANSCRIPTION UNIT

In order to measure transcription of *nif* operons, we made use of a system where a promoterless tetracycline resistance gene is fused by *in vitro* techniques to *nif* promoters, and thus under control of these *nif* promoters. *E. coli* cells harbouring such a modified Tc-gene are only tetracycline-resistant if the *nif* promoter is activated, e.g. in the presence of *nifAgp* at 30 °C. Details of the construction of plasmids carrying Tc-genes under the control of *nif* promoters are reported elsewhere (Pühler *et al.* 1983*b*; Pühler *et al.* 1983*c*). In Fig. 4.3 the

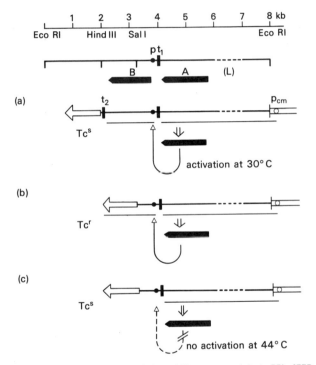

FIG. 4.3. Transcriptional analysis of the *nif*B operon. (a) A *Hind*III-*Eco*RI sub-fragment of fragment A (Fig. 4.1) carrying the genes *nif*B and *nif*A and a dele-tion in the *nif*L-gene was fused to a promoterless Tc-gene (⟵▭). The *nif*A-gene is expressed from a constitutive promoter (p_{Cm}) located on the vector plasmid. (b) A *Sal*I-*Eco*RI subfragment was cloned in front of a promoterless Tc-gene. Activation of the *nif*B promotor (p) results in the expression of the Tc-gene (Tc^r: tetracycline-resistant). (c) Inactivation of the *nifAgp* at 44 °C leads to a tetracycline-sensitive (Tc^s) plasmid phenotype. The resistance pat-terns can be explained by three transcriptional signals on the cloned fragments:　　t_1 : terminator of the *nif*AL-operon
　　　　　　　　t_2 : terminator of the *nif*B-gene
　　　　　　　　p : promoter of the *nif*B-gene

results of experiments using the Tc-gene under the control of the *nifB* promoter are summarized. An additional result of these experiments is that a transcriptional terminator (t_{nifB} designated t_2 in Fig. 4.3(a)) has to be postulated beyond the *nifB*-gene but before the *Hind*III site of fragment A. As shown in Fig. 4.3(b), removal of this terminator leads to expression of the Tc-gene when the *nifB* promoter is activated by *nifAgp*. This transcriptional activation is inhibited at elevated temperature due to the temperature sensitivity of the *nifAgp* (Fig. 4.3(c)). Interestingly, the *nifAgp* can also be supplied *in trans* from another plasmid (data not shown in this article).

From the results of this section it is evident that *nifB* is located on a monocistronic transcription unit which is controlled by a *nifA*-dependent promoter. What should be made clear is that our results do not support the existence of a *nifQ*-gene. First, we could not identify its gene product, and second, there is no space for another gene in the *nifB* transcription unit.

4.5 IDENTIFICATION OF FURTHER *nif*Agp-DEPENDENT PROMOTERS IN THE *K. pneumoniae nif*-REGION

In the previous sections we have demonstrated how *nifAgp* can activate the promoter of the *nifB* transcription unit. The Tc fusion technique has turned out to be an especially powerful tool in the analysis of the organization of *nif* operons. We therefore used this technique to detect further *nifAgp*-dependent promoters in the *Klebsiella nif*-region. We have already reported that a promoter to the right of the *nifH*-gene can be activated by *nifAgp* (Pühler *et al.* 1983*b*; Pühler *et al.*, 1983*c*). This is shown in Fig. 4.4. The *Hind*III-*Eco*RI fragment B of the *Klebsiella nif*-region (Fig. 4.1) has been cloned into pBR322, resulting in an operon fusion between Tc and p_{nifKDH}. The activation of this promoter by the *nifAgp* at 30 °C results in the expression of the Tc-gene. In addition, we now report that the location of p_{nifKDH} can be mapped by a series of Tn5 insertions. The expression of the Tc resistance can no longer be activated by *nifAgp* in plasmids carrying Tn5 inserted into the *nifH*, *nifD*, and *nifK* coding regions. This indicates that the transcription of the Tc-gene evidently starts from a promoter to the right of *nifH*.

By a very similar experiment, a further *nifAgp*-dependent promoter was found between the coding regions of *nifY* and *nifE* (Fig. 4.5). Again, the operon fusion technique was used to link promoterless Tc-genes to different restriction fragments of the *Klebsiella nif*-region. We first worked with the *Cla*I fragment C (Fig. 4.1)), carrying the coding regions of part of *nifK*, *nifY*, *nifE*, and part of *nifN*. The Tc-gene is not activated when Tn5 is inserted into *nifN* or *nifE*. In contrast, Tn5 insertions in *nifY* and *nifK* do not influence the activation of the Tc-gene.

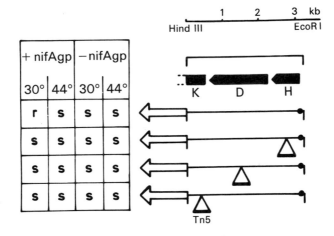

Fig. 4.4. Mapping of a *nifAgp*-dependent promoter which transcribes *nifK*, *nifD*, and *nifH*. A hybrid plasmid was constructed which carries a promoterless Tc-gene (◁▭) fused to the *Hind*III-*Eco*RI fragment B (Fig. 4.1) of the *Klebsiella nif*-region. Plasmids with Tn*5* insertions into *nifK*, *nifD*, and *nifH* were isolated and included in the activation experiments. The tetracyline resistance pattern (r = resistant, s = sensitive) for the different plasmids are listed The *nifA*-gene product (*nifAgp*) was supplied from another plasmid. The location of the indicated promoter (p) p*nif*KDH is the main result of the experiments reported in this figure.

From these findings a *nifAgp*-dependent promoter has to be postulated between the coding regions of *nifY* and *nifE*. In order to answer the question as to whether transcription of *nifX* can also be initiated from the *nifAgp*-dependent promoter to the right of *nifE*, the small *Eco*RI-*Pst*I fragment was fused to the Tc-gene. Subsequent enlargement of this fragment showed that the *nifX* coding region is also transcribed from the above-mapped promoter. Thus the main conclusion of Fig. 4.5 can be summarized as follows: the *nifAgp*-dependent promoter to the right of *nifE* transcribes the coding regions *nifX*, *nifN*, and *nifE*.

The next fragment which we used in our Tc fusion experiments was the *Cla*I-*Eco*RI fragment D (Fig. 4.1) carrying the coding regions *nifX*, *nifU*, and part of *nifS*. In this experiment, we could identify a *nifAgp*-dependent promoter between the coding region of *nifX* and *nifU* (Fig. 4.6). This result is based again on Tn*5* insertions. Tn*5* inserted into *nifX* did not inhibit expression but such insertions into *nifU* and *nifS* inhibited the expression of the Tc-gene. By using the larger *Eco*RI fragment E (Fig. 4.1), we were then able to demonstrate that transcription of *nifV* and *nifM* is initiated at the same *nifAgp*-dependent promoter. This finding is again supported by Tn*5* insertions into *nifV* and *nifM*. The results summarized in Fig. 4.6 are that *nifM*, *nifV*, *nifS*, and *nifU*

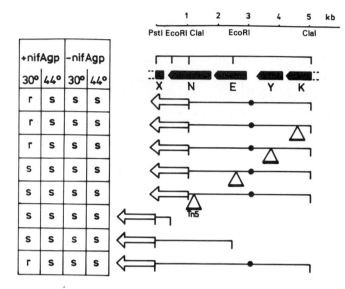

Fɪɢ. 4.5. Mapping of *nifAgp*-dependent promoter which transcribes *nifX*, *nifN*, and *nifE*. The *Cla*I fragment C (Fig. 4.1) was fused to a promoterless Tc-gene (⟵) of an *E. coli* vector plasmid. Specific Tn*5* insertions into the hybrid plasmid were isolated. They map in the following coding regions: *nifN*, *nifE*, *nifY*, and *nifK*. The Tc fusion was also carried out within the coding region of *nifX*. Using first a *Pst*I-*Eco*RI and subsequently enlarged fragments (see Figure) it could be shown that *nifX* is also expressed from the promoter between *nifE* and *nifY*. The conditions which result in tetracycline resistance are again listed (r: resistant, s: sensitive). The location of the *nifA*-dependent promoter p_{nifXNE} (*p*) is the main conclusion of the experiments reported in this figure.

belong to one transcriptional unit which is transcribed from a *nifAgp*-dependent promoter located between *nifX* and *nifU*.

4.6 CONCLUDING REMARKS

In transcriptional analysis of the *K. pneumoniae nif* region, four different *nifAgp*-depending promoters were identified. They are located in each case to the right of *nifB*, *nifH*, *nifE*, and *nifU* and are designated p⁺ (Fig. 4.7). The operons governed by these promoters were also analysed. From our data we conclude that the 16 *nif*-genes of *K. pneumoniae* are organized in 7 operons. The *nifALp* operon is the operon which regulates their transcription in concert with signals from the glutamine pathway (Ow and Ausubel 1983). *nifAgp* acts as an activator and *nifLgp* as an in-activator or repressor of nearly all the other *K. pneumoniae nif* operons. For this operon we were also able to map the transcriptional

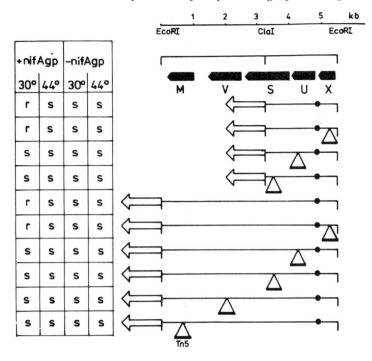

FIG. 4.6. Mapping of a *nifAgp*-dependent promoter which transcribes *nifM*, *nifV*, *nifS*, and *nifU*. The *Cla*I-*Eco*RI fragment D and the *Eco*RI fragment E (Fig. 4.1) were fused to a promoterless Tc-gene (⇐▭). Tn*5* insertions into *nifM*, *nifV*, *nifS*, *nifU*, and *nifX* were isolated. All the resulting plasmids were tested to determine whether they carried a Tc-gene which can be transcribed from a *nifAgp* dependent promoter. The results are listed (r: resistant, s: sensitive). The location of the indicated promoter ($p_{nifMVSU}$ designated p) is the main conclusion to the experiments reported in this figure.

terminator. The other interesting operon is *nifFp*+, which is the only operon to be transcribed in the opposite direction. The promoter of this operon is also *nifAgp*-dependent (data not shown in this article). The operon next to it contains the genes *nifM*, *nifV*, *nifS*, and *nifU* and shows the organization *nifMVSUp*+. This operon is followed by *nifXNEp*+ and *nifYKDHp*+. It should be mentioned that we do not have transcriptional data which clearly show that *nifY* belongs to this operon, but previous minicell experiments support this view. In the analysis of the *nifJp* operon we used different fusions with the Tc-gene, but were not able to demonstrate a *nifAgp*-dpendent promoter.

In summary, there are still some unanswered questions concerning the transcriptional organization of the *K. pneumoniae nif* region. In particular, very little is known about the transcriptional termination signals. These questions have to be answered before the complex regulation of the entire *Klebsiella nif*-region can be understood.

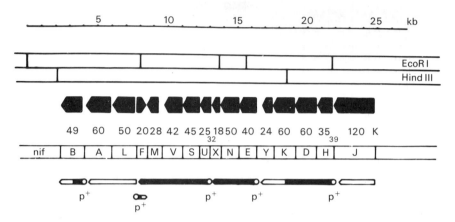

FIG. 4.7. Transcriptional organization of the *Klebsiella nif*-region. In comparison to the *Eco*RI-*Hin*dIII restriction map, the coding regions (big black arrows) and the molecular weights of *nif*-gene products are given. Proteins showing two bands on SDS polyacrylamide gels are indicated with both molecular weights. Promoters which are dependent on the presence of *nifA*-gene product are designated p^+ The direction of transcription is marked by arrows. Regions where *nifAgp*-dependent transcription could be demonstrated are shown in black.

REFERENCES

CHANG, A. C. Y. and COHEN, S. N. (1978). *J. Bact.* **134**, 1141–56.

DIXON, R., KENNEDY, C., and MERRICK, M. (1981). In *Genetics as a tool in microbiology* (ed. S. W. Glover and D. A. Hopwood), pp. 161–85. Cambridge University Press.

KENNEDY, C., CANNON, F., CANNON, M., DIXON, R., HILL, S., JENSEN, J., KUMAR, S., MCLEAN, P., MERRICK. M., ROBSON, R., and POSTGATE, J. (1981). In *Current perspectives in nitrogen fixation* (ed. A. H. Gibson and W. E. Newton), pp. 146–56. Australian Academy of Science, Canberra.

PÜHLER, A., BURKARDT, H. J., and KLIPP, W. (1979*a*). *Molec. & gen. Genet.* **176**, 17–24.

——— (1979*b*). In *Plasmids of medical, environmental and commercial importance* (ed. K. Timmis and A. Pühler), pp. 435–48. Elsevier/North-Holland Biomedical Press.

— KLIPP, W. (1981). In *Biology of inorganic nitrogen and sulfur* (ed. H. Bothe and A. Trebst), pp. 276–86. Springer-Verlag, Heidelberg.

——— (1983*a*). In *Nitrogen fixation: The chemical–biochemical–genetics interface* (ed. Müller and Newton), pp. 111–34. Plenum Press, London.

—— WEBER, G. (1983*b*). *Proc.   4th Int. Symp. Genet. Industrial Micro-organisms*, pp. 320-7.

——— (1983*c*). In *Proc. 1st Int. Symp. Molec. Genet. Bacteria–Plant Interactions* (ed. A. Pühler). Springer-Verlag, Heidelberg.

Ow, D. W. and AUSUBEL, F. M. (1983). *Nature, Lond.* **301**, 307–13.

ZHU, J. and BRILL, W. J. (1981). *J. Bact.* **145**, 116–18.

5 Azospirillum

Claudine Elmerich

5.1 INTRODUCTION

In 1922, Beijerinck described a new bacterium, which he called first *Azotobacter spirillum* and later *Spirillum lipoferum* (see Krieg 1976). Becking (1963) rediscovered the bacterium, whose ability to fix nitrogen in pure culture was established by the ^{15}N isotopic method. However, the potential agronomic importance of the bacterium was raised when Döbereiner and Day (1976) reported its association with the roots of grasses from various geographical origins. *S. lipoferum* isolates were aerobic gram-negative bacteria, curved rod shaped, with a polar flagellum and containing globules of poly-β-hydroxybutyrate. Krieg (1977) observed that the DNA base composition of 11 *S. lipoferum* isolates was 69–71 moles per cent G + C, a value much higher than the average 38 per cent found in the members of the *Spirillum* genus, and he suggested the generic name *Azospirillum*. In addition, *S. lipoferum* isolates were divided on the basis of nutritional properties into two groups (Okon *et al.* 1976*b*) or even three groups (Neyra *et al.* 1977). DNA homology was consistent with the existence of two groups (Krieg 1977). Consequently, Tarrand *et al.* (1978) defined the *Azospirillum* genus and two species *A. brasilense* and *A. lipoferum* which corresponded roughly to the two groups previously described by Okon *et al.* (1976*b*). In 1983, new isolates corresponding to a third species *A. amazonense* were discovered (Magalhaes *et al.* 1983).

5.2 NUTRITIONAL PROPERTIES

As defined by Tarrand *et al.* (1978), *A. brasilense* and *A. lipoferum* can grow on organic acids such as malate, succinate, lactate, or pyruvate. Strains of *A. lipoferum* can utilize a large number of carbohydrates including glucose, mannitol, sorbitol, etc. which are not used by *A. brasilense*, though some strains grow on glucose. *A. brasilense* is prototroph whereas *A. lipoferum* requires biotin. Colonies at old stages of growth develop a pink red pigment and some strains are naturally highly pigmented in young cultures (Eskew *et al.* 1977; Tarrand *et al.* 1978; Nur *et al.* 1980). Pigmentation was shown to be due to carotenoids (Nur *et al.* 1981). No amino acid can be used both as the only carbon and nitrogen source (Gauthier 1978). In the presence of maltate and under aerobic conditions, the nitrogen sources used are ammonia (Okon *et al.* 1976*a*), nitrate (Neyra and Van Berkum 1977), glutamate, glutamine and amino acids of the same family (Asp, Arg, Pro, Lys), histidine and purines (Bani *et al.* 1980). Dinitrogen can be used only under micro-aerobic conditions (Von Bülow and Döbereiner 1975; Okon *et al.* 1976*a,b*).

Denitrification under anaerobic conditions was reported (Krieg 1977; Eskew *et al.* 1977; Neyra *et al.* 1977). It appeared later that all *A. lipoferum* strains were capable of denitrification, whereas *A. brasilense* strains could be divided into two groups, one containing denitrifying organisms, the other containing strains able to dissimilate NO_3^- to NO_2^- but not to produce gas from NO_2^- (Neyra *et al.* 1977; Döbereiner and De Polli 1980).

Studies of hydrogen metabolism in *Azospirillum* revealed the existence of an H_2-uptake hydrogenase activity (Chan *et al.* 1980; Berlier and Lespinat 1980; Yates *et al.* 1981; Volpon *et al.* 1981; Pedrosa *et al.* 1982). This activity was extremely oxygen sensitive, it could support H_2-dependent acetylene reduction by whole cells but only under conditions of carbon-source starvation (Yates *et al.* 1981; Pedrosa *et al.* 1982). Several other aerobic nitrogen-fixing organisms with H_2-uptake hydrogenases were shown to grow autotrophically (Hanus *et al.* 1979; Pedrosa *et al.* 1980). Autotrophy and methylotrophy were also demonstrated in *Azospirillum* (Sampaio *et al.* 1981). Malik and Schlegel (1981) reported autotrophic growth for all *A. lipoferum* strains tested, but none of the *A. brasilense* strains were able to grow under autotrophic conditions.

5.3 NITROGEN METABOLISM

As pointed out by Neyra *et al.* (1977) and Bothe *et al.* (1981), *Azospirillum* spp participate in all steps of the nitrogen cycle except nitrification.

5.3.1 Nitrogen fixation

In nitrogen-free semi-solid media *Azospirillum* behaves as a typical micro-aerobic organism and forms a pellicle below the surface, which moves depending on the oxygen availability (Döbereiner and Day 1976; Barak *et al.* 1982). In liquid cultures analysed in fermenters, optimal dissolved oxygen tensions between 0.003 and 0.007 atm were reported for growth under conditions of nitrogen fixation and the doubling time was 5 to 7 h (Okon *et al.* 1976*b*, 1977; Ahmad 1978; Volpon *et al.* 1981; Gauthier 1978). No respiratory protective mechanism of the nitrogenase activity, as observed in *Azotobacter* (Drozd and Postgate 1970), was detected when the O_2 tension was increased (Nelson and Knowles 1978; Volpon *et al.* 1981). By growing *A. brasilense* strain Sp7 in a fermentor, under nitrogen-fixing conditions, we observed that the specific activity of the nitrogenase complex, assayed in whole cells by the acetylene-reduction test, decreased before the end of the exponential phase. This suggested that either the synthesis was stopped or that the enzyme was inactivated. We therefore looked for the excretion of nitrogenous compounds in the growth medium. Using the very sensitive method of Chaykin (1969), ammonia up to 0.3 mM, was detected in the culture supernatant, as shown in Fig. 5.1(a). The peak of ammonia excretion was transitory (half a doubling time) and apparently ammonia was re-utilized for growth (Gauthier 1978). Excretion of nitrogenous compounds in the growth medium was also reported for strain Br17 (Volpon *et al.* 1981). Nitrogenase biosynthesis is repressed by low amounts of fixed nitrogen (see Fig. 5.1(b) such as 0.5 mM NH_4^+, 1 mM NO^{-3} (Döbereiner and De Polli 1980) and 50 µg ml^{-1} glutamine (Gauthier 1978). No repression was observed when glutamate was used as the nitrogen source (Pedrosa and Yates 1984). In the presence of 0.25 per cent NH_4Cl, methionine sulfoximine (20 mg ml^{-1}) de-repressed nitrogenase biosynthesis (Okon *et al.* 1976*a*) as previously shown for other diazotrophs (Gordon and Brill 1974). In addition, the enzyme is subject to an *in vivo* ammonia or glutamine switch off (Fig. 5.1(c)) (P. Jara, this laboratory, unpublished) as demonstrated for photosynthetic bacteria (Zumft and Castillo 1978; Meyer *et al.* 1978; Zumft *et al.* 1981).

Specific nitrogenase activity, assayed with whole cells during growth in fermentors or under derepression conditions, ranged from 70 to 200 nmoles C_2H_4 min^{-1}mg protein^{-1}. Values obtained with crude extracts were much lower, 8 nmoles C_2H_4 min^{-1}mg protein^{-1}, a difference most probably due to inactivation of the Fe-protein during extraction (Okon *et al.* 1977; C. Elmerich, unpublished). Okon *et al.* (1977) observed a decrease of activity of the extracts upon storage at -18 °C, but activity was restored by addition of pure Fe-protein from *Azotobacter vinelandii* or *Rhodospirillum rubrum*. Nitrogenase components were purified from nitrogen fixing culture of *A. brasilense* Sp7 (Okon *et al.* 1977; Ludden *et al.* 1978; Pedrosa and Yates 1984). Properties of the complex were similar to that of the *Rhodospirillum rubrum* enzyme (Ludden and Burris 1976). The activity was increased by Mn^{++} and by an activating factor purified from NH_4^+ grown cells (Okon *et al.* 1977; Ludden *et al.* 1978). Using antisera prepared against *K. pneumoniae*

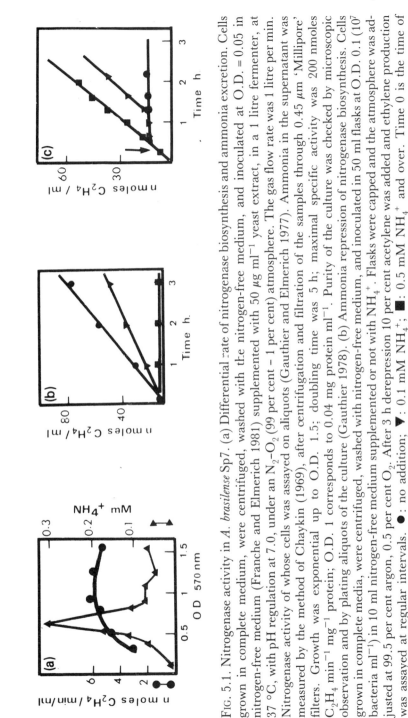

FIG. 5.1. Nitrogenase activity in *A. brasilense* Sp7. (a) Differential rate of nitrogenase biosynthesis and ammonia excretion. Cells grown in complete medium, were centrifuged, washed with the nitrogen-free medium, and inoculated at O.D. = 0.05 in nitrogen-free medium (Franche and Elmerich 1981) supplemented with 50 μg ml^{-1} yeast extract, in a 1 litre fermenter, at 37 °C, with pH regulation at 7.0, under an N_2–O_2 (99 per cent – 1 per cent) atmosphere. The gas flow rate was 1 litre per min. Nitrogenase activity of whose cells was assayed on aliquots (Gauthier and Elmerich 1977). Ammonia in the supernatant was measured by the method of Chaykin (1969), after centrifugation and filtration of the samples through 0.45 μm 'Millipore' filters. Growth was exponential up to O.D. 1.5; doubling time was 5 h; maximal specific activity was 200 nmoles C_2H_4 min^{-1} mg^{-1} protein; O.D. 1 corresponds to 0.04 mg protein ml^{-1}. Purity of the culture was checked by microscopic observation and by plating aliquots of the culture (Gauthier 1978). (b) Ammonia repression of nitrogenase biosynthesis. Cells grown in complete media, were centrifuged, washed with nitrogen-free medium, and inoculated in 50 ml flasks at O.D. 0.1 (10^7 bacteria ml^{-1}) in 10 ml nitrogen-free medium supplemented or not with NH_4^+. Flasks were capped and the atmosphere was adjusted at 99.5 per cent argon, 0.5 per cent O_2. After 3 h derepression 10 per cent acetylene was added and ethylene production was assayed at regular intervals. ● : no addition; ▼ : 0.1 mM NH_4^+; ■ : 0.5 mM NH_4^+ and over. Time 0 is the time of acetylene addition (this laboratory, unpublished). (c) Ammonia inhibition of nitrogenase activity. The same protocol as in Fig. 5.1(b), was used but NH_4^+ was added 20 min after acetylene addition as indicated by the arrow. ■ = no addition; ▼ = 0.1 mM NH_4^+; ● = 1 mM NH_4^+ (P. Jara, unpublished result).

MoFe-protein and Fe-protein, cross-reacting material was precipitated from nitrogen-fixing cultures of *A. brasilense* Sp7. Molecular weights of the polypeptides after SDS gel electrophoresis were estimated at 60 000 and 64 000 for the MoFe-protein and 33 000 and 36 000 for the Fe-protein (Nair *et al.* 1983). The polypeptides were not detected from ammonia or air-grown cultures. The two bands observed for the Fe-protein might correspond to the presence of active and covalently-modified inactive polypeptides in the extracts.

5.3.2 Ammonia assimilation

Ammonia uptake in ammonia-limited culture was described as an energy-dependent mechanism (Hartmann and Kleiner 1982). This result includes *Azospirillum* spp amongst the bacteria that possess a specific ammonia carrier (Kleiner 1981).

Azospirillum contains a glutamine synthetase with features similar to the *E. coli* enzyme (Ginsburg and Stadtman 1973). Its biosynthesis is repressed (five- to tenfold) by high concentrations of ammonia in the growth medium, and its activity is modulated by adenylylation (Okon *et al.* 1976a; Gauthier and Elmerich 1977; Bani *et al.* 1980). A second glutamine synthetase was described in *Rhizobiaceae* (Darrow and Knotts 1977) but it was not detected in *Azospirillum* (Fuchs and Keister 1980). A NADP dependent glutamate synthase and a NAD glutamate dehydrogenase were also characterized (Okon *et al.* 1976a; Bani *et al.* 1980; Gauthier 1978). The presence of an alanine dehydrogenase was also mentioned (Papen and Werner 1982). Mutants impaired in glutamine synthetase activity (Gauthier and Elmerich 1977) and in glutamate synthase activity (Bani *et al.* 1980) were isolated. It is likely that both enzymes are responsible for NH_4^+ assimilation under nitrogen-limiting conditions as previously demonstrated for other diazotrophs (Nagatini *et al.* 1971). The pathway of NH_4^+ assimilation under non-limiting conditions was not established. Bani *et al.* (1980) proposed, according to the behaviour of glutamate-synthase-deficient strains, that glutamate dehydrogenase was the major route. As no glutamate dehydrogenase mutant has yet been isolated, this assumption requires confirmation. Gauthier (1978) found an apparent K_m for ammonia higher than 30 mM which suggests a dissimilatory role for the enzyme, as is the case for most bacterial NAD-dependent glutamate dehydrogenases (Tempest *et al.* 1973).

5.3.3 Regulation of nitrogen fixation and of ammonia assimilation

Mutants deficient in glutamine synthetase and glutamate synthase are impaired in nitrogen fixation (Gauthier and Elmerich 1977; Bani *et al.*

1980). In particular, Nif⁻ and Nif^c (which fix nitrogen in the presence of NH_4^+) phenotypes were observed among the glutamine auxotrophs (Gauthier and Elmerich 1977). These phenotypes were also described in *Klebsiella pneumoniae* mutants (see, e.g., Streicher *et al.* 1974; Leonardo and Goldberg 1980; Espin *et al.* 1981). The glutamate synthase-deficient mutants, which also displayed an Asm⁻ phenotype since they could not utilize a series of nitrogen sources (Nagatini *et al.* 1971), were unable to grow on nitrogen-free medium (Bani *et al.* 1980). None of the *Azospirillum* mutants was genetically or biochemically characterized as impaired in the structural gene for the corresponding enzyme. Thus, the hypothesis suggesting the involvement of glutamine synthetase or glutamate synthase in *nif*-gene regulation should be treated with caution. In *Enterobacteriaceae*, regulation of the assimilation of nitrogen compounds is very sophisticated (see Merrick 1983 for a review). At least three regulatory genes are involved: *glnF*, *glnL*, and *glnG* (also designated *ntrA*, *ntrB*, *ntrC*) (see, e.g., Garcia *et al.* 1977; McFarland *et al.* 1981; Espin *et al.* 1982; Sibold and Elmerich 1982). Recently, Pedrosa and Yates (1984) described *Azospirillum* mutants that behave like *ntrC* mutants of *K. pneumoniae*, and whose nitrogenase activity was restored by a *glnAntrBC* episome of *K. pneumoniae*. Thus, it is likely that in *Azospirillum* nitrogen fixation and ammonia assimilation is regulated through a *ntr* like control.

5.3.4 Nitrate assimilation and dissimilation

All *Azospirillum* spp can utilize nitrate as the only nitrogen source under aerobic conditions (Neyra and Van Berkum 1977; Tarrand *et al.* 1978, reviewed by Döbereiner and De Polli 1980). More recently, it also became apparent that denitrifying strains, too, could use NO_3^- under anaerobic conditions as the only nitrogen source, as well as an electro acceptor (Bothe *et al.* 1981). Under anaerobic conditions, the suggestion that nitrate respiration can support nitrogen fixation was controversial (Neyra and Van Berkum 1977; Nelson and Knowles 1978). It was finally established that nitrate repressed nitrogenase biosynthesis, but that nitrogen fixation could occur for a short period before assimilation of ammonia resulting from nitrate reduction (Bothe *et al.* 1981). Mutants isolated as chlorate resistant clones were impaired in nitrate reductase or nitrite reductase activity, or both, and they could not dissimilate or assimilate NO_3^- under anaerobic or aerobic conditions (Magalhaes *et al.* 1978). Furthermore, the mutants which contained a wild-type nitrogenase activity (Magalhaes *et al.* 1978) were unable to fix nitrogen in anaerobiosis when NO_3^- was the electron acceptor (Scott *et al.* 1979).

5.4 PLASMIDS AND PHAGES

5.4.1 Plasmid content

All *Azospirillum* strains examined contain at least one plasmid and some contain as many as six molecular species, the size of which ranges from 4 Mdal to over 300 Mdal (Franche and Elmerich 1981; Heulin *et al.* 1982; Singh and Wenzel 1982; Wood *et al.* 1982; Plazinsky *et al.* 1983). Plasmids were characterized by electrophoretic migration, in agarose gels, of cell lysates obtained by various techniques such as those described by Meyers *et al.* (1976), Eckhardt (1978), and Casse *et al.* (1979). An example of plasmids detected in a few *Azospirillum* strains is shown in Fig. 5.2(a). Discrepancies between the number and the size of the plasmids carried by a given strain were observed. For example, Singh and Wenzel (1982) found three plasmids of 5.4, 12, and 51 kb in strain *A. lipoferum* Br17 (ATCC29709) whereas Franche and Elmerich (1981) found a single large plasmid of about 200 Mdal in the same strain. It is not clear whether this was due to instability of the plasmids, to technical artefacts, or strain mis-naming.

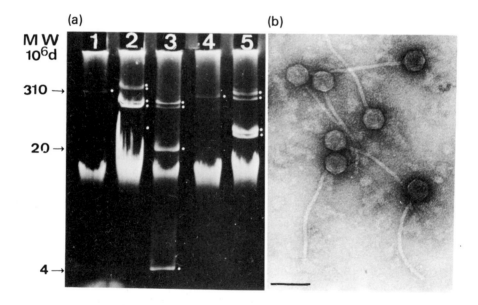

FIG. 5.2. Plasmids and bacteriophage. (a) Plasmid content of *Azospirillum* strains. Plasmids were extracted according to the methods of Casse *et al.* (1979). 1 *Pseudomonas putida* contained a 310 Mdal plasmid. 2: *A. brasilense* Sp7; 3: *A. brasilense* RO7; 4: *A. lipoferum* Br17; 5: *A. lipoferum* S28 [see Franche and Elmerich (1981) for details]. (b) Electron micrograph of bacteriophage Al-1 particles; the bar represents 50 nm (see Elmerich *et al.* (1982) for details).

The two taxonomic groups, *A. brasilense* and *A. lipoferum*, cannot be differentiated on the basis of their plasmid content. Moreover, there is no indication that plasmids with the same apparent molecular weight correspond to the same molecular species. Some plasmids of relatively low molecular weight were purified by CsCl ethidium bromide gradient and restriction analysis was performed (Franche and Elmerich 1981; Singh and Wenzel 1982).

5.4.2 Phenotypes associated to plasmids

No evidence of self-transmissibility of the plasmids was reported. Spontaneous loss (Franche and Elmerich 1981) and temperature or acridine-orange-induced curing (Heulin *et al.* 1982; Wood *et al.* 1982) of some plasmids have been described. Phenotypic changes concomitant with plasmid loss were reported (Heulin *et al.* 1982; Wood *et al.* 1982). However, no specific feature was demonstrated as plasmid-borne. Nevertheless, possible functions for plasmids can be suggested such as: carbon-source utilization, heavy metals resistance, plant hormone production, plant host specificity or recognition, nitrogen fixation, etc.

5.4.3 Introduction of *IncP* plasmids into *Azospirillum*

Plasmids of the IncP group are broad-host-range plasmids (Datta and Hedges 1971) and they can be introduced into *Azospirillum*. Kinetics of transfer of RP4 and R68-45 (Haas and Holloway 1978) suggested that the transfer was maximal after 6 to 8 h mating (10^{-2} per donor) (Franche *et al.* 1981). IncP-plasmids containing Mu were also introduced but the phenomenon of suicide observed in *Rhizobium* (Beringer *et al.* 1978) was not detected (Franche 1981; Elmerich and Franche 1982; Singh 1982). The pRD1 plasmid which contains the *nif*-genes of *K. pneumoniae* (Dixon *et al.* 1976) was transferred to *Azospirillum* at a low frequency and it was found to be stable in the recipient (Polsinelli *et al.* 1980). Introduction of IncP-plasmids into *Azospirillum* did not modify the resident plasmid pattern at least in the case of *A. brasilense* strain Sp7 (Franche 1981).

5.4.4 Lysogenic state

Mitomycin C-induced lysis was observed with most *Azospirillum* spp. *A. brasilense* Sp7 was sensitive, while *A. lipoferum* Br17 was resistant (Franche and Elmerich 1981; Elmerich *et al.* 1982). Conditions of lysis were studied in detail for *A. brasilense* Sp7. Lysis occurred for drug concentrations ranging from 0.1 to 1 μg ml^{-1} whatever the growth conditions. Attempts to obtain plaques or growth inhibition by spotting lysates onto lawns of each of the other mitomycin-C-sensitive or-resistant strains were unsuccessful. This suggested that most strains were lysogen for a defective prophage. Phage-like particles were purified from strain Sp7 lysates and were examined by electron microscopy. They appeared as icosahedric particles devoid of DNA (Franche and Elmerich 1981).

5.4.5 Plaque-forming bacteriophages

Phages titrating *Azospirillum* have been isolated in Brazil. A temperate phage, forming plaques on *A. lipoferum* Br17, was studied (Elmerich *et al.* 1982). The phage isolated in J. Döbereiner's laboratory, was named Al-1. Its morphology and size are similar to that of the coliphage λ (Fig. 5.2(b)). However, it carries five to six spikes at the distal end of the tail. It contains a double-stranded linear DNA molecule of 36 kb with cohesive ends. Stable lysogens were obtained and the prophage DNA was found to be maintained as a plasmid. This is a new example of extrachromosomal lysogeny (Ideka and Tomizawa 1968). Until now, the phage host-range was limited to a few strains of *A. brasilense* and *A. lipoferum* of Brazilian origin. In particular, strain Sp7 is resistant to this phage. No relationship was established between the hypothetical lysogenic state of the hosts and the sensitivity to Al-1. No transducing property of Al-1 has yet been detected.

5.5 GENETICS

5.5.1 Isolation of mutants

Classical techniques of mutagenesis previously described in *E. coli* (Miller 1972) such as ultraviolet irradiation or treatment by ethyl-methane sulfonate (EMS) or N-methyl-N'-nitro-N-nitrosoguanidine (NTG) followed by penicillin or D-cycloserine enrichment can be successfully applied to *Azospirillum*.

Isolation of mutants impaired in various functions was reported. Auxotrophs were obtained from at least four different strains: *A. brasilense* Sp7 (Franche *et al.* 1981), *A. brasilense* 13t (Wood *et al.* 1982), *A. brasilense* Sp6 (Polsinelli *et al.* 1980), *A. lipoferum* Br17 (C. Elmerich, unpublished). When an enrichment procedure for auxotrophs is used the frequency of mutant isolation is roughly 1 to 5×10^{-3}. Antibiotic-resistant mutants can be isolated without mutagenesis (Mishra *et al.* 1979). However, the spontaneous rate is very low and mutagenesis is often required. Mutants of the nitrogen metabolism (see Sections 5.3 and 5.6) were obtained. In addition, antimetabolite-resistant mutants (Hartmann 1982; Barberio *et al.* 1982; Hartmann *et al.* 1983) have been described.

5.5.2 Transposon mutagenesis

Transposon mutagenesis with Tn5, which codes for kanamycine resistance (Km^R), was investigated. The suicide plasmid pJB4JI (IncP,

Mu,Gm,Tn5) used by Beringer *et al.* (1978) to introduce Tn5 in the chromosome of *Rhizobium* was found to be very stable in *A. brasilense* and no spontaneous Tn5 transposition was detected in strains ATCC 29710 (Singh 1982) and Sp7 (Elmerich and Franche 1982). Elmerich and Franche (1982) reported a method of isolation of Tn5-containing mutants that was based on the incompatibility between pJB4JI and pJB3JI (IncP,Tc,Cb). In their method the molecular events which led to the transposition in *Azospirillum* DNA are rather obscure, and it was not demonstrated that the auxotrophic phenotypes were due to a physical insertion of the transposon. An alternative procedure was developed with the use of mobilizable plasmids, containing Tn5, such as pSUP2021 constructed by Simon *et al.* (1983). This plasmid cannot replicate outside the enteric bacteria. It was introduced in *Azospirillum* Sp7 to perform random mutagenesis. The frequency of Tn5 transposition was 10^{-7} and two per cent auxotrophs were found among the Km^R mutants (H. Bozouklian, this laboratory, unpublished).

5.5.3 Genetic exchange

Phages and plasmids of *Azospirillum* cannot be used, as yet, for genetic exchange. Mishra *et al.* (1979) reported DNA-mediated transformation in *A. brasilense* strain Sp7, but until now this technique has not been used to establish genetic linkage. Plasmid R68-45 (Haas and Holloway 1978) can promote chromosome mobilization in various gram-negative bacteria including *Rhizobium* (see, e.g., Kondorosi *et al.* 1980). Similar properties were observed in *A. brasilense* strain Sp7 (Franche *et al.* 1981). Crosses were performed with a series of multiple auxotrophs. The frequency of transfer was about 10^{-6} per recipient, whatever the marker tested. The ratio between recombinants and transconjugants which received R68-45 was always close to 10^{-5} and inheritance of the plasmid was observed in 90 per cent of the recombinants. Apparently no stable R68-45 episome was formed. The marker transfer promoted by R68-45 appeared to be not polarized which suggested the existence of multiple origins of transfer in *Azospirillum* as has been previously reported in *Pseudomonas aeruginosa* (Haas and Holloway 1978). Linkage data between several pairs of markers was established (Franche *et al.* 1981). Similar results were obtained with strain *A. brasilense* Sp6 (Balzicalupo and Gallori 1983).

5.5.4 Construction of partial diploids

By using suitable vectors it is now possible to construct partial diploids in a series of gram-negative bacteria. Plasmids such as RK_2, RP4, RSF1010, or their derivatives, with broad-host-range, can be used as vectors (Ditta *et al.* 1980; Bagdasarian *et al.* 1981). Ditta *et al.* (1980) constructed the pRK290 plasmid which was used to obtain a gene bank of *Rhizobium meliloti*. This vector is a derivative of RK_2 which confers resistance to tetracycline and contains single *Eco*RI and *Bgl*II sites for cloning. It is not self-transmissible, but it can be transferred with the aid of plasmid-pRK2013 (Ditta *et al*. 1980). Transfer of pRK290, using the same methodology, was reproduced in *Azospirillum*. In addition, a *nif* fragment was cloned in the plasmid and *nif* partial diploids were constructed which appeared to be stable (Jara *et al*. 1983). Recently a gene bank of *Azospirillum* Sp7 DNA fragments was constructed in the cosmid vector pVK100 (Knauf and Nester 1982). The gene bank was used to clone the *glnA* gene (which codes for the glutamine synthetase subunit) by complementation of strain 7029 (Gauthier and Elmerich 1977), a glutamine auxotroph of Sp7 (C. Fogher, this laboratory, unpublished).

5.6 GENETICS OF NITROGEN FIXATION

5.6.1 Homology with *Klebsiella pneumoniae* nitrogen fixation (*nif*) genes

In *K. pneumoniae*, 17 *nif*-genes organized in seven transcriptional units have been identified (see, e.g., MacNeil *et al*. 1978; Merrick *et al*. 1980; Pühler and Klipp 1981; Sibold 1982). *NifHDK* are the structural genes for the nitrogenase complex polypeptides; *nifH* codes for the single subunit of component 2 (Fe-protein) and *nifDK* for the two subunits of component 1 (MoFe-protein). The three genes and the *nifY*-gene (Pühler and Klipp 1981) are transcribed in the order *nifHDKY*. This operon and a part of *nifE* are carried by a 6.2 kb *Eco*RI fragment which was cloned in plasmid pSA30 (Cannon *et al*. 1979). Using this fragment as a hybridization probe, homology was detected with total DNA from a large number of diazotrophs (Ruvkun and Ausubel 1980; Mazur *et al*. 1980) including several strains of *Azospirillum* (Elmerich and Franche 1982; Quiviger *et al*. 1982). Two *A. lipoferum* strains and three *A. brasilense* strains were examined and the size of the homologous *Eco*RI and *Hind*III fragments were different from one strain to another (Quiviger *et al*. 1982). With *A. brasilense* strain Sp7, the probe hybridized with a single 6.7 kb *Eco*RI fragment which was contained within a 24 kb *Hind*III fragment and a 22 kb *Bam*HI fragment. With *A. lipoferum* strain Br17 homology was found with two *Eco*RI fragments of 16 and 1.8 kb and with a single 15 kb *Hind*III fragment (Quiviger *et al*. 1982). In the case of strains Sp7 and Br17 homology with a *nifA* probe was also detected (Nair *et al*. 1983).

5.6.2 Cloning of a *nif* cluster of *A. brasilense* Sp7

The 6.7 kb *Eco*RI fragment was cloned in the λ gt7-ara6 vector, then subcloned in plasmid vectors pACYC184 (Quiviger *et al*. 1982) and pRK290 (Jara *et al.* 1983) to yield plasmids pAB1 and pAB35 respectively. The physical map of the plasmids was established (see Fig. 5.3) and derivatives containing *in-vitro*-generated deletions were constructed (Jara *et al.* 1983). Heteroduplex analysis of phages containing the *K. pneumoniae* or *Azospirillum nif* DNA was performed. Homology on a sequence approximately 5 kb long was detected. This was consistent with the existence of a complete *nifHDK*-gene cluster in *Azospirillum*. Moreover, the direction of transcription could be determined from the heteroduplex analysis (from the left to the right in Fig. 5.3). Attempts to identify products encoded by pAB1-plasmid, after segregation in *E. coli* minicells were laborious. Only very faint polypeptide bands could be detected as compared to the result obtained when plasmid pSA30 (where the transcription proceeds from the *cat* promoter) was used (Quiviger *et al.* 1982). It was suggested that the weak expression observed might be due to the lack of a specific *nif*-gene activator.

5.6.3 Physical localization of the *nif* genes

Plasmid location of *nifHD*-genes was demonstrated in several *Rhizobium* spp (see, e.g., Bánfalvi *et al.* 1981; Hombrecher *et al.* 1981; Prakash *et al.* 1981; Rosenberg *et al.* 1981). Using plasmids pSA30 or pAB1 no homology with plasmid DNA of several *Azospirillum* strains (including Sp7 and Br17) was found (C. Elmerich, unpublished).

5.6.4 Characterization of *nif* mutants.

Attempts to isolate Nif⁻ mutants have been carried out in several laboratories including this one. As reported by Bani *et al.* (1980), most of the mutants isolated as non- or poorly-growing on nitrogen-free

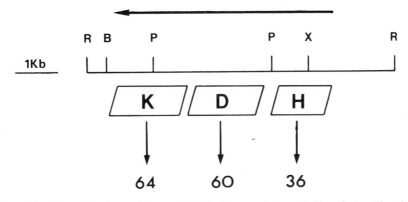

FIG. 5.3. Organization of the *nifHDK* cluster of *Azospirillum* Sp7. The three genes are organized in a single transcription unit transcribed from *nifH* to *nifK*. The molecular weight of the *nif* polypeptides is indicated in kb. Restriction sites : B : *Bgl*II; P : *Pst*I; R : *Eco*RI; X : *Xho*I.

medium had an Asm⁻ phenotype. The other mutants isolated after chemical mutagenesis fall into two classes: regulatory mutants and nitrogenase mutants. Pedrosa and Yates (1984) reported the isolation of Nif⁻ mutants with *nifA* and *ntrC* phenotypes. The *ntrC*-like mutant was complemented by plasmids pGE10 and pCK3, which carry the *glnAntrBC* regulon and the *nifA*-gene of *K. pneumoniae* respectively. The *nifA*-like mutant was complemented by pCK3 only. This is in favour of regulatory mechanisms related to those existing in *K. pneumoniae*. Two nitrogenase mutants impaired in the MoFe-protein were described (Jara *et al.* 1983; Pedrosa and Yates 1984). One of them, strain 7571, was characterized genetically by complementation with pAB35, which contain the wild type *nifHDK* cluster, and by lack of complementation by pAB36, which is depleted of *nifD* and part of *nifK* but still contains *nifH* (Jara *et al.* 1983). The mutant was characterized biochemically by complementation of crude extract with pure MoFe-protein of *K. pneumoniae* (Nair *et al.* 1983).

By directed mutagenesis of the 6.7 kb *Eco*RI fragment carrying the *nifHDK* cluster of strain Sp7, using the procedure of Simon *et al.* (1983), Tn*5* insertions were obtained in the parent strain in *nifH*, *nifD*, and *nifK* respectively. Insertions in *nifH* were polar on *nifD* and *nifK*, as determined by genetic complementation and by gene-product analysis (B. Perroud, this laboratory, unpublished).

5.7 ASSOCIATION WITH PLANTS

Azospirillum spp were isolated from the rhizosphere of a large number of monocotyledon and of some dicotyledon plants from tropical, temperate, or cold climates (see, e.g., Von Bülow and Döbereiner 1975; Döbereiner and Day 1976; Döbereiner *et al.* 1976; Kumari *et al.* 1976; Tarrand *et al.* 1978; Nur *et al.* 1980; Haahtela *et al.* 1981; Rao and Venakateswarlu 1982; see also reviews by Döbereiner and De Polli 1980; Van Berkum and Bohlool 1980; Patriquin 1982).

5.7.1 Site and process of the association

In most cases, the bacteria were isolated after surface sterilization of the roots. Thus, it appeared that *Azospirillum* could invade the cortical and vascular tissues of the host (Döbereiner and Day 1976; Kumari *et al.* 1976; Patriquin and Döbereiner 1978). Studies on adsorption of *A. brasilense* strains to roots of *Pennisetum americanum* (pearl millet) revealed a preferential colonization by this species as compared to other genus (Umali-Garcia *et al.* 1980). Bacteria were located mostly in the mucigel. In addition, protease-sensitive non dialyzable-substances were isolated from the root exudates. These substances promoted association of bacteria to root hairs and might be considered as 'lectin-like' compounds (Umali-Garcia *et al.* 1980). Invasion of root tissues suggested the production of bacterial enzymes with pectinolytic activity (Umali-Garcia *et al.* 1980). A polygalacturonic acid transeliminase activity was characterized from several *Azospirillum* spp. including *A. brasilense* Sp7 (Tien *et al.* 1981). The activity was detected in culture supernatants and was found also associated with the outside of the bacterial cell wall. Polygalacturonic acid and pectin were preferentially degraded and the two compounds were inducers of the enzyme synthesis. However, the activity detected was much lower than that observed with plant pathogens such as *Erwinia* (Tien *et al.* 1981).

5.7.2 Phytohormone production

The effect of *Azospirillum* inoculation on the growth of plant roots has also been studied. A large enhancement of the number of lateral roots and of root hairs was noticed (Tien *et al.* 1979; Umali-Garcia *et al.* 1980; Kapulnik *et al.* 1981). This was most likely due to phytohormone production (Tien *et al.* 1979; Reynders and Vlassak 1979). *A. brasilense* 13t produced auxins (indole-acetic-acid, indole-lactic-acid), giberellin, and cytokinin-like substances (Tien *et al.* 1979). Hartmann *et al.* (1983) examined three strains of *A. brasilense* (including Sp7 and SpCd) and three strains of *A. lipoferum* (including Br17) for indole-acetic-acid and anthranilic acid production. The major difference between the two species was the detection of anthranilate in culture supernatants of *A. lipoferum* strains but not of *A. brasilense*. In addition, tryptophan highly stimulated indole-acetic-acid production by *A. brasilense* strains but the effect was weak in th case of *A. lipoferum* (Hartmann *et al.* 1983). Mutants resistant to fluorotryptophan, which excreted higher amounts of indole-acetic-acid, were isolated from *A. brasilense* strain SpCd (Hartmann *et al.* 1983).

5.7.3 Role of cyst formation

The association of *Azospirillum* with plant callus has also been examined. Bacteria were inoculated directly onto the callus or at its proximity (Child and Kurz 1977) in a similar way to that originally devised to demonstrate nitrogen fixation of *Rhizobium* spp in the free-living state (Kurz and La Rue 1975). Plant cell callus from various origins supported growth of the bacteria and stimulated nitrogenase activity (Child and Kurz 1977; Berg *et al.* 1980; see also review by Giles and Vasil 1980). Moreover, bacteria recovered from sugar cane callus tissue cultures displayed a much higher nitrogenase activity than the original parent strain (Berg *et al.* 1980). This enhancement was tentatively attributed to spherical forms designated C forms, which corresponded to capsulated bacteria (Berg *et al.* 1980). The C forms were also detected in older cultures and may be considered as 'cysts' or 'cocoid bodies' described in the *Spirillum* genus (Krieg 1976). It was not shown that the C forms could fix nitrogen, nor that they were similar to the spherical cells found in old cultures. However, Berg *et al.* (1980) suggested that capsulation regulated oxygen flow to N_2-fixing cells. Other hypotheses were formulated by Lamm and Neyra (1981) and Papen and Werner (1982), who observed cyst formation in old cultures of *Azospirillum*. Cysts were more resistant than vibrioid cells, and consequently could survive more efficiently in the soil (Lamm and Neyra 1981). Papen and Werner (1982) found that cyst formation was concomitant with a decrease of nitrogenase activity and that encystation and de-encystation could account for a biphasic growth in long lasting cultures (Papen and Werner 1980). Thus the process of cyst formation in the soil might play an important role at the level of infection with the host plant.

5.7.4 Host-plant specificity

For the moment no real specificity of association of the bacteria with the plant has been demonstrated but there are indications that some specificity could exist. Baldani and Döbereiner (1980) showed that most of the strains isolated from maize were *A. lipoferum* and that most of the strains isolated from wheat or rice were *A. brasilense* Nir⁻, suggesting a difference of specificity between the two species towards C_4 and C_3 plants (Döbereiner and De Polli 1980). Use of fluorescent antibodies against whole bacteria was in agreement with the two different host-specificity groups (De Polli *et al.* 1980). Hartmann *et al.* (1983) suggested that the different behaviour in auxin metabolism between *A. lipoferum* and *A. brasilense* could reflect different susceptibilities of association with plants. The role of auxins in *Rhizobium*-legume infection was also investigated, and a recent report by Badenoch-Jones *et al.* (1982) denied the involvement of auxin in the process.

Further experiments indeed are required, in particular with bacterial mutants, to determine if specific genetic information is involved in the recognition process and whether stimulation of plant growth is due to nitrogen fixation or to phytohormone production (O'Hara *et al.* 1981).

5.8 CONCLUSION

In less than ten years, a relatively large amount of information has been accumulated on the molecular biology and the genetics of *Azospirillum* as well as on the physiology of its association with plants. Unfortunately, most of the reports are still preliminary. For example, the nitrogenase complex is not well studied and the regulation of its biosynthesis and activity requires further study. No other *nif* product has been identified and even the enzymes involved in the first steps of the nitrogen metabolism are not well characterized. Little information exists on phages and all the endogenous plasmids are cryptic. Although a few Nif⁻ mutants have been described, the reason why they are so difficult to isolate as compared to *K. pneumoniae* or *Azotobacter* is still obscure.

However, the possibility of transferring chromosomal markers by plasmids such as R68-45 and the development of *in vitro* DNA recombinant techniques rendered *Azospirillum* amenable to genetic analysis. From that point of view, it may be expected that, in the next few years, a circular linkage map will be established and that substantial progress will be realized on the functional organization of the *nif*-genes and on their localization.

In spite of the fact that the bacteria were poorly known, the potential agronomic importance of *Azospirillum* gave rise to considerable interest. Inoculations of plants were performed with wild-type strains without really controlling eventual host plant–bacteria specificity, which might be an explanation for the contradictory observations reported. In this area, genetics may also be useful in the future by providing good performing mutants or recombinant strains and by helping to identify the determinants of the plant–bacteria interaction.

ACKNOWLEDGMENTS

Unpublished experiments from this laboratory were carried out in cooperation with Drs. C. Fogher, C. Franche, D. Gauthier, P. Jara, and with Mr. H. Bozouklian and B. Perroud, and I wish to acknowledge their contribution. I also wish to thank Mrs. M. Ferrand for typing the manuscript. Work in the author's laboratory was supported by funds from Elf Aquitaine and Entreprise Minière et Chimique.

REFERENCES

AHMAD, M. H. (1978). *J. gen. appl. Microbiol.* **24**, 271–8.

BADENOCH-JONES, J., SUMMONS, R. E., DJORDJEVIC, M. A., SHINE, J., LETHAM, D. S., and ROLFE, B. G. (1982). *Appl. & environ. biol.* **44**, 275–80.

BAGDASARIAN, R. L., LURZ, R., RÜCKERT, B., FRANKLIN, F. C. H., BAGDASARIAN, M. M., FREY, J., and TIMMIS, K. N. (1981). *Gene* **16**, 237–47.

BALDANI, V. L. D. and DÖBEREINER, J. (1980). *Soil Biol. Biochem.* **12**, 434–44.

BÁNFALVI, Z., ŠAKANYAN, V., KONCZ, C., KISS, A., DUSHA, I., and KONDOROSI, A. (1981). *Molec. & gen. Genet.* **184**, 318–25.

BANI, D., BAZZICALUPO, M., FAVELLI, F., GALLORI, E., and POLSINELLI, M. (1980). *J. gen. Microbiol.* **119**, 239–44.

BARAK, R., NUR, I., OKON, Y., and HENIS, Y. (1982). *J. Bact.* **152**, 643–9.

BARBERIO, C., BAZZICALUPO, M., GALLORI, E., and POLSINELLI, M. (1982). In *Azospirillum genetics, physiology, ecology* (ed. W. Klingmüller) EXS42, pp. 47–53. Birkhaüser, Basel.

BALZICALUPO, M. and GALLORI, E. (1983). In *Azospirillum* II (ed. W. Klingmiller) EXS48, pp. 24–8. Birkhaüser, Basel.

BECKING, J. H. (1963). *Antonie van Leeuwenhock* **29**, 326.

BERG, R. H., TYLER, M. E., NOVICK, N. J., VASIL, V., and VASIL, I. K. (1980). *Appl. & environ. Microbiol.* **39**, 642–9.

BERINGER, J., BEYNON, J. L., BUCHANAN-VOLLASTON, A. V., and JOHNSTON, A. W.B. (1978). *Nature, Lond.* **275**, 633–4.

BERLIER, Y. M. and LESPINAT, P. A. (1980). *Arch. Microbiol.* **125**, 67–72.

BOTHE, H., KLEIN, B., STEPHAN, M. P., and DÖBEREINER, J. (1981). *Arch. Microbiol.* **130**, 96–100

CANNON, F. C., RIEDEL, G. E., and AUSUBEL, F. M. (1979). *Molec. & gen. Genet.* **174**, 59–66.

CASSE, F., BOUCHER, C., JULLIOT, J. S., MICHEL, M., and DÉNARIÉ, J. (1979). *J. gen. Microbiol.* **113**, 229–42.

.CHAN, Y. K., NELSON, L. M., and KNOWLES, R. (1980). *Can. J. Microbiol.* **26**, 1126–31.

CHAYKIN, S. (1969). *Analyt. Biochem.* **31**, 375–82.

CHILD, J. J. and KURZ, W. G. W. (1978). *Can. J. Microbiol.* **24**, 143–8.

DARROW, R. A. and KNOTTS, R. R. (1977). *Biochem. biophys. Res. Commun.* **78**, 554–9.

DATTA, N. and HEDGES, R. W. (1971). *Nature, Lond.* **234**, 222.

DE POLLI, H., BOHLOOL, B. B., and DÖBEREINER, J. (1980). *Arch. Microbiol.* **126**, 217–22.

DITTA, G., STANFIELD, S., CORBIN, D., and HELSINSKI, D. (1980). *Proc. natn. Acad. Sci. U.S.A.* **77**, 7347–51.

DIXON, R., CANNON, F. C., and KONDOROSI, A. (1976). *Nature, Lond.* **260**, 268–71.

DÖBEREINER, J. and DAY, J. M. (1976). In *Proc. 1st Int. Symp. Nitrogen Fixation* (ed. W. E. Newton and C. J. Nyman), pp. 518–36. Washington State University Press, Pullman.

— and DEPOLLI, H. (1980). In *Nitrogen fixation: Proc. Phytol. Soc. Eur. Symp.* (ed. W. D. P. Stewart and J. R. Gallon), pp. 301–33. Academic Press, London.

— MARRIEL, I. E., and NERY, M. (1976). *Can. J. Microbiol.* **22**, 1464–73.

DROZD, J. and POSTGATE, J. R. (1970). *J. gen. Microbiol.* **63**, 63–73.

ECKHARDT, T. (1978). *Plasmid* **1**, 584–8.

ELMERICH, C. and FRANCHE, C. (1982). In *Azospirillum genetics, physiology, ecology.* (ed. W. Klingmüller) EXS42, pp. 9–17. Birkhaüser, Basel.

— QUIVIGER, B., ROSENBERG, C., FRANCHE, C., LAURENT, P., and DÖBER—EINER, J. (1982). *Virology* **122**, 29–37.

ESKEW, D. L., FOCHT, D. D., and TING, I. P. (1977). *Appl. & environ. Microbiol.* **34**, 582–5.

ESPIN, G., ALVAREZ-MORALES, A., and MERRICK, M. (1981). *Molec. & Gen. Genet.* **184**, 213–7.

—— CANNON, F., DIXON, R., and MERRICK, M. (1982). *Molec. & Gen. Genet.* **186**, 518-24

FRANCHE, C. (1981). Thèse de Doctorat de 3ᶦᵉᵐᵉ Cycle, Université de Paris VII, Paris.

— and ELMERICH, C. (1981). *Ann. microbiol. Inst. Pasteur* **132**A, 3-17.

— CANÉLO, E., GAUTHIER, D., and ELMERICH, C. (1981). *FEMS Microbiol. Lett.* **10**, 199-202.

FUCHS, R. L. and KEISTER, D. L. (1980). *J. Bact.* **144**, 641-8.

GARCIA, E., BANCROFT, S., RHEE, S. G., and KUSTU, S. (1977). *Proc. natn. Acad. Sci. U.S.A.* **74**, 1662-6.

GAUTHIER, D. (1978). Thèse de Doctorat de 3ᵉᵐᵉ Cycle, Université de Paris VII, Paris.

— and ELMERICH, C. (1977). *FEMS Microbiol. Lett.* **2**, 101-4.

GILES, K. L. and VASIL, I. K. (1980). In *Perspectives in plant cell and tissue culture* (ed. I. K. Vasil), *Int. Rev. Cytol., Suppl.* 11B, pp. 81-99. Academic Press, New York.

GINSBURG, A. and STADTMAN, E. R. (1973). In *The enzyme of glutamine metabolism* (ed. S. Prusiner and E. R. S. Stadtman), pp. 9-43. Academic Press, New York.

GORDON, J. K. and BRILL, W. J. (1974). *Biochem. biophys. Res. Commun.* **59**, 967-71.

HAAHTELA, K., WARTIOVAARA, T., SUNDMAN, V., and SKUJINS, J. (1981). *Appl. & environ. Biol.* **41**, 203-6.

HAAS, D. and HOLLOWAY, B. W. (1978). *Molec. & gen. Genet.* **158**, 229-37.

HANUS, F. J., MAIER, R. J., and EVANS, H. J. (1979). *Proc. natn. Acad. Sci. U.S.A.* **76**, 1788-92.

HARTMANN, A. (1982). In *Azospirillum genetics, physiology, ecology* (ed. W. Klingmüller) EXS42, pp. 59-68. Birkaüser, Basel.

— and KLEINER, D. (1982). *FEMS. Microbiol. Lett.* **15**, 65-7.

— Singh, M., and Klingmüller, W. (1983). *Can. J. Microbiol.,* **29**, 916-23.

HEULIN, T., BALLY, R., and BALANDREAU, J. (1982). In *Azospirillum genetics, physiology, ecology* (ed. W. Klingmüller) EXS42, pp. 92-9. Birkhaüser, Basel.

HOMBRECHER, G., BREWIN, N. J., and JOHNSTON, A. W.B. (1981). *Molec. & gen. Genet.* **182**, 133-6.

IKEDA, H. and TOMIZAWA, J. (1968). *Cold Spring Harbor Symp. Quant. Biol.* **33**, 791-8.

JARA, P., QUIVIGER, B., LAURENT, P., and ELMERICH, C. (1983). *Can. J. Microbiol.,* **28**, 968-72.

KAPULNIK, Y., OKON, Y., KIGEL, J. NUR, I., and HENIS, Y. (1981). *Pl. Physiol.* **68**, 340-3.

KLEINER, D. (1981). *Biochim. biophys. Acta.* **639**, 41-52.

KNAUF, V. C. and NESTER, E. W. (1982). *Plasmid* **8**, 45-54.

KONDOROSI, A., VINCZE, E., JOHNSTON, A. W. B., and BERINGER, J. E. (1980). *Molec. & gen. Genet.* **178**, 403-8.

KRIEG, N. R. (1976). *Bact. Rev.* **40**, 55-115.

— (1977). In *Genetic engineering for nitrogen fixation* (ed. A. Hollaender), vol. 9, pp. 463-72. Plenum Press, New York.

KUMARI, M. L., KAVIMANDAN, S. K., and SUBBA-RAO, N. S. (1976). *Indian J. exp. Biol.* **14**, 638–9.

KURZ, W. G. and LA RUE, T. A. (1975). *Nature, Lond.* **256**, 407–9.

LAMM, R. B. and NEYRA, C. A. (1981). *Can. J. Microbiol.* **27**, 1320–5.

LEONARDO, J. M. and GOLDBERG, R. B. (1980). *J. Bact.* **142**, 99–110.

LUDDEN, P. W. and BURRIS, R. H. (1976). *Science* **194**, 424–6.

— OKON, Y., and BURRIS, R. H. (1978). *Biochem. J.* **173**, 1001–3.

MACNEIL, T., MACNEIL, D., ROBERTS, G. P., SUPIANO, M. A., and BRILL, W. J. (1978). *J. Bact.* **136**, 253–66.

MCFARLAND, N., MCCARTER, L., ARTZ, S., and KUSTU, S. (1981). *Proc. natn. Acad. Sci. U.S.A.* **78**, 2135–9.

MAGALHAES, I. M. S., NEYRA, C. A., and DÖBEREINER, J. (1978). *Arch. Microbiol.* **117**, 247–52.

— BALDANI, J. I., SOUTO, S. M., DUYKENDALL, J. R., and DÖBEREINER, J. (1983). *A. Acad. Brasil. Cienc.* **55**, 417–30.

MALIK, K. A. and SCHLEGEL, H. G. (1981). *FEMS Microbiol. Lett.* **10**, 63–7.

MAZUR, B., RICE, D., and HASELKORN, R. (1980). *Proc. natn. Acad. Sci. U.S.A.* **77**, 186–90.

MERRICK, M., FILSER, M., DIXON, R., ELMERICH, C., SIBOLD, L., and HOUMARD, J. (1980). *J. gen. Microbiol.* **117**, 509–20.

— (1983). *Eur. molec. Biol. Orgn J.* **3**, 501–7.

MEYER, J., KELLY, B. C., and VIGNAIS, P. M. (1978). *Biochimie* **60**, 245–60.

MEYERS, J. A., SANCHEZ, D., ELWELL, I. P., and FALKOW, S. (1976). *J. Bact.* **127**, 1529–37.

MILLER, J. H. (1972). *Experiments in molecular genetics* (2nd edn). Cold Spring Harbor Laboratory, New York.

MISRHA, A. K., ROY, P., and BHATTACHARYA, S. (1979). *J. Bact.* **137**, 1425–7.

NAGATINI, H., SHIMIZU, M., and VALENTINE, R. C. (1971). *Arch. Mikrobiol.* **79**, 164–75.

NAIR, S. K., JARA, P., QUIVIGER, B., and ELMERICH, C. (1983). In *Azospirillum II* (ed. W. Klingmuller) EXS48, pp. 29–38. Birkhaüser, Basel.

NELSON, L. and KNOWLES, R. (1978). *Can. J. Microbiol.* **24**, 1395–403.

NEYRA, C. A., DÖBEREINER, J., LALANDE, R., and KNOWLES, R. (1977). *Can. J. Microbiol.* **23**, 300–5.

— and VAN BERKUM, P. (1977). *Can. J. Microbiol.* **23**, 306–10.

NUR, I., OKON, Y., and HENIS, Y. (1980). *Can. J. Microbiol.* **26**, 714–718.

— STEINITZ, Y. L., OKON, Y., and HENIS, Y. (1981). *J. gen. Microbiol.* **122**, 27–32.

O'HARA, G. W., DAVEY, M. R., and LUCAS, J. A. (1981). *Can. J. Microbiol.* **27**, 871–7.

OKON, Y., ALBRECHT, S. L., and BURRIS, R. H. (1976a). *J. Bact.* **128**, 592–7.

——— (1976b). *J. Bact.* **127**, 1248–54.

— HOUCHINS, J.P., ALBRECHT, S. L., and BURRIS, R. H. (1977). *J. gen. Microbiol.* **98**, 87–93.

PATRIQUIN, D. G. (1982). In *Advances in agricultural microbiology* (ed. N. S. Subba Rao), pp. 139–190. Oxford University Press.

— and DÖBEREINER, J. (1978). *Can. J. Microbiol.* **24**, 734–42.

PAPEN, H. and WERNER, D. (1980). *Arch. Microbiol.* **128**, 209–14.

—— (1982). *Arch. Microbiol.* **132**, 57–61.

PEDROSA, F. O., DÖBEREINER, J., and YATES, M. G. (1980). *J. gen. Microbiol.* **119**, 547–51.

— STEPHAN, M., DÖBEREINER, J., and YATES, M. G. (1982). *J. gen. Microbiol.* **128**, 161–6.

— and YATES, M. G. (1984). *FEMS Microbiol. Lett.* **29**, 95–101.

PLAZINSKY, J., DART, P., and ROLFE, B. (1983). *J. Bact.* **155**, 1429–33.

POLSINELLI, M., BALDANZI, E., BAZZICALUPO, M., and GALLORI, E. (1980). *Molec. & gen. Genet.* **178**, 709–11.

PRAKASH, R. K., SCHILPEROORT, R. A., and NUTI, M. P. (1981). *J. Bact.* **145**, 1129–36.

PÜHLER, A. and KLIPP, W. (1981). In *Biology of inorganic nitrogen and sulfur* (ed. H. Bothe and A. Trebst), pp. 276–86. Springer-Verlag, Heidelberg.

QUIVIGER, B., FRANCHE, C., LUTFALLA, G., RICE, D., HASELKORN, R., and ELMERICH, C. (1982). *Biochimie* **64**, 495–502.

RAO, A. V. and VENKATESWARLU, B. (1982). *Can. J. Microbiol.* **28**, 778–82.

REYNDERS, L. and VLASSAK, K. (1979). *Soil. biol. Biochem.* **11**, 547–8.

ROSENBERG, C., BOISTARD, P., DÉNARIÉ, J., and CASSE-DELBARD, F. (1981). *Molec. & gen. Genet.* **184**, 326–33.

RUVKUN, G. B. and AUSUBEL, F. M. (1980). *Proc. natn. Acad. Sci. U.S.A.* **77**, 191–5.

SAMPAIO, M. J. A. M., DA SILVA, E. M. R., DÖBEREINER, J., YATES, M. G., and PEDROSA, F. O. (1981). In *Current perspectives in nitrogen fixation* (ed. A. H. Gibson and W. E. Newton), pp. 444. Australian Academy of Sciences, Canberra.

SCOTT, D. B., SCOTT, C. A. and DÖBEREINER, J. (1979). *Arch. Microbiol.* **121**, 141–5.

SIBOLD, L. (1982). *Molec. & gen. Genet.* **186**, 569–71.

— and ELMERICH, C. (1983). *Eur. molec. Biol. Orgn J.* 1982, 1551–8.

Simon, R., Priefer, and Pühler, A. (1984). *Bio/Tecnology* **1**, 784–91.

SINGH, M. (1982). In Azospirillum *genetics, physiology, ecology* (ed. W. Klingmüller) EXS42, pp. 35–43. Birkhaüser, Basel.

SINGH, M. and WENZEL, W. (1982). In Azospirillum *genetics, physiology, ecology* (ed. W. Klingmüller) EXS42, pp. 44–51. Birkhaüser, Basel.

STREICHER, S. L., SHANMUGAM, K. T., AUSUBEL, F., MORANDI, C., and GOLDBERG, R. B. (1974). *J. Bact.* **120**, 815–21.

TARRAND, J. J., KREIG, N. R., and DÖBEREINER, J. (1978). *Can. J. Microbiol.* **24**, 967–80.

TEMPEST, D. W., MEERS, J. L., and BROWN, C. M. (1973). In *enzyme of glutamine metabolism* (ed. S. Prusiner and E. R. Stadtman), pp. 167–82 Academic Press, New York.

TIEN, T. M., DIEM, H. G., GASKINS, M. H., and HUBBELL, D. H. (1981). *Can. J. Microbiol.* **27**, 426–31.

— GASKINS, M. H., and HUBBEL, D. H. (1979). *Appl. & environ. Microbiol.* **37**, 1016–24.

UMALI-GARCIA, M., HUBBEL, D. H., GASKINS, M. H., and DAZZO, F. B. (1980). *Appl. & environ. Microbiol.* **39**, 219–26.

Van Berkum, P. and Bohlool, B. B. (1980). *Microbiol. Rev.* **44**, 491–517.

Volpon, A. G. T., De-Polli, H., and Döbereiner, J. (1981). *Arch. Microbiol.* **128**, 371–5.

Von Bülow, J. F. M. and Döbereiner, J. (1975). *Proc. natn. Acad. Sci. U.S.A.* **72**, 2389–93.

Wood, A. C., Menezes, E. M., Dykstra, C. and Duggan, D. E. (1982). In Azospirillum *genetics, physiology, ecology* (ed. W. Klingmüller) EXS42, pp. 18–34. Birkhaüser, Basel.

Yates, M. G., Walker, C. C., Partridge, C. D. P., Pedrosa, F. O., Stephan, M. and Döbereiner, J. (1981). In *Current perspectives in nitrogen fixation* (ed. A. H. Gibson and W. E. Newton), pp. 97–100. Australian Academy of Science Press, Canberra.

Zumft, W. G., Alef, K., and Mümmler, S. (1981). In *Current perspectives in nitrogen fixation* (ed. A. H. Gibson and W. E. Newton), pp. 190–3. Australian Academy of Science Press, Canberra.

— and Castillo, F. (1978). *Arch. Mirobiol.* **117**, 53–60.

6 Azotobacter biology, biochemistry, and molecular biology

Betty E. Terzaghi and Eric Terzaghi

6.1 INTRODUCTION

In this chapter, the word azotobacters will be used to denote the family Azotobacteraceae in general; specific genera and species will be referred to by name.

The azotobacters have been difficult to analyse genetically, and this problem, combined with the fact that they do not form nitrogen-fixing symbioses, has resulted in their playing a minor role in studies of the molecular biology of nitrogen fixation. They are of historic interest because of their use as soil inoculants for non-symbiotic biological nitrogen fixation and, in the laboratory for studies of an aerobic free-living microbial nitrogenase system. They are of current and future interest for studies of the molecular biology of nitrogen fixation because of their systems to protect nitrogenase from oxygen, because of the development of genetic transfer systems which will make them more amenable to analysis, and, uniquely, because of the recent (controversial) claim that they possess an alternative nitrogen-fixing system.

6.2 BIOLOGY

This section is not intended to be a comprehensive review of the biology of the Azotobacteraceae, but rather a generalized introduction to characteristics associated with the family.

6.2.1 Classification

The first recognition of biological nitrogen fixation by free-living aerobic soil bacteria is credited to Beijerinck in 1901. Today we recognize the Azotobacteraceae as free-living, aerobic, gram-negative, nitrogen-fixing rods. Because there is considerable variation in many traits (e.g. cyst formation, flagella, pigment formation, cell size and shape, polysaccharide production), classification by traditional methods has been

127

controversial. *Bergey's manual of determinative biology* (Buchanan and Gibbons 1974) lists four genera in the Azotobacteraceae as determined by GC content: *Azotobacter*, *Azomonas*, *Beijerinckia*, and *Derxia*. However, this classification has recently been challenged in several ways.

By numerical assessment of a wide range of characters, a reclassification into six genera has been proposed (Thompson and Skerman 1979), adding *Azomonotrichon macrocytogenes* for *A. macrocytogenes* and *Azorhizophilus paspali* for *A. paspali*, the azotobacter believed to form an association (now called a rhizocoenosis) with the tropical grass *Paspalum* (Döbereiner 1966).

Alternatively, the inclusion of a new genus, *Xanthobacter*, has been proposed (Wiegel *et al.* 1978) which would include bacteria capable of microaerobic diazotrophy (*Corynebacterium*; *Mycobacterium*). An entirely different grouping of the Azotobacteraceae has been made on the basis of the degree of rRNA homology (which correlates with the overall phenotypic similarity of bacterial genera) (DeSmedt *et al.* 1980). Of four rRNA super-families, *Azotobacter* and *Azomonas* (and some pseudomonas species) were placed together while *Beijerinckia* was grouped with *Xanthobacter*, *Azospirillum*, and *Rhizobium* in a different super-family, and *Derxia* was placed in a third super-family along with *Alcaligenes*, different pseudomonas species, and other bacteria. The *Azotobacter* species, including *A. paspali*, were all closely related and certain similarities to some pseudomonads were noted.

Finally, Becking (1981) has argued that *Azospirillum* should be included in the Azotobacteraceae, and Tchan and co-workers (1983), using data from immuno-electrophoresis, have suggested another taxonomy. Whatever the outcome of these taxonomic disagreements, for this review the classification in Bergey's (Buchanan and Gibbons 1974) will be used. Virtually all work relevant to molecular biology has been done with *A. vinelandii* and *A. chroococcum*.

6.2.2 Distribution

Azotobacters can be found in a wide variety of habitats: soils, leaves, and both marine and fresh water; at temperatures ranging from tropical to arctic and at a pH range from 3–9; however, there is species specificity in their distribution and in general they are restricted to environments rich in organic matter (see reviews by Jensen 1965; Mishustin and Shil'nikova 1971; Thompson and Skerman 1979; Becking 1981; Gordon 1981). Most isolations are made from neutral soils and it is generally accepted that the optimum pH for nitrogen fixation is around neutrality, with nitrogen fixation ceasing at pHs below 6 (Burk *et al.* 1934). There are exceptions, most notably the *Beijerinckias*, which can grow and fix nitrogen at pHs as low as 3.5 (Becking 1978).

Fragmentary information is available on the nitrogen fixation of these bacteria in aquatic environments (see Jordan 1981), although more is known of their ecological and geographical distribution (see previously-cited reviews). Differences have been reported in the nitrogen-fixing efficiency of salt-tolerant azotobacters at different salinities. In general, decreasing fixation rates have occurred with increasing salinity; however, fixation can still occur at 40 per cent salinity (Dicker and Smith 1981).

6.2.3 Growth characteristics

Investigation into the growth of azotobacters has contributed to the concept of a bacterial life-cycle. Azotobacters pass through various changes of cell form, depending on the species, the stage of growth, and the culture conditions. Reports of very small cellular forms called variously germinal, L-forms, or microgonidia, were generally discounted; recently however, a small (0.3 μm) filterable form was described which grows readily in soil but is difficult to find in cultures grown in defined medium (Gonzalez-Lopez and Vela 1981). This filterable form does not fix nitrogen (on molybdenum-containing medium) (Vela and Gonzalez-Lopez 1982), but can generate the large nitrogen-fixing form, which in turn can produce the small form. This finding raises the question of the true distribution of the Azotobacteraceae in soils and the contribution these bacteria could be making to soil fertility. DNA hydridizations between the large and small cell forms, and with *nif* probes, should be performed to confirm strain identity and the presence or absence of nitrogen-fixation genes.

The most prominent cell forms in laboratory cultures are the comparatively large (2–3 × 3–6 μm) oval, spherical, or rod-shaped cells, frequently in distorted shapes (e.g. peanut, spindle) (Mishustin and Shil'nikova 1971; Post *et al.* 1982). The fragility of azotobacters and the care needed in isolation has recently been emphasized (Postgate 1981), but in light of all the recent findings on azotobacter, perhaps another growth medium is also required. The structure and molecular composition of the outer layers of the cells are beginning to receive some systematic attention (Page and von Tigerstrom 1982; Bingle *et al.* 1983; Emery and Melton 1983). In addition to these forms, many species make thick-walled cysts in stationary phase; an extensive literature on cyst formation and germination has developed (reviewed by Sadoff 1975).

In the laboratory, azotobacters are normally grown on simple nitrogen-free salts containing phosphate, magnesium, calcium, iron, molybdenum, and a carbon source. Azotobacters are capable of metabolizing a very wide range of carbon compounds from organic acids to simple alcohols, but usually sucrose, glucose, or mannitol are used as carbon sources in laboratory media. The nitrogen requirement is met by elemental nitrogen, urea, ammonia, some simple amino acids, nitrate, and nitrite. The energy requirement for nitrogen fixation is met by a very high rate of aerobic metabolism; the resulting high oxygen demand is believed to contribute to the maintenance of a minimal intracellular oxygen tension, a requirement of the oxygen-sensitive nitrogenase (reviewed by Robson and Postgate 1980).

Eady and Robson (1984) grew *A. vinelandii* in both batch and continuous culture under conditions of N- and Mo-limitation and looked at growth and nitrogen fixation. They found that acetylene reduction activity was disproportionately low and also that high concentrations of nitrogenase were not a prerequisite to obtaining adequate rates of nitrogen fixation, observations of importance to assessments of biological nitrogen fixation.

Fekete *et al.* (1983*b*), Page and Huyer (1984), and Knosp *et al.* (1984) imposed iron limitation on *A. vinelandii* OP and a mutant derivative and looked at the effects on growth and siderophore production of adding different iron-containing minerals. changes in proteins associated with the outer membrane and with the siderophore system were documented. In an associated study, the protein composition of ultrastructure of the surface layer were examined by SDS-PAGE and freeze– etch electron microscopy (Bingle *et al.* 1984) suggesting the involvement of a protein of about 65 000 molecular weight in this surface layer.

There is disagreement over whether calcium is essential for growth of *Azotobacter* species since it is not essential for all Azotobacteraceae. Although it was gradually accepted that calcium was necessary, particularly for cyst formation (Norris and Jensen 1958; Page and Sadoff 1975), the requirement might be related to the phosphate concentration in most media, for good growth of a variety of *Azotobacter* strains has been obtained on a medium low in phosphate and not supplemented with calcium (where background calcium contaminant levels were less than 1 p.p.m.) (B. E. Terzaghi, unpublished data). The role of phosphate in azotobacter metabolism is multiple, with effects on cell wall structure, oxygen uptake, and cell viability (Dalton and Postgate 1969; Tsai *et al.*, 1979). Since the above was written more information has appeared on the role of Ca^{2+} in *Azotobacter* growth on Burk's medium. Bingle *et al.* (1984) found differences in surface layer properties related to Ca^{2+} concentration, while Ramos and Robson (1985*a, b*) suggested that Ca^{2+} was involved with PEP-carboxylase; most of their Fos mutants (oxygen-sensitive fixation with sugars) of *A. chroococcum* could be restored to air-tolerant diazotrophy by increasing the Ca^{2+} concentration.

On minimal medium, optimal growth occurs at temperatures between 25 and 30 °C (see Thompson and Skerman 1979). While growth can occur under micro-aerobic conditions, it is improved by aeration and under optimal conditions, nitrogen-fixing growth proceeds with generation times as low as two hours.

Storage of cultures is generally poor at 4 °C or by lyophilization; for long term storage, sealed cultures left at room temperature on agar, or freezing of liquid cultures at low temperature are satisfactory (Becking 1981; B. E. Terzaghi, unpublished). A detailed study of encystment and other changes occurring in *A. vinelandii* upon storage under different conditions has now appeared (Page 1983), confirming that storage at 4°C should be avoided in favour of storage at 20°C on non-acidic medium. Ruppen *et al.* (1983) compared protein turnover during vegetative growth, encystment, and germination.

Although phage infection is not prevalent, phage lysis of azotobacters does occur (Duff and Wyss 1961); hence phage-typing can be used for strain identification. Phages can also be carried in pseudolysogenic cultures (Thompson *et al.* 1980a). Lysogeny perhaps occurs but has not been studied (Thompson *et al.* 1980a; Page 1982). Pseudolysogenic conversion of *A. vinelandii* leads to loss of the polysaccharide coat, acquisition of flagellae and motility, and the colonies acquire a yellow pigmentation (Thompson *et al.* 1980a). The similarity between the isolation of *A. vinelandii* strain UW (Bush and Wilson 1959) and the isolation of a phage-converted cell is described in Thompson *et al.* (1980b); the possibility that some UW cultures might carry phages as pseudolysogens should be considered. Azotobacters are somewhat resistant to bacterial parasitism but can be lysed by pseudomonads (Postgate 1967) and bdellovibrios (Sullivan and Casida 1968).

6.2.4 Compounds produced

Azotobacters produce a wide variety of compounds, depending on the strain. Some species form polysaccharides, occasionally copiously, particularly amongst the *Beijerinckias* (Derx 1950). The exopolysaccharide of *A. vinelandii* has a composition similar to alginic acid, an industrially important polymer, and active research on *A. vinelandii* for production of alginate is being conducted (Horan *et al.* 1981).

Pigments are formed by some strains under certain conditions, and have been used as taxonomic criteria. Formation of a green fluorescent pigment in iron-depleted cultures of *A. vinelandii* has been correlated with the attainment of transformation competence (Page 1978). Other compounds produced by azotobacters include vitamins, an antibiotic, and the hormones indole acetic acid, gibberellin, and cytokinin (Mishustin and Shil'nikova 1971; Apte and Shende 1981). These latter compounds may be significant in plant-microbial interactions.

Azotobacters, particularly *A. chroococcum* and *A. vinelandii*, have long been used as soil inoculants. Extensive study of azotobacters has been conducted in the Soviet Union because of that practice. Agricultural usage of azotobacters continues today, particularly in countries which have warm, alkaline soils such as India, Egypt, and Iraq, and accordingly, a large literature exists on field use of azotobacters. It is generally agreed that yields, for a wide variety of crops, have been improved by inoculation; however, the means by which the plant benefits remains controversial, whether by nitrogen fixation or from bacterial production of hormones which stimulate plant growth. One widely-cited case supporting hormonal involvement was performed at a sufficiently acid pH that nitrogen fixation is not likely to be attributable to aerobic free-living azotobacter species (Brown 1976). We have not been able, using several methods, either to get *A. paspali* strains to grow on nitrogen-free media at those pHs cited or to obtain mutants with normal growth at such pHs (B. E. Terzaghi, unpublished).

Beneficial non-symbiotic associations have been reported between azotobacters and certain plants, such as *Beijerinckia* and rice or sugar cane (Becking 1978; Döbereiner and Boddey 1981), but the only such association receiving more thorough study has been that between *A. paspali* and the tropical C_4 grass *Paspalum notatum* (Döbereiner 1966). Using the ^{15}N isotopic dilution technique Boddey *et al.* (1983) found that the *Paspalum notatum* batatais cultivar obtained about 11 per cent of its nitrogen from *A. paspali*. Associations have also been reported with a few other paspalum species, all tropical and all with C_4 metabolism (Döbereiner 1970; Kass *et al.* 1971; Day *et al.* 1975). The nature of the association has not been elucidated, but microscopic examination showed bacteria-like objects associated with the outer mucigel layer of the root (Döbereiner *et al.* 1972). Most work has been done with plants taken from the field and laboratory duplication of the association in sand showed nitrogen gains in roots but not leaves (Kass *et al.* 1971). Incorporation of ^{15}N has since confirmed transfer of nitrogen to the entire plant (De Polli *et al.* 1977).

Non-specific associations have also been reported, claiming that biological nitrogen fixation is greatly enhanced when azotobacters are grown in association with other micro-organisms rather than in pure culture. The list of synergistic organisms includes other bacteria, actinomycetes, fungi, yeast, amoebae, ciliates, and blue-green algae (Jensen and Holm 1975).

Nitrogen-fixing associations have also been developed in the laboratory. Introduction of *A. vinelandii* into carrot callus cell cultures on nitrogen-free medium resulted in intercellular establishment of the bacteria (Carlson and Chaleff 1974). A fusion hybrid between *Azotobacter* and *Agrobacterium* has been reported which can fix nitrogen in galls on tomato (Qian-you and Chen-ying 1981) and uptake of *A. vinelandii* by protoplasts of *Saccharomyces cerevisiae* was successful but nitrogen fixation was not reported (Yamada and Sakaguchi 1981). The reported endosymbiotic association claimed between *A. vinelandii* and the fungus *Rhizopogon rubescens* (Giles and Whitehead 1975) must be regarded as unproven inasmuch as a fungal contaminant with identical properties to their 'modified' fungus was subsequently found in the parental *A. vinelandii* culture (Terzaghi 1980*a*). The original bacterial and fungal materials fixed by Giles and Whitehead (1975) for electron microscopy have recently been re-examined. Fungal contamination was detected in the *A. vinelandii* culture and, by the criteria of pore morphology and conidial formation, there was no evidence of the *Rhizopogon*-type basidiomycete fungal presence, only ascomycetous fungi, in the 'modified' fungal material (Terzaghi and Christensen 1985).

A different approach to developing an azotobacter association is in progress and is aimed at breeding corn with nitrogen-fixing, ammonium-excreting *A. vinelandii* bound to the roots (Brill 1981). Since this anticipated association is not naturally-occurring, modification of both the plant and the bacterium is planned. Development of bacterial mutants is described here and in the Molecular Biology section. Gordon and Jacobson (1983) found that methyl alanine resistant *A. vinelandii* mutants produce 23 per cent of the normal level of nitrogenase when grown in the presence of ammonium, while methyl-ammonium-resistant mutants fix more N_2 than wild type and excrete that excess into the medium. These mutants are being evaluated for potential use as fertilizers but plant benefits have not yet been reported. Plant modifications include breeding for corn plants which produce carbon compounds on their roots to supply the growth requirements of the bacteria. Claims of satisfying 1 per cent of the plant's nitrogen supply by such modifications have been made (Brill 1981), leaving a further 99 per cent to be obtained.

6.3 BIOCHEMISTRY

In recent published symposia and reviews of nitrogen fixation (Sprent 1979; Stewart and Gallon 1980; Gibson and Newton 1981; Postgate 1982), a wealth of information on both the genetic determination of those genes directly implicated in the process of nitrogen fixation and ammonia assimilation, and on the chemical properties of those molecules responsible for the splitting of dinitrogen, has been presented. However, there has been a surprising lack of information about the metabolic context within which this remarkable process takes place. Accordingly, we shall briefly survey what is currently known of the general biochemical properties of the azotobacters, and in addition, present in Table 6.1 a list of those proteins of metabolic significance that have been at least partially purified and characterized. The nitrogenase complex itself, and its regulation, we regard as a black box into which are fed energy, reducing power, and dinitrogen, and out of which comes NH_4^+. The reader is referred to Chapter 1 for details concerning the contents of that box. Also, the reader is warned that much of the work has been done with the species *A. vinelandii*, *A. beijerinckia*, and *A. chroococcum*, and that extrapolation to other genera must be done with caution.

Perhaps the most striking biochemical feature of the Azotobacteraceae is the fact that they are obligate aerobes, yet are effective in nitrogen fixation in spite of the notorious oxygen sensitivity of the nitrogenase enzyme complex. This immediately raises the interesting question of how these conflicting physiological requirements are resolved. As outlined by Gallon (1981), under normal conditions of aerobic growth and nitrogen fixation, 'respiratory protection', in which reducing power is generated metabolically or by H_2 scavenging, is utilized (or wasted) for converting O_2 to H_2O and thus maintaining a low internal pO_2. This strategy results in one of the highest (and most inefficient) rates of O_2 utilization known in the microbial world (Gutschick 1980; Robson and Postgate 1980), and this general point will be expanded upon in the discussion of metabolism below. Post *et al.* (1982) have pointed out that this metabolic protection of nitrogenase is accompanied by a striking increase of intracytoplasmic membrane area. An examination of superoxide dismutase, a potential element of the respiratory protection system, has very recently been initiated (Moore *et al.* 1983). Should the pO_2 level exceed the capacity of the respiratory system, *de novo* nitrogenase synthesis is repressed (Robson and Postgate 1980) and the existing nitrogenase complex is endowed with 'conformational protection' through the binding of a supplementary Fe–S protein (Haaker and Veeger 1977). Monitoring nitrogenase activity of cells grown in a chemostat under various regimens of nutrient and oxygen supply yielded results consistent with this interpretation (Post *et al.* 1983).

6.3.1 Intermediary metabolism
6.3.1.1 *Pathways*
In azotobacters, glucose is metabolized via the Entner–Douderoff pathway (Mortenson *et al.* 1955) and there is a functional tricarboxylic acid (TCA) cycle (Jackson and Dawes 1976). An early report of a 6-phosphogluconate dehydrogenase (EC 1.1.1.44) (Senior and Dawes 1971), a key enzyme in the oxidative pentose phosphate cycle, was subsequently shown to be an artefact (Stephenson *et al.* 1978). Pentoses are presumed to be synthesized from fructose-6-phosphate and

TABLE 6.1
Azotobacter proteins which have been characterized

Protein	Reference
Intermediary metabolism	
Glucose-6-phosphate dehydrogenase (EC 1.1.1.49)	Stephenson *et al.* (1978)
Pyruvate dehydrogenase (EC 1.2.4.1)	Stephenson *et al.* (1978)
6-Phosphogluconate dehydratase (EC 4.2.1.12)	Brester *et al.* (1975)
	Stephenson *et al.* (1978)
2-Keto-3-deoxy-6-phosphogluconate aldolase (EC 4.1.2.1)	Stephenson *et al.* (1978)
Isocitrate dehydrogenase (EC 1.1.1.42)	Chung and Franzen (1969)
	Barrera and Jurtshuk (1970)
	Slobbe and Voordouw (1981)
Phosphoenol carboxylase (EC 4.1.1.31)	Liao and Atkinson (1971)
Pyruvate kinase (EC 2.7.1.40)	Liao and Atkinson (1971)
Glutamine synthetase (EC 6.3.1.2)	Kleinschmidt and Kleiner (1978)
	Lepo *et al.* (1979)
Acetyl-CoA reductase (EC 1.1.1.36)	Ritchie *et al.* (1971)
Electron transport	
Ferredoxin	Shethna (1970)
	Furey *et al.* (1982)
	Howard *et al.* (1983)
Flavodoxin	Van Lin and Bothe (1972)
	Van Schagen and Muller (1981)
Cytochrome C_4	Tissières (1956)
	Swank and Burris (1969)
	Yang (1981)
Cytochrome C_5	Tissières (1956)
	Swank and Burris (1969)
Cytochrome O	Yang and Jurtshuk (1978)
Cytochrome b_{559}	Yang (1981)
Cytochrome b_{561}	Yang (1981)
Cytochrome d	Yang (1981)
Cytochrome a_1	Yang (1981)
Pyridine nucleotide transhydrogenase (EC 1.6.1.1)	Voordouw *et al.* (1980)
	Voordouw *et al.* (1982)
Cytochrome oxidase	Jurtshuk *et al.* (1981)
Others	
Glucohydrolase (NMN specific)	Imai (1979)
Nitrogenase complex	See Chapter 1
Nitrogenase 'protection factor' (Fe–S protein)	Shethna *et al.* (1968)
Bacterioferritin	Chen and Crichton (1982)
	Li *et al.* (1980)
Fe storage complex-bacterioferritin cytochrome b 577.5	Bulen *et al.* (1973)
Mo storage protein	Pienkos and Brill (1981)
5S rRNA	Dams *et al.* (1983)

glyceraldehyde-3-phosphate, via transketolase and transaldolase, both of which activities have been demonstrated. The absence of phospho-fructokinase activity indicates that glycolysis is not operative (Stephenson *et al.* 1978).

It has been known for some time that azotobacters are able to utilize acetate as the sole carbon and energy source to support growth and nitrogen fixation (Bergey *et al.* 1923). However, it has only recently been reported that there is an active isocitrate lyase (George and Melton 1982), a key enzyme in the glyoxalate cycle which is the usual pathway for acetate utilization.

Azotobacters share with a number of other groups of bacteria the ability to accumulate poly-β-hydroxybutyrate (PHB). Up to 70 per cent of the cell dry weight may be in the form of this storage product (Jackson and Dawes 1976), which is synthesized from the pivotal inter-mediate acetyl-CoA. PHB is believed to play a key role in 'respiratory protection' of the nitrogenase by serving as a sink for reducing power under low O_2 conditions and as a supplementary endogenous source of reducing power under high O_2 conditions.

Although it is known that a variety of sources of fixed nitrogen (beyond that produced by endogenous nitrogenase) may be utilized (Jensen 1954), neither the pathway of NH_4^+ assimilation nor of the metabolism of amino acids or other fixed nitrogen sources has been examined in detail. Kleiner and Kleinschmidt (1976) demonstrated the presence of both glutamine synthetase (GS) and glutamate synthase (GOGAT), but were unable to find significant levels of glutamate dehydrogenase in either actively growing or stationary phase cells. This suggests that NH_4^+ assimilation proceeds via amination of glutamate. Neither GS nor its substrate appear to be directly involved in the regulation of nitrogenase (Kleinschmidt and Kleiner 1981) as may be the case in other nitrogen-fixing systems (Tyler 1978). However, although GS synthesis appears to be unregulated, its activity is modulated by adenylylation in response to ammonium ion (Lepo *et al.* 1982). Razaaq and Leach (1978) have demonstrated that arginine is synthesized via a transacetylase pathway, which appears to be con-stitutive in one particular wild-type isolate (Razaaq and Leach 1979).

Hardisson *et al.* (1969) studied the metabolism of aromatic com-pounds and demonstrated that para-hydroxybenzoate and shikimate are metabolized via the β-ketoadipate pathway, in common with other bacteria (Durham and Ornston 1980), but that benzoate is handled by a novel and unique pathway via catechol and a meta cleavage to yield α-hydroxy muconic semialdehyde. A novel pathway for the synthesis of isoprenoid compounds (such as the ubiquinone side chain), via an acetolactate intermediate, has been demonstrated by Pandian *et al.* (1981).

Synthesis of alginic acid, a copolymer of guluronic acid and mannuronic acid, has been studied by several groups. It requires the derepression of phosphomannose isomerase (epimerase) (Horan *et al.* 1983) and utilizes starting materials produced by the Embden–Meyerhof pathway (Lynn and Sokatch 1984). The role of Ca^{2+} in this biosynthetic process was examined by Annison and Couperwhite (1984).

6.3.1.2 *Regulation of the primary pathways*

In an extensive series of experiments, Dawes and his collaborators have utilized chemostats in order to assess the effects of C-, O-, and N-limitation upon the activities of key enzymes of the Entner–Douderoff pathway, the TCA cycle, and the accumulation of PHB. These experiments were supplemented by studies of metabolite and co-factor regulation of purified enzyme preparations (Carter and Dawes 1979; Ward *et al.* 1977; Jackson and Dawes 1976). Their results show that the flow of electrons and carbon is regulated such that under nitrogen-fixing conditions and oxygen limitation, PHB is accumulated, acting as an effective electron and carbon sink. As the pO_2 rises, PHB synthesis is repressed and its utilization, via acetyl-CoA and the TCA cycle is initiated, serving not only as an energy and electron source for cellular activities, but also for O_2-reduction and hence 'respiratory protection' of nitrogenase. Growth on a fixed nitrogen source led to similar results, although co-limitation of growth by nitrogen and oxygen yielded an anomalous PHB response which could be attributed to the conflicting demands of incipient nitrogen-fixation, respiratory protection, and PHB synthesis/degradation. Liao and Atkinson (1971*a,b*) have studied the regulation of the catabolic and anabolic pathways, in the presence of fixed nitrogen, leading from phosopho-enol-pyruvate (PEP). They observed that, as in other microbial systems, a high energy charge (high ATP) favours the anabolic pathways [via PEP carboxylase (EC 4.1.1.31)] and inhibits the catabolic [via pyruvate kinase (EC 2.7.1.40)].

Bosma *et al.* (1984, 1985) have devised an improved purification procedure for pyruvate dehydrogenase (PDC) and have proposed a model for its quaternary structure which corresponds to the monomer of the cubic octomeric structure that has been proposed for the *Escherichia coli* PDC.

Diauxic growth on glucose and acetate or several TCA cycle intermediates has recently been observed, with the latter compounds being the preferentially utilized substrates (George and Melton 1982; Terzaghi 1980*c*, and unpublished results). The level of isocitrate lyase was highest during the acetate phase of growth and ATP generation and nitrogen fixation (as measured by acetylene reduction) were highest during the glucose-utilizing phase.

There is little or no question that the availability of fixed nitrogen (Shah *et al.* 1972) or the lack of molybdenum, respectively (Nagatani and Brill 1974; Bishop *et al.* 1982) repress the synthesis of one component or both of nitrogenase. Furthermore, the introduction of NH_4^+ results in the immediate cessation of nitrogenase activity (Laane *et al.* 1980) though the enzyme itself seems not to be sensitive to inhibition by NH_4^+ (Gordon *et al.* 1981). The cellular mechanism (s) of regulation of the activity or the synthesis of nitrogenase remains obscure (see Chapter 1). However, it does not follow from these observations that azotobacters cannot be net exporters of fixed nitrogen. Gordon *et al.* (1981) report that under carefully contrived conditions, the bulk of ^{15}N supplied as dinitrogen can be recovered from the medium. Perhaps more significantly from the ecological point of view, as much as 40 per cent of the nitrogen fixed by azotobacters can be recovered in the form of extracellular, mostly high molecular weight material (Jensen 1954).

6.3.2 Hydrogenase

The evolution of H_2, representing a substantial loss of energy and reducing power as well as inhibiting nitrogenase (Hill 1978; Robson and Postgate 1980), is an inevitable consequence of the activity of nitrogenase; however, the amount produced seems to depend on the physiological state of the cell (Yates *et al.* 1980; Hageman and Burris 1980). In order to cope with this potentially wasteful side product, an apparently Ni^{2+}-dependent synthesis (Partridge and Yates 1982) of a unidirectional hydrogenase (Hup) is induced (Yates *et al.* 1980). The level of hydrogenase activity appears to be modulated by physiological factors such as the growth rate or level and nature of carbon source. Although the mechanism of action of hydrogenase is not yet clear, there is evidence that it operates via an independent electron transport chain and that it can contribute substantially to the nitrogen fixing economy of the cell (Yates *et al.* 1980). The contribution to 'respiratory protection' of nitrogenase from the reducing power generated by hydrogenase has not been determined, but the energy-efficiency of hydrogenase activity is reduced with increasing pO_2 (Laane *et al.* 1979). One group of azotobacters (*Derxia*) are able to grow chemo-autotrophically on H_2 and CO_2, utilizing the energy and reducing equivalents supplied by the hydrogenase (Pedrosa *et al.* 1980).

6.3.3 Electron transport

The very high rate of respiration of azotobacters early attracted interest in their electron transport system (Robson and Postgate 1980; Tissières 1956), and has provided insight into 'respiratory protection' of nitrogenase. Although this topic is covered in Chapter 3, particularly as it relates to nitrogen fixation and membrane integrity, we wish to briefly discuss the electron transport system in the special context of 'respiratory protection'.

A system of converging and branching pathways of electron flow was proposed (Haddock and Jones 1977) in which only one common energy conserving step is shared (ubiquinone Q to a cytochrome b). Only one of the electron sources (NADH) is believed to enter the system via energy conserving steps, and one of the principal pathways from the b-type cytochrome to O_2, via cytochrome d, is not coupled to ATP synthesis. Under conditions of O_2 excess, cytochrome d becomes the predominant cytochrome species and there is an uncoupling of the first step in electron transport from NADH. More recently, Yang (1981) has made a detailed study of the redox properties of the main components of the electron transport system solubilized from membrane preparations and has suggested modifications to some of the details of the original proposal.

Most current work focuses on the structure and function of various elements of the electron transport, hydrogen evolution, and oxygen management systems. Ferredoxin iron-sulphur clusters were examined spectroscopically by Lutz *et al.* (1983) and by Morgan *et al.* (1984*b*). Anaerobic assembly of the molecule yields only the species bearing the 4Fe–4S cluster (Morgan *et al.* 1984*a*). An improved flavodoxin preparation technique was described by Carlson and Laugenman (1983). The amino-acid sequence of the dihaem cytochrome C4 was determined and sequence similarities between the two halves was noted, along with a detectable homology with the monohaem cytochrome c of unrelated species (Ambler *et al.* 1984). Spectral studies of cytochrome d suggest that it is a major pathway of energy nonconservative NADH reduction (Yang 1984) to which is generally attributed metabolic protection of nitrogenase from oxygen. Further support for this idea is offered by the isolation of TCA cycle mutants (citrate synthase negative) in which nitrogen fixation is unusually oxygen-sensitive (Ramos and Robson 1985*a,b*). Cytochrome oxidase and superoxide dismutase of *A. vinelandii* have been compared to those of a variety of other species and genera of free-living bacteria (Jurtshuk *et al.* 1984; Moore *et al.* 1984).

The amino-acid sequence of the nitrogenase iron protein of *A. vinelandii* has been determined and compared with that of other free-living nitrogen-fixing organisms (Hausinger and Howard 1982). This group has also begun a chemical dissection of this molecule by examining thiol reactivity (Hausinger and Howard 1983) and iron removal (Anderson and Howard 1984). Removal of up to 16 electrons from the entire nitrogenase complex did not alter its activity (Lough *et al.* 1983) Scherings *et al.* (1983) report the *in-vitro* reconstitution of an oxygen-tolerant three-component nitrogenase complex with a sedimentation constant of 34S. The molybdenum–iron protein has been reported to catalyse the reduction of a number of substrates by molecular hydrogen (Wang and Watt 1984). These authors suggested that should molecular oxygen be reduced in this reaction, a degree of physiological oxygen protection would be realized. Wong and Maier (1984) examined the molecular hydrogen oxidizing capacities of several of the known electron transport cytochromes and Kow and Burris (1984) purified and characterized a membrane-bound hydrogenase.

The interaction of ammonium and methyl ammonium with glutamine synthetase and ammonium transport was examined (Barnes *et al.* 1983; Jayakumar and Barnes 1984). The latter molecule appears to inhibit only ammonium transport, a point relevant to the phenotype of the methyl-ammonium-resistant mutants reported by Gordon and Jacobson (1983). It was pointed out that use of the latter molecule as an ammonium analogue can give misleading results as it is converted by glutamine synthetase into the impermeant metabolite, gamma-glutamyl methyl amide. The effect of ammonium (Cefudo *et al.* 1984; Klugkist and Haaker 1984) and of molecular oxygen (Post *et al.* 1983*a*; Tsygankov *et al.* 1984) has been further examined under a variety of natural and artificial physiological conditions. Phosphate limitation, under conditions of nitrogen sufficiency, led to an increase in respiratory rate (Haldenwang and Behrens 1983) which the authors suggest is due to a shift from oxidative phosphorylation to substrate phosphorylation. Goyal *et al.* (1984) determined the effect of water stress upon several characteristics of the phosphate-scavenging alkaline phosphatase.

6.3.4 Permeation

The membrane system has received a great deal of attention, particularly in the context of electron transport and the cellular energy economy (see Chapter 3), so we shall discuss only the transport of ions (including Fe and Mo) and organic molecules.

Because of its role in regulation of nitrogenase activity, there has been some interest in the mechanism of transport of NH_4^+ (Laane *et al.* 1980; Barnes and Zimnial 1981). While there is agreement that the ions enter the cell along an electric gradient facilitated by a specific carrier which is not coupled to ATP, the identity of the carrier is uncertain. Ca^{2+}, which can be maintained at levels below that of the medium, is believed to be transported by means of a specific calcium-proton antiport (Barnes 1980). Molybdenum and iron are essential for nitrogen fixation, yet molybdenum is rare and iron is generally inaccessible to the cell. Page and von Tigerstrom (1982), in the course of studies on induction of transformation competence, have described several membrane proteins induced by iron-starvation and several others produced under molybdenum-starvation (although at least one of the latter is produced constitutively at reduced levels). Simultaneously, small fluorescent molecules with extraordinarily high binding constants for $Fe^{2+,3+}$ and MoO_2^{2-} (Stiefel *et al.* 1980; Fekete 1983*a*), are produced which are perhaps transported specifically by the newly-appearing membrane components. The later authors have also studied sidero-phore reductase activity in *A. vinelandii* (Lodge *et al.* 1983). Both iron and molybdenum are stored intracellularly by specific binding proteins: bacterioferritin cytochrome $b_{557.5}$ for iron (Stiefel *et al.* 1980) and molybdenum storage protein (Pienkos and Brill 1981).

Glucose uptake has been examined by Barnes (1972) and shown to require an inducible L-malate/FMN-dependent carrier which is not coupled to ATP or PEP. Energy-dependent transport of pyruvate and TCA cycle intermediates has been studied by Postma and Van Dam (1971). One constitutive carrier for pyruvate was found, as were four inducible carriers for: citrate and isocitrate; succinate, fumarate, L-malate and oxaloacetate; 2-oxo-glutarate; and malonate.

The amount of cytoplasmic membrane and intracytoplasmic membrane (Post *et al.* 1983*b*) and the localization of cytochrome oxidase (Payne and Socolovsky 1984) and of nitrogenase, glutamine synthetase, and glutamate (Röckel *et al.* 1983) on these membranes was studied.

6.3.5 Encystment

In common with a number of other groups of bacteria, the azotobacters are able to assume a metabolically quiescent and physically resistant state, termed a cyst. Following exposure to certain molecules, such as β-hydroxy-butyrate (BHB) or crotonate, the cell passes through an orderly sequence of events including an increase in the level of BHB dehydrogenase, accumulation of granules of PHB, cessation of DNA duplication, and nitrogen fixation and finally cessation of cell division (Sadoff *et al.* 1971; Stevenson and Socolofsky 1966). In the presence of adequate levels of calcium, a resilient external double layer is laid down, comprised of several types of uronic acids (Page and Sadoff 1975) as well as proteins and lipids. Su *et al.* (1981), Su and Sadoff (1981), and Reusch and Sadoff (1983) have characterized a number of novel lipids with resorcinolic, pyronic, phenolic, and polymethylene side chains, whose syntheses accompany encystment. In the later paper referred to, a model of an inner/outer membrane structure is proposed.

Upon return to favourable conditions, germination of the cysts similarly entails an orderly sequence of events. Respiration, CO_2 production, RNA and protein synthesis start immediately at a modest rate and increase some hours later upon initiation of nitrogen fixation and DNA synthesis (Loperfido and Sadoff 1973).

6.4 MOLECULAR BIOLOGY AND GENETICS OF NITROGEN FIXATION IN AZOTOBACTER

Proceeding from this basic biological and biochemical background, we can now consider the molecular biology of azotobacters. We shall discuss sequentially the location, organization and regulation of genetic information affecting nitrogen fixation. This information relates principally to *A. vinelandii.* *

6.4.1 DNA
6.4.1.1 *Physical characterization*

DNA hyridization studies with the Azotobacteraceae (DeLey and Park 1966) resulted in the classification of the family into four genera, as already mentioned. In all genera there is a preponderance of GC base pairs, ranging from 53–59 per cent for *Azomonas*, 55–59 per cent for *Beijerinckia*, 63–66 per cent for *Azotobacter*, to 70 per cent for *Derxia*.

A. *vinelandii* DNA has an average GC content of 65 per cent and contains no unusual bases (Sadoff *et al.* 1979). Within the cell, nucleoids (bodies which take up standard nuclear stains) have been detected (Dondero and Zelle 1953; Sadoff *et al.* 1971). By a comparison of DNA renaturation kinetics ($Cot_{1/2}$) and sedimentation values, Sadoff *et al.* (1979) showed that *A. vinelandii* and *Escherichia coli* folded chromosomes are similar in respect to unique sequence length and chromosome size. However, using their figure of 15×10^{-14}g of DNA per mid-exponential phase *A. vinelandii* cell (Sadoff *et al.* 1971) and the mass of a single *E. coli* genome as 3.5×10^{-15}g, they calculated that each *vinelandii* cell must contain 40 chromosomes.

Several groups have measured the cellular DNA content, with varying results. Belozerski *et al.* (1957) claimed the DNA was 0.7–0.8 per cent of the cell dry weight (less than the normal bacterium), while Sadoff *et al.* (1971) reported that encysting cells are 2.5 per cent dry weight DNA, and Vela and Gonzalez-Lopez (1982) indicated that the very small filterable cell type of *A. vinelandii* has 'twice as much DNA as the large *A. vinelandii* cell'. Robson (Postgate 1981) found that growing *A. chroococcum* cells have fifty times more DNA than *E. coli*. In part these differences reflect variation in the cell types or the age of the cultures examined, but the variation reported does not always correspond to the changes in cell volume. The decrease in DNA content with culture age (Sadoff *et al.* 1971) implies a very large increase in DNA in cells which are just commencing growth.

---

* The following terms and abbreviations will be used: Fe protein = component II or dinitrogenase reductase; MoFe protein = component I or dinitrogenase; FeMoco = iron-molybdenum cofactor; *nif*H = gene coding for the subunits of the Fe protein of nitrogenase in *K. pneumoniae*; *nif*D = gene coding for the α-subunit of the MoFe protein; *nif*K = gene coding for the β-subunit of the MoFe protein.

The estimates of 40 chromosomes combined with a Cot$_{1/2}$ value comparable to *E. coli* DNA are difficult to reconcile with the one-hit ultraviolet inactivation curve for *A. vinelandii* (Terzaghi 1980*b*) and the ease with which Nif⁻ mutants are isolated or transformed to Nif⁺ (see below). These results suggest that there is only one functional genomic unit per cell. Thus, if *A. vinelandii* does contain 40 chromosomes, as suggested by Sadoff *et al.* (1979), it is easiest to reconcile the information by concluding that there are a great many which are inactive. Alternatively, the cell counts may be low and the chromosome number accordingly in error.

Since the above was written, several techniques have now been employed to look at the question of how many functional chromosomes and *nif* genes there are in azotobacters. Medhora *et al.* (1983) constructed a gene library of *A. vinelandii* which was then tested for *leu* and *nif*KDH genes. Their data suggest that *leu* is present in 40 copies while only two *nif* regions were detected. Robson *et al.* (1984*b*) found variation in plasmid number and size in eight *A. chroococcum* strains, but no detectable genetic markers on the plasmids. Genome size, as determined by restriction enzyme digestion and two-dimensional electrophoresis, produced smaller estimates of total cellular DNA than that found by Sadoff *et al.* (1979) but still an average of 20–5 chromosome equivalents per cell. Ultraviolet dose-response curves for these strains fell into two classes, one with exponential kinetics and the other with multi-hit killing curves (ca. 10 hits). More information is needed to understand the differences in these responses.

6.4.1.2 *Plasmids*

Azotobacter DNA need not all be chromosomal. Plasmids have been found in *A. chroococcum, A. vinelandii, A. beijerinckii* and *A. paspali* (Kennedy *et al.* 1981; Robson 1981; Slot *et al.* 1982), ranging in size from 7 to 260 Mdal, but as yet there is no evidence to link nitrogen fixation genes with plasmids. In fact, in the *A. vinelandii* strain (UW) from which virtually all Nif⁻ mutant strains studied have been derived, no plasmids of any size have been found in several independent investigations (Bishop *et al.* 1977*a*; Kennedy *et al.* 1981; Robson 1981). An *A. vinelandii* strain carrying a large plasmid which has DNA sequences hybridizing to a Kp *nif* probe has now been reported (Yano *et al.* 1984). In strains with plasmids, plasmid curing has not yet resulted in a Nif⁻ phenotype (Kennedy *et al.* 1981; Robson 1981) though further investigations are underway (Slot *et al.* 1982). We may conclude that to date, there is no evidence to suggest other than a chromosomal location for *nif* genes except in the strain of Yano *et al.* (1984).

6.4.2 Conservation of *nif* genes

Within the genome, Southern blot hybridization was observed between a 6 kb fragment of *Klebsiella pneumoniae* DNA carrying nitrogenase structural genes *nifKDH* (Ruvkun and Ausubel 1980) and five fragments of EcoR1-digested total DNA (12.5, 8.1, 3.5, 2.2, and 0.8 kb) from *A. vinelandii* Nif⁻ mutant UW 10 (see Section 6.4.5.1). The region of homology was, in fact, restricted almost exclusively to the *nifDH* genes as the three largest fragments hybridized only to a smaller probe carrying that region and not to fragments carrying other portions of the *K. pneumoniae nif* region (Fig. 6.1). The 2.2 kb fragment hybridized to a fragment carrying parts of *nifK* and *D*, and the 0.8 kb fragment hybridized to both the *nifKD* and *nifDH* fragments. None of the *A. vinelandii* DNA hybridized to a 3.6 kb *Rhizobium meliloti* fragment which showed homology to the initial 6 kb *Klebsiella* probe. No explanation was suggested for either the latter observation or for the high number of *A. vinelandii* fragments which hybridized, but considering the possibility that there could be an alternative nitrogen fixation system in *A. vinelandii* (Bishop *et al.* 1980) and that nitrogen fixation gene sequences are reiterated in *R. phaseoli* (Quinto *et al.* 1982), perhaps the multi-fragment hybridization can be explained by additional copies of *nifD* and/or *nifH* carried in the genome. Alternatively, the structural genes, which are adjacent in *K. pneumoniae*, may have become dispersed in *A. vinelandii*.

Evidence is accumulating that, as suggested, there are two *nifH* genes in *Azotobacter* species. Premakur, Lemos, and Bishop (1984) ran two-

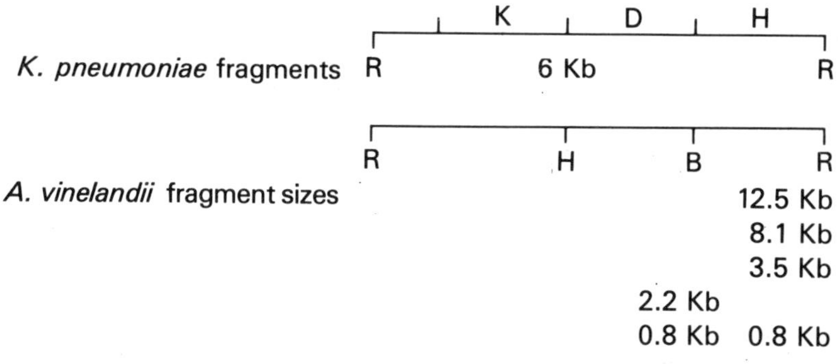

FIG. 6.1. Sizes of EcoRl fragments in *A. vinelandii* total DNA which hybridize to *Klebsiella pneumoniae* restriction fragments *nif KDH* (6kb), *nifKD*, and *nifDH*. Data from Ruvkun and Ausubel (1980). R = EcoRl site; H = HindIII site; B = BamHl site.

dimensional electrophoretic gels of Nif⁻ mutants grown in Mo-deficient medium, and reported the presence of a protein with similar mobility to the conventional dinitrogenase reductase but with a more basic pI. From *A. chroococcum* gene libraries, recombinant plasmids carrying *nif* genes have been characterized to show a *nifKDH* cluster and an additional unlinked *nifH*-like gene sequence (Jones *et al.* 1984). A region corresponding to the *nifHDK* promoter was also identified, exhibiting multicopy inhibition of acetylene reduction activity and growth in N₂ which could be alleviated by multiple copies of Kp *nifA*.

Further confirmation of homology of *Klebsiella* and *Azotobacter nif* information is given by the cloning of two EcoR1 fragments of *A. vinelandii* DNA, a 2.6 kb fragment which shows sequence homology to (*Klebsiella*) *nifD*- and *nifK*-genes and a 1.4 kb fragment with homology to *nifD* and possibly *nifH* (Bishop and Bott 1983). Interestingly, from 3 248 transformants screened, only these two fragments were found which hybridized to the *nifDK* probe. There are discrepancies between the data of Bishop and Bott (1983) and Ruvkun and Ausubel (1980) which must be resolved so that an EcoR1 map of the structural gene region of *A. vinelandii* can be constructed.

Similarity of *nif* regulation between *K. pneumoniae* and *A. vinelandii* and *A. chroococcum* has also recently been demonstrated (Kennedy and Robson 1983). Introduction of the *nifA*-gene from *K. pneumoniae* into *Azotobacter* mutants deficient in both components I and II restored the Nif⁺ phenotype to the presumed regulatory mutants but not to a nitrogenase structural gene mutant. These results are interpreted to suggest that *nifA* activation of *nif* genes might also be conserved among diazotrophs.

6.4.3 Mutant isolation

Difficulty in isolating mutants and subsequent mutant instability have continually plagued genetic analysis of azotobacters (see Mishra and Wyss 1968; Page and Sadoff 1976). To date only the Nif⁻ structural gene mutants have been easy to isolate and maintain stably, though instability also exists here for iron metabolism (W. J. Page, personal communication) and for ammonium derepression (to be discussed). Investigations into instability are beginning, but as yet, the underlying cause is unknown.

The failure to obtain mutants and thereby construct a genetic map has been attributed to the postulated high chromosome number, with consequent problems in segregation and mutant expression (Sadoff *et al.* 1979). As already noted, this explanation is not completely satisfactory. We can only conclude that more research on the organization of the azotobacter genome is needed to determine the reasons why non-mutability, instability, and variable DNA content have been observed.

A process for selection of (mostly Trp⁻) auxotrophic mutants of *A. chroococcum* has been reported (Kashyap and Chopra 1983). Although it was not possible to obtain *nif* mutants of *A. chroococcum* using either Tn5 or Tn7, the aforementioned Fos mutants were isolated using nitrosoguanidine (Robson *et al.* 1984*a*). These mutants fix N_2 with certain sugars as carbon source only at low oxygen partial pressures. One type of mutant can be corrected by introduction of the Kp *nifA* gene, while the other type is complemented by the *E. coli* citrate synthase gene.

6.4.4 Gene transfer

Analysis of those mutants which have been isolated is no longer limited by methods of gene transfer but rather by a lack of scientists to use the methodologies available. Possibilities exist for transfer and expression of genes from other genera, perhaps because *A. vinelandii* appears to lack a restriction system and hence readily accepts DNA (David *et al.* 1981).

6.4.4.1 *Transformation with total DNA*

Successful transformation with total DNA was first claimed in *A. chroococcum* (Sen and Sen 1965), but another decade passed before a reproducible transformation system was developed in *A. vinelandii* (Page and Sadoff 1976*a, b*). Page and co-workers (Page and von Tigerstrom 1978, 1979; Page and Doran 1981; Page 1982) have systematically investigated the factors leading to induction of competence and hence transformation by varying the components of the system and they are now able reliably to achieve transformation frequencies of 10^{-2} to 10^{-3} in liquid media. Intergeneric transformation of nitrogen fixation genes into *A. vinelandii* was also initiated by this group (Page 1977, 1978). They are currently determining the mechanism for DNA transport through the cell envelope (Doran and Page 1981, 1982) and they have further defined the physiology of the transformation system, in particular, factors affecting competence and transport of DNA across the cell envelope (Doran and Page 1983).

The major limitation of the transformation system is now not bacterial receptiveness, but the paucity of mutants. Those mutants which have been mapped were done so when the system was not well developed and therefore the linkage studies (to be discussed) must be regarded as tentative for several reasons. In general, unpurified DNA has been used and transformation frequencies have often been very low. We have little understanding of the size, uptake, and recombination of the DNA; a high frequency of recombination might equally be explained by a low percentage of cells being competent, but those competent being able to accept many DNA molecules. This is particularly a concern when transformation frequencies are very low. In addition, no information is available on the size of the DNA molecules in the transforming mixtures and no linkage studies have yet been done with limiting concentrations of DNA. Furthermore, in many cases only large colony formation on nitrogen-free medium was measured and no further characterization of suspected transformants was performed. To illustrate this potential problem, we have found instability and persistent segregation in some transformants (manuscript in preparation).

Transformation data regarding linkage could be improved by using DNA cut by restriction endonucleases and separated into size classes. At limiting DNA concentrations, only genes on the same fragment should be jointly transformed. This method could also be used to identify fragments encoding specific genes by transformation of Nif⁻ mutants to Nif⁺ with given fragment sizes. Finally, the transformation procedure used successfully for the introduction of plasmid DNA (David *et al.* 1981) should be tried with linear DNA.

6.4.4.2 *Plasmid transfer: transformation and conjugation*

Gene transfer using plasmids has more recently come into vogue. Cannon and Postgate (1976), David *et al.* (1981), and Kennedy and Robson (1983) incorporated *nif* or antibiotic resistance genes into wide-host-range cloning vectors (RP41; RP4; RSF1010; pKT230) which were then introduced into mutants of *A. vinelandii* or *A. chroococcum* by conjugation or transformation (using calcium) and the genes expressed. This general methodology is just beginning to be exploited for analytical purposes, and should be of value for complementation, cloning, and transposon mutagenesis of azotobacter genomes in a similar fashion to that which has been employed for analysis in the rhizobia.

6.4.4.3 *Transduction*

The third means of gene transfer is by transduction; it was first reported in azotobacter in 1962 (Wyss and Nimeck) but that work was not continued. The development of a generalized transduction system has now been reported for *A. chroococcum* (Du *et al.* 1979); transduction frequencies of 10^{-5} have been observed.

6.4.5 **Mutant analysis**

Azotobacter mutants we have found described in the literature which we have surveyed are listed in Table 6.2; it should not be considered an exhaustive listing. It is dominated by mutants affected in nitrogen fixation and contains few amino-acid-requiring mutants. Recent claims of isolating such auxotrophs (Leach and Battikhi 1978; Du *et al.* 1979) indicate that selection procedures rather than failure to transport amino acids (Roberts and Brill 1981) have been at fault.

TABLE 6.2

Types of mutants isolated in azotobacter strains.
Those mutants marked with an asterisk are discussed further in the text

Mutant phenotype	Strain	Reference
Mannitol non-fermenting	A. chroococcum	Smith (1935)
Nif⁻ (nitrate induced)	A. vinelandii	Stumbo and Gainey (1938)
Leucine requiring, glucose non-fermenting, growth factor⁻	Azm. agilis	Karlsson and Barker (1948)
Nif⁻ (pH conditional)	A. vinelandii; Azm. agilis	Wyss and Wyss (1950)
Nif⁻	A. vinelandii; Azm. agilis	Green et al. (1953)
Mucoid-deficient	A. vinelandii	Bush and Wilson (1959)
Nif⁻ (citric acid cycle auxotroph)	A. vinelandii	Mumford et al. (1959)
Nif⁻, Adenine⁻, Antibiotic-resistant (Str, Ery, Pen)	A. vinelandii	Mishra and Wyss (1968)
*Nif⁻	A. vinelandii UW	Fisher and Brill (1969) Shah et al. (1973)
Nif⁻	A. vinelandii	Sorger and Trofimenkoff (1970)
Nif⁻ (temperature sensitive)	A. vinelandii	Benemann et al. (1971)
*Nif⁺ ammonium derepressed	A. vinelandii UW	Gordon and Brill (1972)
Nif⁻, Ad⁻, Uracil⁻, Hypoxanthine⁻, Rif-resistant	A. vinelandii	Page and Sadoff (1976)
*Ammonium excreting	A. vinelandii	Gordon et al. (1977)
Nif⁻ (oxygen-conditional)	A. vinelandii	Schenk and Wyss (1977)
Amino acid and vitamin auxotrophs	A. vinelandii	Leach and Battikhi (1978)
Met⁻, Gln⁻, Cys⁻, Arg⁻, Nif⁻	A. chroococcum	Du et al. (1979)
Tetra-methyl-P-phenylenediamine oxidase⁻	A. vinelandii	Hoffman et al. (1979)
Sulfite reductase	A. vinelandii	Sadoff et al. (1979)
*Nif⁻ (Mo-storage⁻)	A. vinelandii UW	Peinkos et al. (1980)
Ammonium derepressed	A. vinelandii	Terzaghi (1980c)
Tungsten tolerant	A. vinelandii	Riddle et al. (1981)
Nif⁻	A. chroococcum	Sadisvam and Gouri (1981)
Hup⁻ (hydrogen uptake)	A. chroococcum	Postgate et al. (1982)
Glucose metabolism	A. vinelandii OP	McKenney and Melton (1983)

Studies of the molecular biology of mutants defective in nitrogen fixation began in 1969–70 when two groups successfully isolated Nif⁻ mutants in *A. vinelandii* (Fisher and Brill 1969; Sorger and Trofimenkoff 1970). The nitrosoguanidine-induced mutants isolated from a mucoid-deficient mutant (strain UW or OP) of strain O became the first characterized Nif⁻ mutants and have formed the foundation for most subsequent work on the nitrogen fixation system in azotobacters. Virtually no mutants have yet been isolated in the oxygen protection system for nitrogenase.

6.4.5.1 *Structural gene mutants*

The Nif⁻ mutants isolated by Fisher and Brill (1969) were initially characterized biochemically and serologically for the presence and activity of the completed MoFe and Fe proteins. On this basis, a limited number of categorizations could be made, and with the very fast advances then occurring in *Klebsiella*, analysis of *A. vinelandii* Nif mutants stagnated for a while. More recently, two-dimensional gel electrophoresis profiles (Bishop *et al.* 1980, 1982; Page and Collinson 1982; Riddle *et al.* 1982) have given further information and shown the value of not depending solely on an antibody reaction as a measure of synthesis of a protein; two-dimensional gels should be performed on all mutants not yet analysed.

The first mutant category (UW 91) includes mutants with MoFe protein activity *in vitro* and also antigenic cross-reactivity to the complete Fe protein but which lack Fe-protein activity *in vitro* (Shah *et al.* 1973). Mutant UW 91 cannot be activated by the *K. pneumoniae nifA*-gene product (Kennedy and Robson 1983) and has long been considered defective in the structural gene for the Fe protein of nitrogenase (Bishop *et al.* 1977). It phenotypically resembles *nifH* mutants of *K. pneumoniae* (see Section 6.3); however, it is able to grow on molybdenum-deficient nitrogen-free medium (see Section 6.4.6) (Bishop *et al.* 1980). In plate transformations with other Nif mutants, Nif⁺ recombinants were recovered in all crosses but little recombination was observed with MoFe-protein mutants (UW 6, 10, and 38), suggesting close linkage (Bishop and Brill 1977; Fig. 6.2).

The second category contains mutants (UW 6, 38) which have Fe-protein activity *in vitro* but which lack both *in vitro* activity and antigenic cross-reactivity to the complete MoFe protein (Shah *et al.* 1973). Mutant UW 38 is unusual in that it hyper-produces the Fe protein but, because revertants regain (half) the wild-type activity of both MoFe and Fe proteins, it is thought to be mutant at a single site (Shah *et al.* 1974). Two-dimensional gel analysis shows that the mutants produce the β- (but not the α-) subunit of the MoFe protein (Bishop *et al.* 1980; Bishop, unpublished, reported in Page and Collinson 1982). By marker rescue, Bishop and Bott (1983) found that the 2.6 kb fragment of wild-type DNA could correct both the UW 6 and UW 38 mutations. Transformation data also suggest close linkage between these mutations (Bishop and Brill 1977). As with UW 91, both strains can grow in molybdenum-deficient nitrogen-free medium (Bishop *et al.* 1980; Page and Collinson 1982).

Mutant UW 10 has Fe-protein activity *in vitro* and antigenic cross-reactivity to the MoFe protein, but no MoFe-protein activity *in vitro* (Shah *et al.* 1973). That it is mutated in one of the structural genes for the MoFe protein (Bishop *et al.* 1977*b*) is supported by its marker rescue with the aforementioned 1.4 kb cloned fragment of wild-type DNA (Bishop and Bott 1983). A low number of Nif$^+$ transformants are obtained in crosses with the previously discussed mutants (Bishop and Brill 1977), and the strain grows in molybdenum-deficient nitrogen-free medium (Bishop *et al.* 1980).

6.4.5.2 *Other Nif mutants*

Although there are similarities between the phenotypes of the remaining mutants and *Klebsiella* mutants, more information must be obtained before gene- and functional-assignments can be made.

Mutant UW 45 has an active Fe protein but only 12 per cent of the normal serological reactivity to the functional MoFe protein; it can be fully activated *in vitro* by the addition of FeMo cofactor, including cofactor extracted from mutant UW 10 or even from other nitrogen-fixing genera (Nagatani *et al.* 1974). *Klebsiella* genes *nifB*, *N*, and *E* have all been implicated in FeMoco synthesis, and possibly *nifQ*, *J*, and *C* are also involved (Roberts and Brill 1981; Chapter 1). The tentative mapping data place the UW 45 mutation between the structural genes and the UW 1 mutation (Bishop and Brill 1977; Fig. 6.2).

Mutant UW 3 lacks *in vitro* activity for both nitrogenase proteins but has serological cross-reactivity to the MoFe protein (Shah *et al.* 1973). No Fe protein is detectable in two-dimensional gels from bacteria transferred to molybdenum-containing nitrogen-free media, but the strain can grow in molybdenum-deficient nitrogen-free media and under such conditions, produces a protein of the same size and same pI as the Fe protein (Bishop *et al.* 1980; Page and Collinson 1982). The mutant can be transformed to Nif$^+$ by DNA from Nif$^-$ strains UW 1, 2, 45, 6, 38, 10, and 91, but not by the 2.6 or 1.4 kb cloned fragments (Bishop and Bott 1983). The mutation maps (tentatively) close to the structural genes (Bishop and Brill 1977; Fig. 6.2), but could be on either side of those genes.

Mutant UW 71 (Pienkos *et al.* 1980) is derived from UW 10; it has an active Fe protein and material that is antigenically cross-reactive with the MoFe protein, but its lack of *in vitro* activity for the protein is attributed to a defect in molybdenum storage. It does not grow in molybdenum-deficient nitrogen-free medium (Page and Collinson 1982) and has not been mapped.

Several mutants used in early tests have not been further characterized. Mutant UW 112 has antigenic cross-reactivity to both the Fe and MoFe proteins but lacks both protein activities *in vitro*, and mutant UW 118 also lacks both protein activities *in vitro* but cross-reacts with antiserum against the MoFe protein; mutant UW 120 has MoFe-protein activity but lacks both *in vitro* activity of, and antigenic cross-reactivity to, the Fe protein (Shah *et al.* 1973). Mutant UW 100 has an active Fe protein but an inactive MoFe protein; the protein is detectable antigenically but fails to give a staining response for Fe (Brill 1976).

Mutant UW 1 lacks both antigenic cross-reactivity and *in vitro* activity of the Fe and MoFe proteins (Shah *et al.* 1973) but because it can revert to Nif⁺ (Fisher and Brill 1969), it was thought to be a point mutant in a regulatory gene. This hypothesis is reinforced by the recent demonstration that it can be activated to the Nif⁺ phenotype by the *nifA* gene product of *K. pneumoniae* (Kennedy and Robson 1983). It can grow (slowly) in molybdenum-deficient nitrogen-free medium and under these circumstances makes a protein with the properties of the Fe-protein (Bishop *et al.* 1980). Nif⁺ revertants are NH_4^+-repressible (Bishop and Brill 1977) and UW 1 is transformable to Nif⁺ by DNA from Nif⁻ mutants UW 6, 38, 45, 91, and 3, and also by the DNA of two cowpea *Rhizobium* strains, but not by DNA from mutant UW 2, by the 2.6 kb and 1.4 kb cloned fragments, or by DNAs from six other Rhizobium strains, whether *Rhizobium* or *Bradyrhizobium* (Bishop and Brill 1977; Page 1978; Bishop and Bott 1983). Its tentative map position is distant from the structural genes and close to the UW 2 mutation (Fig. 6.2). Mutant UW 1 differs from UW 2 in the types of revertants produced.

Evaluation of mutant UW 2 is difficult, in part because much of the data has not been fully published or has come from reviews. The strain lacks antigenic cross-reactivity to both of the completed structural proteins (Gordon and Brill 1972) but can be transformed to Nif⁺ with DNA from mutants UW 3, 6, 38, 45, and 91, but not UW 1 (Bishop and Brill 1977). Because it was suspected to be mutant in a regulatory function, reversion analysis was performed to try to obtain revertants with derepressed synthesis of the nitrogenase structural genes (Gordon and Brill 1972). Three distinct phenotypes were found in 21 spontaneous revertants; seven had detectable acetylene reduction activity in the presence of normally nitrogenase-repressing levels of ammonium,

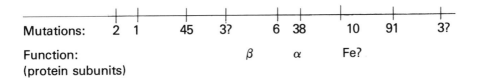

FIG. 6.2. Tentative map of Nif⁻ mutations of *A. vinelandii* derived from plate transformation data of Bishop and Brill (1977) and Page (1978). Distances are not quantitative.

of which three had about 25 per cent of wild-type acetylene reduction activity and four about 0.5 per cent (at the same cell density).

Characterization of one such revertant, strain UW 59, showed that it grew much more slowly on nitrogen-free medium and had half the wild-type specific acetylene reduction activity. Another NH_4^+-derepressed strain, UW 590, with better growth and acetylene reduction activity with and without NH_4^+, was subsequently isolated, presumably from this strain (Gordon *et al.* 1977). Activities of all the NH_4^+-assimilatory enzymes were normal in all these strains (Gordon 1978). All Nif$^+$ transformants of strain UW 2 using DNA from strain UW 59 were NH_4^+-derepressed, whereas only about 16 per cent were NH_4^+-derepressed when strain UW 590 was the DNA donor (Gordon *et al.* 1977; Gordon 1978). Thus, while it is clear that strain UW 590 carries at least two mutations, it is not clear that it is derived from strain UW 59 in a single step. Ammonium-excreting mutants have also been obtained from these NH_4^+-derepressed strains for use in the development of plant associations (Bishop *et al.* 1977*b*; Gordon *et al.* 1977; Brill 1981).

On the basis of the simultaneous loss and re-appearance of both nitrogenase components in strain UW 59, presumably by a single mutational event, and the fact that several classes of revertants were isolated, control by a positive effector was proposed (Gordon and Brill 1972). Roberts and Brill (1981) suggested that the effector may be analogous to the *nifA* product of *Klebsiella*, which we have seen is the case for mutant UW 1 (Kennedy and Robson 1983). A role in molybdenum-processing should also be investigated (see Section 6.4.6).

The discovery that both the UW 2 strain and a derivative NH_4^+-derepressed Nif$^+$ strain (possibly UW 590) are unstable (Terzaghi 1980*a*) has further complicated the characterization of mutant UW 2. Two kinds of Nif$^-$ colonies were found in the UW 2 culture, only one of which (strain PPD 82) was transformable to Nif$^+$. With wild-type (UW) DNA, Nif$^+$ transformants of PPD 82 were all NH_4^+-repressible; in contrast, the NH_4^+-depressed strain produced Nif$^+$ variants which did not express NH_4^+-derepressed activity but whose DNA transformed the PPD 82 strain to both Nif$^+$ and (about 5 per cent) Nif$^+$ NH_4^+-derepressed colonies. Instability of some transformants was observed. Thus, characterization of mutant UW 2 could be complicated by whatever factors lead to this instability.

6.4.5.3 *Genetic mapping*

A tentative map, constructed from the combined transformation data (Bishop and Brill 1977; Page 1978), is given in Fig. 6.2. The proposal that there are two gene clusters in *A. vinelandii*, with regulatory genes unlinked to structural genes (Bishop and Brill 1977), must still be considered speculative.

Transcriptional studies are consistent with a physical contiguity of the three nitrogenase structural genes in *A. vinelandii*. Krol *et al.* (1982) found a large mRNA (1.4×10^6 MW) appearing in bacteria grown a minimum of 34 min in nitrogen-free medium which could hybridize to a *Klebsiella* DNA fragment carrying *nifDH*. A mRNA with similar properties was present in *Klebsiella*, and was considered to be the transcript for the *nifYKDH* operon. An *A. vinelandii* band of 0.2×10^6 MW, not found in *Klebsiella* RNA, also hybridized strongly to the *nifDH* DNA. This band was thought to be a degradation product, but considering the possibility of redundancy in nitrogen fixation genes in *A. vinelandii*, there may be another explanation.

Krol *et al.* (1982) also reported that the Fe and MoFe proteins of *A. vinelandii* are synthesized independently and at different rates, with MoFe protein detectable before Fe protein. To correlate the difference in the onset of synthesis with a single polycistronic message for the (presumed) region, they concluded there must be further regulation at the level of translation. Their finding is in contradiction to the report of coordinate synthesis of the two proteins in *A. vinelandii* (Gordon and Brill 1972; Shah *et al.* 1972).

A restriction map for a region of the *A. chroococcum* genome carrying *nif* genes (Jones *et al.* 1984) confirms close linkage and expression of *nifKDH* as a single transcript, and places *nifV* 15 kb away from *nifKDH*.

6.4.5.4 *Intergeneric gene transfer*

Transfer and expression of genetic information affecting nitrogen fixation in *A. vinelandii* has also been possible with DNA from other azotobacter species and genera, with *Klebsiella* DNA, and with rhizobial DNAs.

Rhizobium DNA from both fast- and slow-growing strains was able to transform Nif$^-$ mutants in another (non-UW) *A. vinelandii* strain to Nif$^+$ at low frequency (Page 1978); these mutants were presumably structural since they were not transformable to Nif$^+$ by DNA from mutants UW 6 and UW 10. The transformants were stable (though no analysis to exclude plasmid introduction was performed). Those transformations correcting the the β-subunit of the MoFe protein were indistinguishable from normal *vinelandii* Nif$^+$ transformants, whereas transformants of mutations in the α-subunit reduced acetylene at a lower rate and had a lag in the outset of activity. This result raises the question of why no homology was detected between *R. meliloti nif*DNA and *vinelandii* DNA (Ruvkun and Ausubel 1980).

The aforementioned transformation of UW 1 to Nif$^+$ with some cowpea *Rhizobium* DNAs was also at lower frequency than *vinelandii* DNA transformations. The transformants were stable, had full acetylene reduction activity but showed a delay in onset of activity. Nitrogen fixation genes might be chromosomally located in *R. loti* (Pankhurst *et al.* 1983), whereas *nif* genes are plasmid encoded in *R. meliloti*, *R. trifolii*, and *R. leguminosarum* (see Chapters 8–11). Thus, although it has been suggested that plasmid-encoded *nif* rhizobial genes might be more efficacious in transforming *A. vinelandii* mutants (Kennedy *et al.* 1981), the data available do not support this suggestion.

Plasmid instability was observed when mutants defective in either the Fe or MoFe protein were complemented with plasmid-encoded *K. pneumoniae nif* genes (Cannon and Postgate 1976). About 50 per cent of wild-type acetylene-reduction activity was restored to (presumed structural gene) mutants UW 10, 100, 38, and 91 by transfer of RP41 (an RP4 derivative carrying sufficient *K. pneumoniae nif*-information to confer acetylene reduction activity to *E. coli*). In contrast to the results of Kennedy and Robson (1983), the regulatory mutant UW 1 did not regain acetylene-reduction activity upon receipt of the plasmid (cf. results with cowpea Rhizobium DNA). The activity with the complemented UW 10 mutant was also half that obtained when UW 10 was transformed with Rhizobium DNA. However, Cannon and Postgate (1983) have recently reported that they are unable to repeat their 1976 claim of correction of structural *nif* mutations in *A. vinelandii* by *Klebsiella nif*, and that therefore their original results must be viewed with caution.

Transformation of the Nif$^-$ isolate PPD 82 for Nif characteristics with DNA from *A. paspali* and *Azm. agilis* has also been successful (Terzaghi, unpublished).

6.4.5.5 *Symbiotic gene transfers*

Intergeneric transfer of symbiotic genes and other kinds of genetic information has also been observed. Antibiotic resistance and nutrition markers from *Rhizobium* were transformed into *Azotobacter* (Sen *et al.* 1969). The claimed (reciprocal) transformation of (fast-growing) *R. trifolii* with azotobacter DNA to a form which could grow on nitrogen-free medium, reduce acetylene, and still nodulate clover (Venkataraman *et al.* 1975), has not yet been repeated.

Symbiotic genes involved in nodulation were transformed from *Rhizobium* into Nif$^-$ *A. vinelandii* strains by selecting first for transformation to Nif$^+$, confirming that Nif$^+$ colonies were *Azotobacter*, and then screening them for the presence of the putative gene product. This technique assumes either linkage between genes or multiple entry of DNA molecules into competent bacteria. In the first experiment, DNA coding for trifoliin binding sites was transferred from *R. trifolii* into mutants UW 1, 6, and 10 as determined by agglutination of 13 per cent of the Nif⁺ transformants by trifoliin or by antiserum to clover root antigens (Bishop *et al.* 1977*a*). No plasmids were found in the transformants, suggesting incorporation of the DNA rather than complementation.

Similarly, the O-antigen-related polysaccharide present on the cell surface of nodulating *R. japonicum* was transferred to three of fifty Nif⁺ transformants of Nif⁻ strain UW 10 (Maier *et al.* 1978). Positive transfer of the gene was assumed if transformants were agglutinated by antiserum against whole cells of *R. japonicum*. By use of such inter-generic transfers, one hopes that it will be possible to move genes into *A. vinelandii* which will be of use for developing novel plant-microbial associations (Brill 1981).

6.4.6 Alternative nitrogen fixation system?

The most controversial topic in the molecular biology of azotobacters at the moment is the question of whether *A. vinelandii* has an alternative nitrogen-fixation system or whether some other explanation can be found for the ability of tungsten-resistant mutants and some, but not all, Nif⁻ mutants to grow and reduce acetylene in a modified nitrogen-free medium. The tungsten-resistant mutants (Riddle *et al.* 1982) give similar results to the Nif⁻ mutants; since the latter have been more extensively characterized, they will be described in detail.

In 1980, Bishop *et al.* reported that Nif⁻ mutants UW 1, 3, 6, 10, 38, 91, and also wild-type, could grow in nitrogen-free medium if it was also deficient in molybdenum (1.3 p.p.b.). The low nitrogenase activity previously detected when *Azotobacter* was inoculated into molybdenum-deficient media was attributed to molybdenum contamination (Esposito and Wilson 1956; Nagatani and Brill 1974; Pienkos and Brill 1981*a*), but molybdenum contamination cannot explain the growth, albeit poor, of Nif⁻ mutants in nitrogen-free media and, in particular, by mutants 'conventionally' defective in or lacking a structural protein of nitrogenase.

The system governing molybdenum-deficient growth, termed Nif B versus Nif A for the conventional molybdenum-containing system (Bishop *et al.* 1980), arose out of work on Nif⁺ revertants of (α-subunit) mutants UW 6 and 10. The 'revertants' (pseudo-revertants) obtained carried the original Nif⁻ mutation (as shown by back-crossing), grew at about half the wild-type rate in nitrogen-free medium, had low levels of acetylene-reduction activity, and under such growth conditions did not have a low-temperature paramagnetic resonance signal at $g = 3.65$. The absence of this signal (considered diagnostic for the molybdenum-containing component of nitrogenase but originating from unpaired electrons in iron) led to the supposition of a possible involvement of molybdenum, the testing of the Nif⁻ strains, and the surprising growth response. To the extent that they have undergone similar testing, the pseudo-revertants resemble strain UW 59.

Two-dimensional gel electrophoresis has been performed on a variety of mutants (Bishop *et al.* 1980, 1982; Page and Collinson 1982; Riddle *et al.* 1982). A consistent change in the protein pattern in growth under molybdenum-deficiency has been claimed, with the replacement of the relevant MoFe-protein subunits by two new proteins of approximately 57 and 50 kdal (Table 6.3). Upon addition of NH_4^+ to molybdenum-deficient cultures, these proteins (and acetylene reduction activity) disappear; the same result is obtained upon addition of molybdenum to most mutants, but with wild-type, acetylene reduction activity increases and the MoFe subunit proteins are formed. Mutant UW 3 is unusual in that, upon addition of molybdenum, nitrogenase activity increases for a short time, then disappears; these changes are paralleled by the disappearance of the 57 and 50 K proteins and the appearance of a protein resembling the β-subunit of the MoFe protein (Page and Collinson 1982). In some conditions, four new proteins were observed (Bishop *et al.* 1982).

In all cases of molybdenum-deficient growth, a protein corresponding in size and pI to the Fe protein, and with Fe-protein activity (P. E. Bishop, personal communication) was observed on the gels. This is not surprising for those mutants affected in the MoFe protein, but there are two unusual cases. Extracts of mutant UW 3 grown in molybdenum-sufficient medium did not contain this Fe protein but it was present in molybdenum-deficient cultures. When molybdenum was added to molybdenum-deficient cultures it disappeared (and nitrogenase activity was lost) (Page and Collinson 1982). Mutant UW 91, which has an active MoFe protein but a functionally inactive Fe protein, grew in molybdenum-deficient medium and had the two new proteins and an unchanged Fe-type protein. Thus, a model for this system must accommodate synthesis of the Fe protein and correction of the functionally-inactive protein in addition to the synthesis of the claimed two new proteins.

There are additional data difficult to explain by a conventional nitrogenase system. The pseudo-revertant strains of UW 6 and UW 10 grew on nitrogen-free medium containing tungstate and produced the four new proteins, while wild-type produced an inactive MoFe protein under these conditions and did not grow (Bishop *et al.* 1980). Since tungsten inhibits molybdenum uptake (see Chapter 1), this result was explained by suggesting that the molybdenum-uptake system had been inactivated in the pseudo-revertants.

Secondly, vanadium or rhenium can stimulate acetylene-reduction activity in all molybdenum-starved cultures, whether Nif⁻ mutant or wild-type (Bishop *et al.* 1982; Page and Collinson 1982). If these metals were substituting for molybdenum, as suggested (see Pienkos *et al.* 1980), no effect would be expected with conventional Nif⁻ mutants.

When molybdenum is added to molybdenum-starved cultures, protein synthesis is required for the expression of nitrogenase activity, but there is a lag before activity is detected; activity is lost in the presence of streptomycin or rifampicin (Page and Collinson 1982). While both growth and acetylene reduction activity are measurably less than normal in molybdenum-deficient medium, the two are not similarly affected; acetylene reduction activity is always decreased more than growth (Bishop *et al.* 1982).

Bishop *et al.* (1980) explain their results by proposing an alternative nitrogen fixation system which is repressed by Mo, and have isolated mutants to test this system (Bishop, personal communication). However, by their model it is difficult to understand why a mutant defective in molybdenum-storage (UW 71) does not grow in molybdenum deficient medium (Page and Collinson 1982). This is particularly

TABLE 6.3

Generation times, acetylene reduction activities, and proteins present in two-dimensional electrophoretic gels of A. vinelandii *wild-type and Nif⁻ strains grown with and without nitrogen and molybdenum. Data assembled from Bishop et al.* (1980, 1982) *and from Page and Collinson* (1982).

Strain	Growth conditions							
	-N, +Mo			-N, -Mo			→ + Mo*	
	(1)	(2)	(3)	(1)	(2)	(3)	(2)	(3)
UW (wild-type)	3	823	60 K 55 K 30 K	3.5	32	57 K 50 K 30 K	244	60 K 55 K 30 K
UW 6	0	0	60 K 0 30 K	3.5-4.5	30	57 K 50 K 30 K	7	60 K 0 30 K
UW 10	0	0	60 K 55 K 30 K	3.5-4.5	21	57 K 50 K 30 K	8	60 K 55 K 30 K
UW 3	0	0	60 K 55 K 0	3-4	39	57 K 50 K 30 K	221	60 K 55 K 30 K
Pseudo revertants (UW 6; UW 10	5-6.2	33-42	2 basic 2 acidic			2 basic 2 acidic		

(1) = generation time, h at 28 °C. (2) = specific activity (n moles C_2H_4 h^{-1} 10^8 cells^{-1}). (3) = MW of new protein seen on gels. *Proteins observed upon addition of Mo to Mo-deficient cultures.

worrying because *A. vinelandii* can normally store large quantities of molybdenum (Pienkos and Brill 1981*b*) and might be an efficient molybdenum scavenger (see Chapter 1). Page and Collinson (1982) have proposed that a 'hybrid' nitrogenase is formed between a normal Fe protein and β-subunits of the MoFe protein. This model does not explain the growth of strain UW 91 which is defective in the Fe protein.

To further complicate the issue, Terzaghi *et al.* (1984) found that their wild-type UW strain and the Nif⁻ isolate PPD 82 did not grow at all in nitrogen-free molybdenum-deficient medium. In contrast, two NH_4^+-derepressed strains derived from strain PPD 82 grew to normal levels, reduced acetylene, and produced proteins corresponding in size and pI to normal nitrogenase component I polypeptides.

The existence of an alternative nitrogen fixation system is not yet proven, but attention has switched from the MoFe protein to the presence of two *nifH*-like proteins (Premakur *et al.* 1984; Robson *et al.* 1984*a*; Jones *et al.* 1984). How this second *nifH*-like protein might be regulated (by Mo deficiency) and whether Fe protein activity is sufficient to provide alternative nitrogen fixation remain to be demonstrated.

6.5 CONCLUDING REMARKS

The key element of nitrogen fixation, nitrogenase, seems to be a constant and conserved feature of all nitrogen-fixing systems. This demanding biological process has become established in a wide variety of ecological niches, each presenting a special set of problems to be solved.

The azotobacters are distinguished by their ability to fix nitrogen in an aerobic environment accompanied by a voracious and catholic appetite for organic carbon and energy. Work has just begun on the genetic determination of the system, and particularly with the application of physical (restriction fragment) mapping and transposon mutagenesis, we shall soon learn whether there are fundamental departures from the *Klebsiella* scheme. That the conserved *Klebsiella* probe hybridizes to a number of fragments of an *Azotobacter* restriction enzyme digest, and that there may be an 'alternative nitrogen-fixing system', suggest that there are differences between these two organisms beyond the obvious aerobic/anaerobic nitrogen-fixing capability. Whether or not these are fundamental differences at the nitrogenase level remains to be seen.

Our still incomplete knowledge of the intermediary metabolism of this group of organisms suggests substantial modulation, if not reorganization, to accommodate the rigorous demands of nitrogen fixation. Much work remains to be done on the regulation of key metabolic crossroads, on the establishment of the peripheral pathways, in understanding the partitioning of electron flow through the various energy-conserving and wasting transport chains, and in clarifying the various active, facilitated, and passive membrane transport systems.

Finally, on quite a different level, the optimistic reports concerning the extracellular and possibly inter- or intracellular association of some azotobacters with specific plant species demands further investigation. This area seems particularly promising for the fruitful collaboration of the molecular biologist and the field biologist to produce bacterial strains of practical utility.

REFERENCES

AMBLER:, R. P., DANIEL, M., MELIS, K., and STOUT, C. D. (1984). *Biochem. J.* **222**, 217–27.

ANDERSON, G. L. and HOWARD, J. B. (1984). *Biochem.* **23**, 2118–22.

ANNISON, G. and COUPERWIIITE, I. (1984). *Appl. Microbiol.* **19**, 321–5.

APTE, R. and SHENDE, S. T. (1981). *Zentbl. Bakt. ParisitKde* Abt. II **136**, 548–54.

BARNES, Jr., E. M. (1972). *Arch. Biochem. Biophys.* **152**, 795–9.

— (1980). *J. Bact.* **143**, 1086–9.

— and ZIMNIAK, P. (1981). *J. Bact.* **146**, 512–16.

—— and JAYAKUMAR, A. (1983). *J. Bact.* **156**, 752–7.

BARRERA, C. R. and JURTSHUK, P. (1970). *Biochim. biophys. Acta* **220**, 416–29.

BECKING, J. H. (1978). In *Environmental role of nitrogen-fixing blue-green algae and asymbiotic bacteria* (ed. V. Granhall), *Ecol. Bull. (Stockholm)* **26**, 116–29.

— (1981). In *The prokaryotes: A handbook on habitats, isolation, and identification of bacteria*, Vol. 1, (ed. M. P. Starr, H. Stolp, N. G. Prüper, A. Balows, and H. G. Schlegel), pp.795–817. Springer-Verlag, Heidelberg.

BEIJERINCK M. W. (1901). *Zentbl. Bakt. ParasitKde* Abt. II **7**, 561–82.

BELOZERSKI, A. N., ZAITSEVA, V., GAVRILOVA, L. P., and MINEYEVA, L. V. (1957). *Mikrobiologiya* **26**, 409.

BENEMANN, J. R., SHEU, C. W., and VALENTINE, R. C. (1971). *Arch. Microbiol.* **79**, 49–58.

BERGEY, D. H., HARRISON, F. C., BREED, R. S., HANMER, B. W., and HUNTOON, F. M. (eds) (1923). *Bergey's manual of determinative bacteriology.* Williams & Wilkins, Baltimore.

BINGLE, W. H., DORAN, J. L., and PAGE, W. J. (1983). *Abstr. A. Mtg. Am. Soc. Microbiol.,* K155.

—— (1984). *J. Bact.* **159**, 251–9.

BISHOP, P. E. and BOTT, K. F. (1983). *Abstr. A. Mtg. Am. Soc., Microbiol.,* K70.

— and BRILL, W. J. (1977). *J. Bact.* **130**, 954–6.

— DAZZO, F. B., APPELBAUM, E. R., MAIER, R. J., and BRILL, W. J. (1977a). *Science* **198**, 938–40.

— GORDON, J. K., SHAH, V. K., and BRILL, W. J. (1977b). In *Genetic engineering for nitrogen fixation* (ed. A. Hollaender), pp. 67–80. Plenum Press, New York.

— Jarlenski, D. M. L., and Hetherington, D. R. (1980). *Proc. natn. Acad. Sci. U.S.A.* **77**, 7342–6.

— (1982). *J. Bact.* **150**, 1244–51.

Boddey, R. M., Chalk, P. M., Victoria, R. L., Matsui, E., and Döbereiner, J. (1983). *Can. J. Microbiol.* **29**, 1036–45.

Bosma, H. J., de Kok, A., van Markwijk, B. W., and Veeger, C. (1984). *Eur. J. Biochem.* **140**, 273–80.

——— Westphal, A. H., and Veeger, C. (1985). *Eur. J. Biochem.* **142**, 541–9.

Bresters, T. W., De Abreu, R. A., De Kok, A., Visser, J., and Veeger, C. (1975). *Eur. J. Biochem.* **59**, 335–45.

Brill, W. J. (1976). In *Symbiotic nitrogen fixation in plants* (ed. P. S. Nutman), pp. 39–47. Cambridge University Press.

— (1981). *Scientific American* **245**, 146–56.

Brown, M. E. (1976). *J. appl. Bact.* **40**, 341–8.

Buchanan, R. E. and Gibbons, N. E. (eds.) (1974). *Bergey's manual of determinative bacteriology* (8th edn), pp. 253–5. Williams & Wilkins, Baltimore.

Bulen, W.A., Le Comte, J. R., and Lough, S. (1973). *Biochem. biophys. Res. Commun.* **54**, 1279–81.

Burk, D., Lineweaver, H., and Horner, C. K. (1934). *J. Bact.* **27**, 325–40.

Bush, J. A. and Wilson, P. W. (1959). *Nature, Lond.* **184**, 381–2.

Cannon, F. C. and Postgate, J. R. (1976). *Nature, Lond.* **260**, 271–2.

— — 1983). *Nature, Lond.* **306**, 290.

Carlson, P. S. and Chaleff, R. S. (1974). *Nature, Lond.* **252**, 393–4.

Carlson, R. and Laugenman, N. (1983). *Prep. Biochem.* **13**, 489–505.

Carter, J. S. and Dawes, E. A. (1979). *J. gen. Microbiol.* **110**, 393–400.

Cefudo, F. J., de la Torre, A., and Paneque, A. (1984). *Biochim. biophys. Res. Commun.* **123**, 431–7.

Chen, M. and Crichton, R. R. (1982). *Biochim. biophys. Acta* **707**, 1–6.

Chung, A. G. and Franzen, J. S. (1969). *Biochem.* **8**, 3175–84.

Dalton, H. and Postgate, J. R. (1969). *J. gen. Microbiol.* **54**, 463–73.

Dams, E., Vandenberghe, A., and De Wachter, R. (1983). *Nucl. Acids Res.* **11**, 1245–52.

David, M., Tronchet, M., and Dénarié, J. (1981). *J. Bact.* **146**, 1154–7.

Day, J. M., Neves, M. C. P., and Döbereiner, J. (1975). *Soil Biol. Biochem.* **7**, 107–12.

De Ley, J. and Park, I. W. (1966). *Antonie van Leeuwenhoek* **32**, 6–16.

De Polli, H., Matsui, E., Döbereiner, J., and Salati, E. (1977). *Soil Biol. Biochem.* **9**, 119–23.

de Smedt, J., Bauwens, M., Tytgat, R., and De Ley, J. (1980) *Int. J. Syst. Bact.* **30**, 106–22.

Derx, H. G. (1950). *Ann. Bogorienses* **1**, 1–11.

Dicker, H. J. and Smith, D. W. (1981). *Appl. & environ. Microbiol.* **42**, 740–4.

Döbereiner, J. (1966). *Pesq. Agropec. Brasil* **1**, 357–65.

— (1970). *Zentbl. Bakt. ParasitKde.* Abt. II **124**, 224–30.

— and Boddey, R. M. (1981). In *Current perspectives in nitrogen fixation* (eds A. N. Gibson and W. E. Newton), pp. 305–12. Australian Academy of Science, Canberra.

— Day, J. M., and Dart, P. J. (1972). *J. gen. Microbiol.* **71**, 103–16.

DONDERO, N. C. and ZELLE, M. R. (1953). *Science* **118**, 34–6.

DORAN, J. L. and PAGE, W. J. (1981). *Abstr. A. Mtg. Am. Soc. Microbiol.*, 85.

—— (1983). *J. Bact.* **155**, 159–68.

— (1982). *Abstr. 13th Int. Congr. Microbiol.*, pp. 51–3.

DU, Q., CHEN, K., and FAN, C. (1979). *Acta bot. sin.* **21**, 449.

DUFF, J. T. and WYSS, O. (1961). *J. gen. Microbiol.* **24**, 273–89.

DURHAM, D. R. and ORNSTON, N. L. (1980). *J. Bact.* **143**, 834.

EADY, R. R. and ROBSON, R. L. (1984). *Biochem. J.* **224**, 853–62.

EMERY, C. and MELTON, T. (1983). *Abstr. A. Mtg. Am. Soc. Micrbiol.* K25.

ESPOSITO, R. G. and WILSON, P. W. (1956). *Proc. Soc. exp. Biol. Med.* **93**, 564–7.

FEKETE, F. A., LODGE, J. S., and EMERY, T. F. (1983*a*). *Abstr. A. Mtg. Am. Soc. Microbiol.*, K178.

— SPENCE, J. T., and EMERGY, T. (1983*b*). Appl. & environ. Microbiol. **46**, 1297–3001.

FISHER, R. J. and BRILL, W. J. (1969). *Biochim. biophys. Acta* **184**, 99–105.

FUREY, W., O'DONNELL, S., and STOUT, C. D (1981). *J. biol. Chem.* **256**, 4185–95.

GALLON, J. R. (1981). *Trends biochem. Sci.* **6**, 19–23.

GEORGE, S. E. and MELTON, T. (1982). *Abstr. A. Mtg. Am. Soc. Microbiol.* K13.

GILES, K. L. and WHITEHEAD, H. C. M. (1975). *Cytobios* **14**, 49–61.

GONZALEZ-LOPEZ, J. and VELA, G. R. (1981). *Nature, Lond.* **289**, 588–90.

GORDON, J. K. (1978). *Proc Steenbock-Kettering Int. Symp. Nitrogen Fixation*, D-21.

— (1981). In *The prokaryotes: A handbook on habitats, isolation and identification of bacteria*, Vol. 1 (ed. M. P. Starr, H. Stolp, N. G. Prüper, A. Balows, and H. G. Schlegel), pp. 781–94. Springer-Verlag, Heidelberg.

— and BRILL, W. J. (1972). *Proc. natn. Acad. Sci. U.S.A.* **69**, 3501–3.

—— SHAH, V. K., and BRILL, W. J. (1981). *J. Bact.* **148**, 884–8.

— WACEK, T. J., MAIER, R. J., BISHOP, P. E., and BRILL, W. J. (1977). In *Agriculture and energy* (ed. W. Lockeretz), pp. 353–9. Academic Press, New York.

— and JACOBSON, M. R. (1983). *Can. J. Microbiol.* **29**, 973–8.

GOYAL, V., CHETAL, S., and NAINWATEE, H. S. (1984). *Folia Microbiol.* **29**, 228–32.

GREEN, M., ALEXANDER, M., and WILSON, P. W. (1953). *Proc. Soc. exp. Biol. Med.* **82**, 361–3.

GUTSCHIK, B. P. (1980). In *Nitrogen fixation*, Vol. 1 (ed. W. E. Newton and W. H. Orme-Johnson), pp. 17–27. University Park Press, Baltimore.

HAAKER, V. and VEEGER, C. (1977). *Eur. J. Biochem.* **77**, 1–10.

HADDOCK, B. A. and JONES, C. W. (1977). *Bact. Rev.* **41**, 47–99.

HAGEMAN, R. V. and BURRIS, R. H. (1980). *Biochim. biophys. Acta* **591**, 63–75.

HALDENWANG, L. and BEHRENS, U. (1983). *Z. allgemeine Microbiol.* **23**, 491–4.

HARDISSON, C., SALA-TREPAT, J. M., and STANIER, R. Y. (1969). *J. gen. Microbiol.* **59**, 1–11.

HAUSINGER, R. P. and HOWARD, J. B. (1982). *J. biol. Chem.* **257**, 2483–7.

—— (1983). *J. biol. Chem.* **258**, 13486–92.

HILL, S. (1978). In *Environmental role of nitrogen-fixing blue-green algae and asymbiotic bacteria* (ed. U. Granhall), *Ecol. Bull. (Stockholm)* **26**, 130–6.

HOFFMAN, P. S., MORGAN, T. V., and DERVATANIAN, D. V. (1979). *Eur. J. Biochem.* **100**, 19–28.

HORAN, N. J., DAWES, E. A., and JARMAN, T. R. (1981). *Proc. Soc. gen. Microbiol.* **8**, 251.

— JARMAN, T. R., and DAWES, E. A. (1983). *J. gen. Microbiol.* **129**, 2985–90.

HOWARD, J. B., LORSBACH, T. W., GHOSH, D., MELIS, K., and STOUT, C. D. (1983). *J. biol. Chem.* **258**, 508–22.

IMAI, T. (1979). *J. Biochem.* **85**, 887–90.

JACKSON, F. A. and DAWES, E. A. (1976). *J. gen. Microbiol.* **97**, 303–12.

JAYAKUMAR, A. and BARNES, Jr., E. M. (1984). *Arch Biochem. Biophys.* **231**, 95–101.

JENSEN, V. and HOLM, E. (1975). In *Nitrogen fixation by free-living micro-organsims* (ed. W. D. P. Stewart), pp. 101–19. Cambridge University Press.

JENSON, H. L. (1954). *Bact. Rev.* **18**, 195–214.

— (1965). In *Soil nitrogen* (ed. W. V. Bartholomew and F. E. Clark), pp. 440–86. American Society of Agronomy, Madison.

JONES, R., WOODLEY, P., and ROBSON, R. (1984). *Mol. Gen. Genet.* **197**, 318–27.

JORDAN, D. C. (1981). In *Current perspectives in nitrogen fixation* (ed. A. H. Gibson and W. E. Newton), pp. 317–20. Australian Academy of Science, Canberra.

JURTSHUK, P., MUELLER, T. J., and WONG, T. Y. (1981). *Biochim. biophys. Acta* **637**, 374–82.

— LIU, J.-K., and MOORE, E. R. B. (1984). *Appl. & environ. Microbiol.* **47**, 1185–7.

KARLSSON, J. L. and BARKER, H. A. (1948). *J. Bact.* **56**, 671–7.

KASHYAP, L. R. and CHOPRA, V. L. (1983). *Ind. J. exp. Biol.* **21**, 485–6.

KASS, D. L., DROSDOFF, M., and ALEXANDER, M. (1971). *Soil Sci. Am. Proc.* **35**, 286–9.

KENNEDY, C., CANNON, F., CANNON, M., DIXON, R., HILL, S., JENSEN, J., KUMAR, S., MCLEAN, P., MERRICK, M., ROBSON, R., and POSTGATE, J. (1981). In *Current perspectives in nitrogen fixation* (ed. A. H. Gibson and W. E. Newton), pp. 146–56. Australian Academy of Science, Canberra.

— and ROBSON, R. (1983). *Nature, Lond.* **301**, 626–28.

KLEINER, D. and KLEINSCHMIDT, J. A. (1976). *J. Bact.* **128**, 117–22.

KLEINSCHMIDT, J. A. and KLEINER, D. (1978). *Eur. J. Biochem.* **89**, 51–60.

— (1981). *Arch. Microbiol.* **128**, 412–15.

KLUGKIST, J. and HAAKER, H. (1984). *J. Bact.* **157**, 148–51.

KNOSP, O., VON TIGERSTROM, M., and PAGE, W. J. (1984). *J. Bact.* **159**, 341–7.

KOW, Y. W. and BURRIS, R. H. (1984). *J. Bact.* **159**, 564–9.

KROL, A. J. M., HONTELEZ, J. G. J., ROOZENDAAL, B., and VAN KAMMEN, A. (1982). *Nucl. Acids Res.* **10**, 4147–57.

LAANE, C., HAAKER, H., and VEEGER, C. (1979). *Eur. J. Biochem.* **97**, 369–77.

— KRONE, W., KONINGS, W., HAAKER, H., and VEEGER, C. (1980). *Eur. J. Biochem.* **103**, 39–46.

LEACH, C. K. and BATTIKHI, M. (1978). *Proc. Soc. gen. Microbiol.* **5**, 110.

LEPO, J. E., WYSS, O., and TABITA, F. R. (1982). *Biochim. biophys. Acta.* **704**, 414–21.

— STACEY, G., WYSS, O., and TABITA, F. R. (1979). *Biochim. biophys. Acta.* **568**, 428–36.

LI. J., WANG, J., ZHONG, Z., TU, Y., and DONG, B. (1980). *Scientia sin.* **23**, 897–904.

LIAO, C. L. and ATKINSON, D. E. (1971a). *J. Bact.* **106**, 31–6.

— (1971b). *J. Bact.* **106**, 37–44.

LIN, L. P. and SADOFF, H. L. (1968). *J. Bact.* **95**, 480–6.

LODGE, J. S., FEKETE, F. A., and EMERY, T. F. (1983). *Abstr. A. Mtg. Am. Soc. Microbiol.* K. 173.

LOPERFIDO, B. and SADOFF, H. L. (1973). *J. Bact.* **113**, 841–6.

LOUGH, S., BURNS, A., and WATT, G. D. (1983). *Biochem.* **22**, 4062–6.

LUTZ, M., MOULIS, J. M., and MEYER, J. (1983). *FEBS Lett.* **163**, 212–16.

LYNN, A. R. and SOKATCH, J. R. (1984). *J. Bact.* **158**, 1161–2.

MAIER, R. J., BISHOP, P. E., and BRILL, W. J. (1978). *J. Bact.* **134**, 1199–201.

MEDHORA, M., PHADNIS, S. H., and DAS, H. K. (1983). *Gene* **25**, 355–60.

MISHRA, A. K. and WYSS, O. (1968). *The Nucleus* **11**, 96–105.

MISHUSTIN, E. N. and SHIL'NIKOVA, V. K. (1971). In *Biological fixation of atmospheric nitrogen*, pp. 184–250. Macmillan, London.

MCKENNY, D. and MELTON, T. (1983). *A. Mtg Am. Soc. Microbiol.*, Abstr. I104.

MOORE, E. R. B., NORROD, E. P., and JURTSHUK, P. (1983). *Abstr. A. Mtg Am. Soc. Microbiol.*, K33.

——— (1984). *FEMS microbiol. Lett.* **24**, 261–5.

MORGAN:, T. V., STEPHENS, P. J., BURGESS, B. K., and STOUT, C. D. (1984a). *FEBS Lett.* **167**, 137–41.

—— DEVLIN, F., STOUT, C. D., MELLIS, K. A., and BURGESS, B. K. (1984b). *Proc. natn. Acad. Sci. U.S.A.* **81**, 1931–5.

MORTENSON, L. G., HAMILTON, P. B., and WILSON, P. W. (1955). *Biochim. biophys. Acta* **16**, 238–44.

MUMFORD, F. E., CARNAHAN, J. E., and CASTLE, J. E. (1959). *J. Bact.* **77**, 86–90.

NAGATANI, H. H. and BRILL, W. J. (1974). *Biochim. biophys. Acta* **362**, 160–6.

— SHAH, V. K., and BRILL, W. J. (1974). *J. Bact.* **120**, 697–701.

NORRIS, J. R. and JENSEN, H. L. (1958). *Arch. Microbiol.* **31**, 198–205.

PAGE, W. J. (1977). *Abstr. A. Mtg Am. Soc. Microbiol.* **77**, K82.

— (1978). *Can. J. Microbiol.* **24**, 209–14.

— (1982). *Can. J. Microbiol.* **28**, 389–97.

— and COLLINSON, S. K. (1982). *Can. J. Microbiol.* **28**, 1173–80.

— and DORAN, J. L. (1981). *J. Bact.* **146**, 33–40.

— and SADOFF, H. L. (1975). *J. Bact.* **122**, 145–51.

— (1976a). *J. Bact.* **125**, 1080–7.

— (1976b). *J. Bact.* **125**, 1088–95.

— and VON TIGERSTROM, M. (1978). *Can. J. Microbiol.* **24**, 1590–4.

— (1979). *J. Bact.* **139**, 1058–61.

— (1982). *J. Bact.* **151**, 237–42.

— (1983). *Can. J. Microbiol.* **29**, 1110–18.

— and HUYER, M. (1984). *J. Bact.* **158**, 496–502.
PANDIAN, S., SAENGCHJAN, S., and RAMAN, T. S. (1981). *Biochem. J.* **196**, 675–81.
PANKHURST, C. E., BROUGHTON, W. J., and WIENEKE, U. (1983). *J. gen. Microbiol.* **129**, 2535–43.
PARTRIDGE, C. D. P. and YATES, M. G. (1982). *Biochem. J.* **204**, 339–44.
PAYNE, H. R. and SOCOLOFSKY, M. D. (1984). *J. Bact.* **159**, 946–50.
PEDROSA, F. O., DÖBEREINER, J., and YATES, M. G. (1980). *J. gen. Micrbiol.* **119**, 547–51.
PIENKOS, P. T. and BRILL, W. J. (1981). *J. Bact.* **145**, 743–51.
— KLEVICKIS, S., and BRILL, W. J. (1981). *J. Bact.* **145**, 248–56.
— SHAH, V. K., and BRILL, W. J. (1980). In *Molybdenum and molybdenum-containing enzymes* (ed. M. Coughlan) 385–401. Pergamon Press, New York.
POST, E., GOLECKI, J. R., and OELZE, J. (1982). *Arch. Microbiol.* **133**, 75–82.
— KLEINER, D., and OELZE, J. (1983*a*). *Arch. Microbiol.* **134**, 68–72.
— VAKALOPOULOU, E., and OELZE, J. (1983*b*). *Arch. Microbiol.* **134**, 265–9.
POSTGATE, J. R. (1967). *Antonie van Leeuwenhoek* **33**, 113–20.
— (1981). In *Current perspectives in nitrogen fixation* (ed. A. H. Gibson and W. E. Newton), pp. 217–28. Australian Academy of Science, Canberra.
— (1982). *Phil. Trans. R. Soc.* B **296**, 375–85.
— PARTRIDGE, C. D. P., ROBSON, R. L., SIMPSON, F. B., and YATES, M. G. (1982). *J. gen. Microbiol.* **128**, 905–8.
POSTMA, P. W. and VAN DAM, K. (1971). *Biochim. biophys. Acta* **249**, 515–27.
PREMAKUR, R., LEMOS, E. M., and BISHOP, P. E. (1984). *Biochim. biophys. Acta* **797**, 64–70.
QIAN-YOU, D. and CHENG-YING, F. (1981). *Acta bot. sin.* **23**, 453–8.
QUINTO, C., DE LA VEGA, H., FLORES, M., FERNÁNDEZ, L., BALLADO, T., SOBERÓN, G., and PALACIOS, R. (1982). *Nature, Lond.* **299**, 724–6.
RAMOS, J. L. and ROBSON, R. L. (1985*a*). *J. gen. Microbiol.* In press.
— — (1985*b*). *J. Bact.* **162**, 746–51.
RAZAAQ, A. and LEACH, C. K. (1978). *Proc. Soc. gen. Microbiol.* **6**, 35–6.
— (1979). *Biochem. Soc. Trans.* **7**, 409–11.
REUSCH, R. W. and SADOFF, H. L. (1983). *Nature, Lond.* **302**, 268–270.
RIDDLE, G. D., SIMONSON, J. G., HALES, B. J., and BRAYMER, H. D. (1981). *Abstr. A. Mtg Am. Soc. Microbiol.*, K91.
— (1982). *J. Bact.* **152**, 72–80.
RITCHIE, G. A. F., SENIOR, P. J., and DAWES, E. A. (1971). *Biochem. J.* **121**, 309–16.
ROBERTS, G. P. and BRILL, W. J. (1981). *A. Rev. Microbiol.* **35**, 207–35.
ROBSON, R. L. (1981). *Proc. Soc. gen. Microbiol.* **8**, 136.
— and POSTGATE, J. R. (1980). *A. Rev. Microbiol.* **34**, 183–207.
— JONES, R., KENNEDY, C. K., DRUMMOND, M., RAMOS, J., WOODLEY, P. R., WHEELER, C., CHESSHYRE, J., and POSTGATE, J. (1984*a*). In *Advances in nitrogen fixation research* (ed. C. Veeger and W. E. Newton), pp. 643–52. Nijhoff and Junk, The Hague.
— CHESSHYRE, J. A., WHEELER, C., JONES, R., WOODLEY, P. R., and POSTGATE, J. R. (1984*b*). *J. gen. Microbiol.* **130**, 1603–12.
RÖCKEL, D., HERNANDO, J. J., VAKALOPOULOU, E., POST, E., and OELZE, J. (1983). *Arch. Microbiol.* **136**, 74–8.

RUPPEN, M. E., GARNER, G., and SADOFF, H. R. (1983). *J. Bact.* **156**, 1243–8.

RUVKUN, G. B. and AUSUBEL, F. M. (1980). *Proc. natn. Acad. Sci. U.S.A.* **77**, 191–5.

SADASIVAM, S. and GOWRI, G. (1981). *Indian J. biochem. Biophys.* **18**, 86.

SADOFF, H. L. (1975). *Bact. Rev.* **39**, 516–39.

— BERKE, L. E., and LOPERFIDO, B. (1971). *J. Bact.* **105**, 185–9.

— SHIMEI, B., and ELLIS, S. (1979). *J. Bact.* **138**, 871–7.

SCHENK, S. P. and WYSS, O. (1977). *J. Bact.* **130**, 1382–6.

SCHERINGS, G., HAAKER, H., and VEEGER, C. (1977). *Eur. J. Biochem.* **77**, 621–30.

—— WASSINK, H., and VEEGER, C. (1983). *Eur. J. Biochem.* **135**, 591–9.

SEN, M. and SEN, S. P. (1965). *J. gen. Microbiol.* **41**, 1–6.

— PAL, T. K., and SEN, S. P. (1969). *J. Microbiol. Serol.* **35**, 533–40.

SENIOR, P. J. and DAWES, E. A. (1971). *Biochem. J.* **125**, 55–66.

SHAH, V. K., DAVIS, L. C., and BRILL, W. J. (1972). *Biochim. biophys. Acta* **256**, 498–511.

—— GORDON, J. K., ORME-JOHNSON, W. H., and BRILL, W. J. (1973). *Biochim. biophys. Acta* **292**, 246–55.

—— STIEGHORST, M., and BRILL, W. J. (1974). *J. Bact.* **117**, 917–9.

SHETHNA, Y. T., DER VAITANIAN, D. V., and BEINERT, H. (1968). *Biochim. biophys. Res. Commun.* **31**, 862–83.

— (1970). *Biochim. biophys. Acta* **205**, 58–62.

SLOBBE, W. and VOORDOUW, G. (1981). *Eur. J. Biochem.* **120**, 15–20.

SLOT, R. L., REUSCH, R. N., and SADOFF, H. L. (1982). *Abstr. A. Mtg Am. Soc. Microbiol.* H22.

SMITH, L. A., HILL, S. and YATES, M. G. (1976). *Nature, Lond.* **262**, 209–10.

SMITH, N. R. (1935). *J. Bact.* **30**, 323.

SORGER, C. J. and TROFIMENKOFF, D. (1970). *Proc. natn. Acad. Sci. U.S.A.* **65**, 74–80.

SPRENT, J. I. (1979). *The biology of nitrogen-fixing organisms*, pp. 51–75. McGraw-Hill, New York.

STEPHENSON, M.P., JACKSON, F. A., and DAWES, E. A. (1978). *J. gen. Microbiol.* **109**, 89–96.

STEVENSON, L. H. and SOCOLOFSKY, M. D. (1966). *J. Bact.* **91**, 304–10.

STEWART, W. D. P. and GALLON, J. R. (eds.) (1980). *Nitrogen fixation, Proc. Phytochem. Soc. Eur. Symp. 18* Academic Press, New York.

STIEFEL, E. I., BURGESS, B. K., WHERLAND, S., NEWTON, W. E., CORBIN, J. L., and WATT, G. D. (1980). In *Nitrogen fixation*, Vol. 1 (ed. W. E. Newton and W. H. Orme-Johnson), pp. 211–22. University Park Press, Baltimore.

STUMBO, G. R. and GAINEY, P. I. (1983). *J. agric. Res.* **57**, 217–27.

SU, C. J., REUSCH, R. N., and SADOFF, H. L. (1981). *J. Bact.* **147**, 80–90.

— and SADOFF, H. L. (1981). *J. Bact.* **147**, 91–6.

SULLIVAN, L. W. and CASIDA, L. E. (1968). *Antonie van Leeuwenhoek* **34**, 188–96.

SWANK, R. T. and BURRIS, R. H. (1969). *Biochim. biophys. Acta* **180**, 473–89.

TCHAN, Y. T., WYSZOMIRSKA-DREHER, Z., NEW, P. B., and ZHOU, J. C. (1983). *Int. J. Systematic Bacteriol.* **33**, 147–56.

TERZAGHI, B. E. (1980*a*). *New Zealand DSIR Plant Physiology Divn. Rept.* 1980.
— (1980*b*). *J. gen. Microbiol.* **118**, 271–3.
— (1980*c*). *J. gen. Microbiol.* **118**, 275–8.
— SHAW, B. D., and PATERSON, A. D. (1984). In *Advances in nitrogen fixation and research* (ed. C. Veegen, W. E. Newton), p. 163. Nijhoff and Junk, The Hague.
— and CHRISTENSEN, M. (1985). *J. Plant Physiol.* In press.
THOMPSON, A. D. and SKERMAN, V. B. D. (1979). *Azotobacteraceae: The taxonomy and ecology of the aerobic nitrogen-fixing bacteria.* Academic Press, London.
THOMPSON, B. J., DOMINGO, E., and WARNER, R. C. (1980*a*). *Virology* **102**, 267–77.
— WAGNER, M. S., DOMINGO, E., and WARNER, R. C. (1980*b*). *Virology* **102**, 278–85.
TISSIÈRÈS, A. (1956). *Biochem. J.* **64**, 582–9.
TSAI, J. C., ALADEGBAMI, S. L., and VELA, G. R. (1979). *J. Bact.* **139**, 639–45.
TSYGANKOV, A. A., YAKUNIN, A. F., and GOGOTOV, I. N. (1984). *Microbiology* **52**, 419–23.
TSYGANKOV, A. A., YAKUNIN, A. F., and GOGOTOV, I. N. (1984). *Microbiology* **52**, 419–23.
TYLER, B. (1978). *A. Rev. Biochem.* **47**, 1127–62.
VAN LIN, B. and BOTHE, H. (1972). *Arch. Microbiol.* **82**, 155–72.
VAN SCHAGEN, C. G. and MULLER, F. (1981). *Eur. J. Biochem.* **120**, 33–39.
VELA, G. R. and GONZALEZ-LOPEZ, J. (1982). *Abstr. A. Mtg. Am. Soc. Microbiol.* J16.
VENKATARAMAN, G. S., ROYCHAUDHURY, P., HENRIKSSON, L. E., and HENRIKSSON, E. (1975). *Curr. Sci.* **44**, 520–1.
VOORDOUW, G., DE HAARD, H., TIMMERMANS, J. A. M., VEEGER, C., and ZABEL, P. (1982). *Eur. J. Biochem.* **127**, 267–74.
— VAN DER VIES, S. M., EWEG, J. K., VEEGER, C., VAN BREEMAN, J. F. L., and VAN BRUGGEN, E. F. J. (1980). *Eur. J. Biochem.* **111**, 347–55.
WANG, Z.-C. and WATT, G. D. (1984). *Proc. natn. Acad. Sci. U.S.A.* **81**, 376–9.
WARD, A. C., RAWLEY, B. I., and DAWES, E. A. (1977). *J. gen. Microbiol.* **102**, 61–8.
WIEGEL, J., WILKE, D., BAUMGARTEN, J., OPITZ, R., and SCHLEGEL, H. G. (1978). *Int. J. syst. Bact.* **28**, 573–81.
WONG, T.-Y. and MAIER, R. J. (1984). *J. Bact.* **159**, 348–52.
WYSS, O. and NIMECK, M. (1962). *Fed. Proc.* **21**, 456.
— and WYSS, M. B. (1950). *J. Bact.* **59**, 287–91.
YAMADA, T. and SAKAGUCHI, K. (1981). *Agric. biol. Chem.* **45**, 2301–9.
YANG, T. Y. (1981). *Can. J. Biochem.* **59**, 137–44.
— and JURTSHUK, P. (1978). *Biochem. biophys. Res. Commun.* **81**, 1032–9.
— (1984). *Curr. Microbiol.* **10**, 309–12.
YANO, K., ANAZAWA, M., MURAI, F., and FUKUDA, M. (1984). In *Advances in nitrogen fixation research* (ed. C. Veeger and W. E. Newton), p. 751. Nijhoff/ Junk, The Hague.

YATES, M. G., PARTRIDGE, C. C. P., WALKER, C. C., VAN DER WARF, A. N., CAMPBELL, F. D., and POSTGATE, J. R. (1980). In *Nitrogen fixation, Proc. Phytochem. Soc. Eur. Symp. 18* (eds. W. D. P. Stewart and J. R. Gallon), pp. 161–76. Academic Press, London.

7 Cyanobacterial nitrogen fixation

Robert Haselkorn

INTRODUCTION

The genetic approach to the elucidation of biochemical pathways has been consistently powerful ever since its introduction by Beadle and Tatum nearly forty years ago. The ability to obtain mutants blocked in each step of a pathway, whether the pathway be the synthesis of a complex molecule or the flow of electrons in a simple reduction, makes it possible to accumulate intermediates, to analyse individual steps in detail, and to study regulation of the pathway. Thus, in the case of nitrogen fixation, application of the methods of genetic analysis developed for *Escherichia coli* to its nitrogen-fixing cousin *Klebsiella pneumoniae* has made the latter the best-understood azotroph even though its contribution to global nitrogen fixation is minute. On the other hand, the absence of a laboratory gene transfer system has seriously hampered comparable studies of the cyanobacteria, even though their contribution to the balance of fixed nitrogen is significant. In the sections that follow, the biochemistry and genetics of nitrogen fixation in *Klebsiella* will be reviewed first, to be followed by a survey of the modes of nitrogen fixation in cyanobacteria. Next, the progress in determining the organization and regulation of the nitrogen fixation (*nif*)-genes in *Anabaena*, exploiting recombinant DNA methods to mitigate the missing gene transfer system, will be examined in detail. Brief sections on anaerobic nitrogen fixation, on the recently discovered aerobic azotrophs, and on symbiotic associations will be followed by a concluding glimpse at the future.

7.1 BACKGROUND

7.1.1 Biochemistry

This section is intended to describe those features of the biochemical conversion of N_2 to NH_3 in *Klebsiella* that are likely to be common to all azotrophs. For detailed discussion of the enzyme mechanism, several reviews may be consulted (Mortenson and Thorneley 1979; Eady, Chapter 1, this volume). A simplified view of the flow of electrons and protons from reduced pyridine nucleotide to N_2 is shown in Fig. 7.1. Some of the steps are still uncertain; for example, the relative positions of F and J may have to be switched. For each component of the pathway a name of convenience has been included along with a catalogue of the known inorganic components of the active protein and the latter's polypeptide constituents. The polypeptides define five genes (F, J, H, D, and K) needed for the polypeptide components of the electron pathway (see Roberts and Brill 1981 and Ausubel and Cannon 1980 for reviews). In addition, four genes are needed for the synthesis of the $MoFe_6S_8$ centres of nitrogenase and four more genes for the maturation of nitrogenase and nitrogenase reductase. The term 'maturation' hides ignorance of what the process actually is; it could involve the insertion or attachment of the Fe–S clusters to the apoprotein.

The key protein in the pathway is H. After H is reduced, it binds ATP and changes conformation, which results in lowering its redox potential 160 mV. In that state its potential is sufficiently low to reduce the KD protein. ATP hydrolysis is somehow coupled to that reduction. When the products of ATP hydrolysis are released from the H protein, it relaxes to the higher potential state from which it can be reduced by J. Thus, H is the muscle which, by coupling ATP hydrolysis with a conformational change, drives nitrogen reduction (Mortenson and Thorneley 1979).

It should not be inferred from the diagram that the stoichiometry of individual steps is known. For example, the molar ratio of H protein to KD protein in *Klebsiella* is thought to be approximately 5 : 1 and optimal

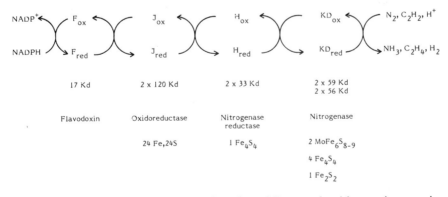

FIG. 7.1 Flow of electrons from reduced pyridine nucleotide to nitrogen in *Klebsiella*. The order of transfer between F and J proteins may be reversed. The H, D, and K proteins are found in all nitrogen-fixing systems to date, including those of cyanobacteria. F and J have only been demonstrated directly in *Klebsiella*.

acetylene reduction *in vitro* requires an excess of H over KD protein (Mortenson and Thorneley 1979; McLean and Dixon 1981).

Nitrogenase activity can be assayed by acetylene reduction *in vitro*, either in crude extracts or by using purified components. The simplest system uses only the H and KD proteins with dithionite as the reductant acting directly on H. The only other requirements are for Mg^{++} and ATP. Dithionite can be replaced by pyruvate in crude extracts of wild type *Klebsiella*. Using extracts prepared from *Klebsiella* mutants it was shown that both the F and J proteins are needed for pyruvate-dependent acetylene reduction. Extracts prepared from H, D, or K mutants are defective in dithionite-dependent acetylene reduction. Such extracts can be repaired *in vitro* by the appropriate addition of purified H or KD protein. The purified nitrogenase component need not come from *Klebsiella*. Many bacterial nitrogenases have been purified and shown to complement *Klebsiella* components *in vitro*. In particular, the H protein from several cyanobacteria will interact with *Klebsiella* KD protein to yield active enzyme *in vitro* (Nagatani and Haselkorn 1978). Purified nitrogenase components from *Anabaena* cross-react with components from *Azotobacter*, *Rhodospirillum*, and *Plectonema* (Hallenbeck *et al.* 1979).

7.1.2 Nif-gene organization and regulation in *Klebsiella*

While this topic will be covered in depth elsewhere in other chapters in this volume, it is necessary to review it briefly here to provide a basis for comparison with *Anabaena* and other cyanobacteria later. The *nif*-gene region of the *Klebsiella* chromosome is shown in the upper part of Fig. 7.2 (Ausubel and Cannon 1980, and many references therein). The roles of genes H, D, and K, which code for the nitrogenase complex, and genes F and J, which code for electron carriers, have been mentioned. Genes Q, B, N, and E are needed for synthesis of the molybdenum cofactor. Genes M, V, S, and U are required for unknown steps in the maturation of nitrogenase. Genes X and Y have totally unknown functions because mutations created by inserting transposons into them have a very weak Nif⁻ phenotype, or none at all. Finally, genes A and L are regulatory, in the following way: gene A codes for an activator and gene L for a repressor of transcription of the other *nif*-genes (Ow and Ausubel 1983; Drummond *et al.* 1983). The *nif* transcription units, defined by polarity of insertion mutations, are shown by horizontal arrows. Initiation of each of these transcripts requires a functional *nifA*-gene product. If *nifA* is missing, or shut down, the other *nif*-genes cannot be turned on. Normally, transcription of *nifA* itself requires another gene, *ntrC*, that is linked to *glnA*, the structural gene for glutamine synthetase. Both *ntrC* and *nifA* require, in order to

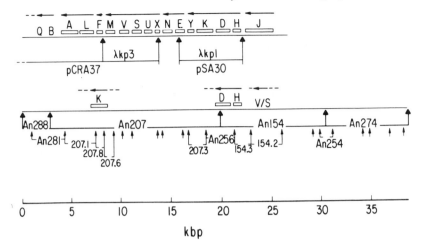

Fɪɢ. 7.2. Physical map of the *nif* genes of *Klebsiella* (above) and of *Anabaena* 7120 (below). Restriction sites for *Eco*RI and *Hind*III are shown as *thick* and *thin arrowheads*, respectively. Horizontal arrows indicate direction and extent of transcription units; dashed lines mean that the ends of the units are not yet defined. *Boxes* indicate gene product sizes where they are known. Numbers below each map (*pCRA*37, *An*288, etc.) refer to recombinant clones containing the indicated fragment. The numbered *Hind*III fragments of *Anabaena* DNA are the ones transcribed during nitrogenase induction.

function, the product of yet another gene, *ntrA* (*glnF*). That product appears to be made constitutively (Merrick 1982).

Let us trace the molecular events that follow induction of this system by withdrawal of ammonia and/or amino acids. The *ntrC*-gene is turned on. Its product, activated by NtrA, promotes transcription of the *nifLA* operon. The *nifA* protein, also activated by NtrA, promotes transcription of all the other *nif* genes. Nitrogenase is synthesized.

The small molecule effectors of this control circuit have not been identified. Ammonia itself is not regulatory because functional glutamine synthetase is required for regulation by ammonia. Whether glutamine or a further metabolite, such as glutaminyl-tRNA, is an effector is still unknown.

Requirement for a functional *ntrC*-gene can be bypassed in several ways. The *nifA*-gene has been cloned into a plasmid in which it is transcribed from a plasmid promoter (Ow and Ausubel 1983; Drummond *et al.* 1983). Alternatively, chromosomal mutants of *Klebsiella* have been isolated in which a small duplication has brought the *nifA*-gene under control of a *his*-gene promoter (Sibold *et al.*1981). In either case, transcription of the other *nif*-genes still requires NtrA. As a framework for consideration of the work on *Anabaena nif*-genes, it will be useful to keep in mind the following possibility: the *nif*-gene promoters are not recognized by unmodified RNA polymerase. Both NtrA and NifA proteins are modifiers of RNA polymerase (σ factors?) required for recognition of *nif*-gene promoters.

7.1.3 Modes of nitrogen fixation in cyanobacteria

Before describing the organization and regulation of the *nif*-genes in those cyanobacteria where detailed comparisons with *Klebsiella* are possible, it will be useful to review what is known in general about the distribution of nitrogen-fixation capability among cyanobacterial species. Much of this information is collected in the classification of cyanobacterial strains of the Pasteur Institute published by Rippka and co-workers in 1979. For detailed information about the strains, their review should be consulted (Rippka *et al.* 1979). A summary of the results with respect to nitrogen fixation is presented in Table 7.1.

The strains have been broadly divided into three classes: unicellular, filamentous, and filamentous with heterocysts. Among the unicellular strains, 22 of a total of 84 are capable of reducing acetylene under anaerobic conditions in the laboratory. Five of these strains, all of the genus *Gloeothece*, can fix nitrogen aerobically as well. Among the remaining 17 strains that can fix anaerobically, ten are marine. The collection includes 36 filamentous strains that do not form heterocysts; 18 of these can reduce acetylene anaerobically. One marine strain of *Oscillatoria* described subsequent to the Rippka *et al.* publication grows on nitrogen-free medium and reduces acetylene aerobically, with no apparent morphological differentiation (Stal and Krumbein 1981). This particular strain grows very slowly, with or without combined nitrogen, but fast-growing strains with similar properties have been isolated (A. Mitsui, personal communication). Finally, the Pasteur collection contain 34 strains, among eight genuses, of heterocyst-forming organisms. These strains will all reduce acetylene aerobically or, in the absence of heterocysts, under anaerobic conditions.

To the extent that the DNA of all these strains has been characterized, there is no simple correlation of a measured property with ability to fix nitrogen. In general, that ability is found in strains with larger genome complexity, 3.6×10^9 or greater, but there are exceptions. Two *Synechococcus*, one *Dermocarpa*, and one *Oscillatoria* strain have genome sizes as low as 3×10^9, near the lower end of the range for all cyanobacteria. Most of the nitrogen-fixing organisms have G-C per cent in the range of 40–46 per cent, but again, there are several *Plectonema* strains with 58 per cent G-C and one *Pseudoanabaena* strain with 52 per cent G-C. Perhaps a more significant variable would be plasmid content, but even that determination would be uninformative in most cases. In some species of *Rhizobium*, the *nif*-genes are located on a

TABLE 7.1
*Acetylene reduction activity in cyanobacterial strains
of the Pasteur collection*

Format	Genus	No. of strains with activity	Remarks
Unicellular	Gloeobacter	0/1	
	Gloeothece	5/5	aerobic activity
	Synechococcus	3/28	anaerobic only*
	Gloeocapsa	0/4	
	Synechocystis	0/17	
	Chamaesiphon	0/2	
	Dermocarpa	2/6, both marine	anaerobic only
	Xenococcus	1/3, marine	anaerobic only
	Dermocarpella	0/1	
	Myxosarcina	1/2, marine	anaerobic only
	Chroococcidiopsis	3/3	anaerobic only
	Pleurocapsa	7/12, 5 marine	anaerobic only
Filamentous, non-heterocystous	Spirulina	0/2	
	Oscillatoria	3/5	anaerobic only*
	Lyngbya, Plecto-nema, Phormidia	12/21, 2 marine	anaerobic only
	Pseudoanabaena	3/8	anaerobic only
Heterocystous	Anabaena, Nodularia, Cylindrospermum, Nostoc, Scytonema, Calothrix, Chlorogloeopsis, and Fischerella		all have aerobic activity

* Strains of these genus have been found to fix nitrogen aerobically (Stal and Krumbein 1980; H. Pearson, personal communication; A. Mitsui, personal communication).

plasmid of such high molecular weight that conventional methods of plasmid preparation do not reveal it (Casse *et al.* 1979).

This survey reveals several interesting paradoxes. The most obvious is the existence of strains such as *Gloeothece* and the marine *Oscillatoria* that fix nitrogen aerobically without heterocysts. This question will be taken up in detail in a later section. A second point is the widespread occurrence of anaerobic acetylene reducing capability. Of what use is this ability? In the laboratory, it is revealed by loading the cells with a carbon reserve, removing combined nitrogen, and then switching to an argon atmosphere and adding dichlorophenyldimethyurea (DCMU) to inhibit O_2 evolution internally (Rippka and Waterbury 1977; Rippka and Stanier 1978). Are these cells likely to encounter analogous circumstances in nature?

7.2 *Nif* GENE ORGANIZATION AND REGULATION IN *Anabaena*

Under aerobic conditions, deprived of combined nitrogen, *Anabaena* differentiates specialized cells at regular intervals along each filament. These cells, called heterocysts, are the exclusive site of nitrogen fixation under aerobic conditions (Haselkorn 1978; Adams and Carr 1981; Neuer *et al.* 1983). Biochemical changes known to occur when a vegetative cell differentiates into a heterocyst are listed in Table 7.2. These changes result in the exclusion of O_2 from the cell, first by limiting diffusion of air into the cell to that possible through the slender connections to adjacent vegetative cells, and second by removal of the O_2-evolving photosystem II reaction centre and accessory proteins. Photosystem I is retained so that ATP can be generated by cyclic photophosphorylation. Photosystem I may also provide some of the reductant for nitrogen fixation (Lockau *et al.* 1978). There is no competition for reductant because CO_2 fixation ceases, due in turn to the destruction of RuBP carboxylase.

TABLE 7.2

Biochemical changes accompanying heterocyst differentiation

Heterocyst features not found in vegetative cells (genes turned on)
    polysaccharide and glycolipid envelope
    isocitrate dehydrogenase
    $NADP^+$: ferredoxin oxidoreductase
    pyruvate: ferredoxin oxidoreductase (oxygen-sensitive)
    glucose-6-phosphate dehydrogenase
    6 phosphogluconate dehydrogenase
    nitrogenase complex
    hydrogenase?
    glutamine transporter?
    protease(s)

Vegetative cell features not found in heterocysts (genes turned off)
    RuBP carboxylase
    Ribulose-P kinase
    glutamate synthase
    glutamate transporter (?)
    phycobiliproteins
    photosystem II reaction center
    sulphate reduction

Vegetative cell features found in heterocysts
    DNA, RNA polymerase, ribosomes, tRNA, etc.
    glutamine synthetase
    most enzymes of glycolysis
    Krebs cycle enzymes *except* pyruvate dehydrogenase and
        α-ketoglutarate dehydrogenase

The path of combined nitrogen following its fixation is shown in Fig. 7.3. This drawing is based in large part on the studies of C. P. Wolk and colleagues, who used $^{13}[N_2]$ to study the fate of newly-fixed N_2 (Wolk *et al.* 1976; Thomas *et al.* 1977). In the heterocyst, newly-made ammonia is rapidly added to glutamate by glutamine synthetase. Glutamine is then transported to adjacent vegetative cells where the amide group is transferred to α-ketoglutarate by the enzyme glutamate synthase (GOGAT). Part of the resulting glutamate is returned to the heterocyst to participate again in the synthesis of glutamine. Glutamine, therefore, is the ammonia transporter from heterocyst to vegetative cell. In addition to the gradients of glutamate and glutamine, there is another important one; reduced carbon flows from vegetative cell to heterocyst to provide reductant for nitrogenase (Wolk 1968). The major carbon compound is thought to be a disaccharide, but it has not yet been identified.

Recent results from Bothe's laboratory on isocitrate dehydrogenase are particularly interesting in this regard (Neuer *et al.* 1983). The

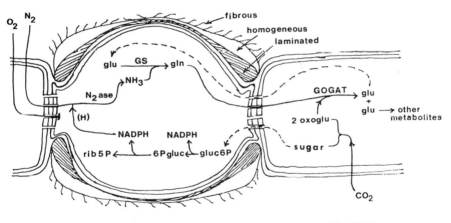

FIG. 7.3. Model of the flow of carbon and nitrogen between heterocyst and vegetative cell in nitrogen-fixing cyanobacteria. Enzymes: N_2ASE = nitrogenase, GS = glutamine synthetase, GOGAT = glutamine-oxoglutarate amido transferase. The mechanism of sugar transport is unknown, as are the steps from that sugar to glucose-6-phosphate. NADPH probably reduces a ferredoxin. An alternate pathway for ferredoxin reduction by pyruvate is not shown. The three layers of the heterocyst envelope (fibrous, homogeneous, laminated) are believed to be impermeable to N_2, CO_2, and O_2. Microplasmodesmata are shown connecting the two cells but the polar bodies are omitted. The heterocyst, lacking photosystem II, neither fixes CO_2 nor produces O_2.

activity of this enzyme, which decarboxylates isocitrate to yield α-ketoglutarate, concomitant with the reduction of $NADP^+$, is readily demonstrated in extracts of heterocysts. The activity cannot be demonstrated in extracts of vegetative cells from differentiated filaments but it is found in cells grown on NH_4^+. It turns out that vegetative cells from N_2-grown cultures contain an inhibitor of the enzyme. This inhibitor has the properties of a protein but it can be replaced by oxidized glutathione or antagonized, in part, by dithiothreitol. One implication of these results is that the heterocyst may be the source of α-ketoglutarate as well as of glutamine, i.e. *both* substrates of GOGAT may be transported out of the heterocyst.

It will be shown later that the *nif* genes of *Anabaena* are not transcribed in the presence of oxygen. In *Klebsiella*, one of the *nif* regulatory genes, L, mediates oxygen repression, but its mechanism is not known. In both cases, the critical protein might be a thioredoxin, since such molecules have been shown to be involved in certain nucleic acid polymerization reactions. A thioredoxin, induced under N^- conditions, might also be the factor that inhibits isocitrate dehydrogenase in vegetative cells of differentiated filaments.

There is, in addition, another likely gradient between heterocysts and vegetative cells. Giddings *et al.* (1981) have shown that heterocysts cannot reduce $SO_4^=$, although $^{35}SO_4^=$ added to whole differentiated filaments is incorporated into protein in heterocysts. Therefore, reduced sulphur, either as $S^=$ or as cysteine, methionine, or other compounds, must flow from vegetative cells into the heterocysts.

The source of reductant for nitrogen fixation in heterocysts was mentioned briefly previously. In the light, it is possible that photosystem I generates some reductant that is coupled to ferredoxin. In the dark, and probably in the light as well, heterocysts reduce $NADP^+$ using glucose-6-phosphate, 6-phosphogluconate, isocitrate, or pyruvate (Lockau *et al.* 1978; Neuer *et al.* 1983). In every case it is believed that a ferredoxin is the eventual donor to nitrogenase. Whether there are specific *nif* gene products that participate in that transfer is still unknown.

The multilayered envelope surrounding the heterocyst makes that cell resistant to lysozyme. For that reason it was possible to separate heterocysts from lysed vegetative cells and to compare the populations of proteins in the two cell types (Fleming and Haselkorn 1973). Even though this was done by rather antiquated one-dimensional gel analysis it could be shown that many of the proteins in the heterocyst are different from the proteins in vegetative cells. The components of nitrogenase are found exclusively in heterocysts, while RuBP carboxylase and the phycobiliproteins, among many others, are missing from heterocysts. During the course of heterocyst differentiation, pulse-labelling and pulse-chase experiments with ^{35}S showed that sets of proteins appeared at different times and some disappeared (Fleming and Haselkorn 1974). The nitrogenase components were a relatively late-appearing set. Thus, whereas nitrogenase induction in *Klebsiella* requires merely the expression of the *nif* genes, in *Anabaena* aerobic nitrogenase induction requires as well the entire programme of gene expression leading to heterocyst differentiation.

In order to understand that programme it would be useful to have cloned genes representative of each of the sets regulated during differentiation. Thus far we have cloned from *Anabaena* a number of *nif*-genes, to be described below (Rice *et al.* 1982); the structural gene for glutamine synthetase (Fisher *et al.* 1981); and genes coding for the large subunit of RuBP carboxylase (Curtis and Haselkorn 1983), the small subunit of RuBP carboxylase (Nierzwicki-Bauer, Curtis, and Haselkorn 1984), the β-subunit of ATPase (Curtis and Haselkorn 1983), and the 32 kdal thylakoid membrane protein (Curtis and Haselkorn 1984). It is important to note that a number of genes acting early in differentiation, such as those coding for specific proteases and the enzymes of glycolipid biosynthesis, have not been cloned yet.

The elegant way to identify and to isolate genes active in *Anabaena* heterocyst differentiation would be to isolate mutants defective in differentiation and then to complement those mutants with cloned DNA introduced by transformation with plasmids. Many suitable mutants of *Anabaena* have been isolated (Wilcox *et al.* 1975; Currier *et al.* 1977) but, until recently, there was not a working system of plasmid transformation or conjugation in any nitrogen-fixing cyanobacterium. This being so, it was necessary to resort to hybridization procedures to identify *Anabaena* DNA fragments carrying those genes for which suitable probes were available from other sources. Of the 17 *nif*-genes of *Klebsiella* shown in Fig. 7.2, only the genes coding for nitrogenase, nitrogenase reductase, and one of the maturation factors were sufficiently conserved in nucleotide sequence to hybridize with *Anabaena* DNA (Mazur *et al.* 1980).

The general strategy of these experiments is straightforward. First, cloned DNA fragments from *Klebsiella*, carrying known *nif*-genes, were hybridized to restriction enzyme digests of total *Anabaena* DNA. Based on the number and size of the fragments identified by the *Klebsiella* probes, appropriate lambda phage vectors were chosen to prepare libraries of *Anabaena* DNA. Thus, to cite one example, the *Klebsiella nifK, D*, and *H* probe identified *Eco*RI fragments of *Anabaena* DNA of 10.5 and 17 kbp. The smaller of these fragments was then sought in a library prepared in the vector lambda gt-7, which accommodates *Eco*RI fragments between 2 and 12 kbp long. The larger fragment was sought, and found, in a lambda Charon 4A library.

Mapping these fragments and extending the physical map was accomplished by re-cloning the *Eco*RI fragments isolated from the lambda vectors into plasmids. The plasmids were then used to isolate, from libraries of *Anabaena* DNA cloned in lambda *Hind*III vectors, the *Anabaena Hind*III fragments that include the ends of the original *Eco*RI fragments. One of the *Hind*III fragments, pAn256, was shown to overlap both the 10.5 kbp and the 17 kbp *Eco*RI fragments, which meant that the latter two are adjacent on the *Anabaena* chromosome. Similar reasoning led to the cloning of two more *Eco*RI fragments on either side of the original two, for a total of 39 kbp. The *Eco*RI and *Hind*III map of this 39 kbp region is shown in the lower part of Fig. 7.2 (Rice *et al.* 1982).

Within the cloned region, the *Anabaena nif* genes were located precisely by detailed Southern hybridization and by measurement of heteroduplex DNA molecules in the electron microscope. The latter were formed by mixing, denaturing, and re-annealing recombinant lambda phage DNA molecules half of which contained *Anabaena* DNA fragments and the other half of which contained *Klebsiella* DNA fragments on which the *nif* genes were accurately mapped. An example of such a heteroduplex DNA molecule, with its interpretation, is shown in Fig. 7.4.

Measurements such as these permitted the approximate location of the *nif* genes. More precise location, with respect to the restriction map, was made possible by subcloning the various *Hind*III fragments shown in Fig. 7.2 and determining the nucleotide sequence of some of them. The first of these was the *Anabaena nifH* gene, coding for nitrogenase reductase, which was found entirely within the *Hind*III fragment in pAn154.3 (Mevarech *et al.* 1980). That same fragment also contained the start of the α-subunit of nitrogenase (*nifD*). The *nifD* sequence was completed by determining the sequence of pAn256 from the *Hind*III site leftward to a point 500 bp beyond the *Eco*RI site that divides An207 from An154 (Lammers and Haselkorn 1983). This arrangement of *nifH* and *nifD* is consistent with the map of *Klebsiella* for these genes. Of great surprise, however, was the finding that the *nifK* gene is around 11 kbp away from *nifD* in *Anabaena*. Originally located by Southern hybridization and by examination of heteroduplex DNA molecules in the electron microscope, the *nifK* gene was unambiguously mapped by sequencing the fragments shown in Fig. 7.2; they clearly code for the β-subunit of nitrogenase (Mazur and Chui 1982).

The map shown in Fig. 7.2 was determined using fragments cloned from vegetative cell DNA. We have recently determined that the *nif* gene region of *Anabaena* is rearranged during heterocyst differentiation (J. W. Golden, S. J. Robinson, and R. Haselkorn, to be published). One aspect of the rearrangement is the excision of 11 kbp between *nifK* and *nifD*. This is accomplished by conservative recombination between two

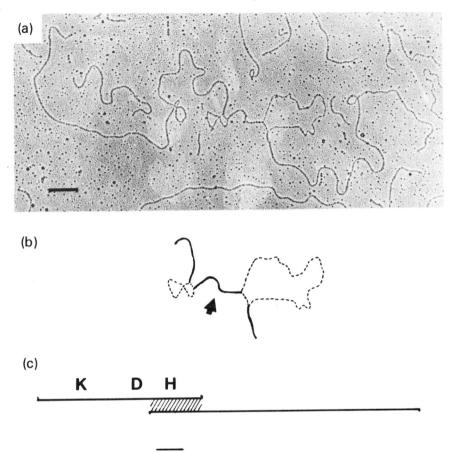

Fig. 7.4. Electron micrograph of a heteroduplex DNA molecule of which one strand contains a cloned fragment of *Klebsiella* DNA with the *nifK*-, *D*-, and *H*-genes and the other strand contains a cloned fragment of *Anabaena* DNA (An154, see Fig. 7.2). The drawing shows the duplex region where the two cloned fragments are homologous. The lowest part of the Figure interprets the duplex to mean that ∿2 kbp at the left end of An154 is homologous to the right end of the *Klebsiella* fragment, i.e. to *nifH* and part of *nifD*. This interpretation was confirmed by sequencing the DNA.

11-bp directly repeated sequences, one copy of which is located approximately 320 bp 5′ of *nifK*, the other copy located 72 bp inside the 3′ end of the *nifD* coding region. The result of this recombination event is the formation of an 11-kbp circular DNA molecule, containing one copy of the 11-bp repeat, and a fused chromosome. In the latter, the *nifD* coding region contains a new 43-amino acid C-terminal sequence and the space between *nifD* and *nifK* is reduced to 199 bp. The rearrangement generates a *nifHDK* operon, which is transcribed as a unit, just as in *Klebsiella*.

The sequence of the *nifD* protein, predicted from the gene sequence following rearrangement, is shown in Fig. 7.5 along with the sequences of the *nifH* and *nifK* gene products. Sequences are now available for the corresponding proteins from many bacteria. Residues that are underlined in Fig. 7.5 are highly conserved and probably involved in critical aspects of the protein's function.

The diagram in Fig. 7.2 shows one more region of sequence homology between the *Klebsiella* probes and *Anabaena* DNA. This region consists of ∿600 bp which correspond, based on Southern hybridization, to either the *nifV* or *nifS* genes of *Klebsiella*. The corresponding gene in *Anabaena* maps to the right of the *nifH* gene, rather than to the left of *nifK*, as in *Klebsiella*. These genes play a poorly-understood role in maturation of the enzyme in *Klebsiella*. The relevant DNA sequences have not yet been determined from either source.

Hybridization of *Klebsiella nif* DNA to total *Anabaena* DNA produced one additional unexpected result. In the *Eco*RI digest of total *Anabaena* DNA, a third band of ∿20 kbp was detected in addition to the 17 and 10.5 kbp bands mapped and shown in Fig. 7.2. This band was recovered from a lambda Charon 4A library and shown to contain a second sequence related to the *Anabaena nifH* gene. The restriction map of this gene differs from the one shown in Fig. 7.2 and there does not appear to be a *nifD*-gene adjacent to it. However, determination of the nucleotide sequence of the region of homology shows an open reading frame nearly identical in amino acid sequence to the product of the original *nifH*-gene (S. J. Robinson, personal communication). This gene is transcribed when nitrogenase is induced.

This brings us to the question of the regulation of *nif*-gene expression which, as will be seen, means transcription. The lower portion of Fig. 7.2 shows the distribution of *Hind*III sites along the 39 kbp of cloned *Anabaena* DNA containing the known *nif*-genes. Each of the *Hind*III fragments was cloned separately into plasmids and used for a quick test of transcription, as follows: total RNA was extracted from *Anabaena* cells grown on ammonia or from cells induced anaerobically for nitrogenase. Each RNA preparation was bound to filters and then each pair of filters (N⁺ RNA, N⁻ RNA) was annealed with nick-translated plasmid. Those plasmids bound by N⁻ RNA, but not by N⁺ RNA, were judged to have contained sequences transcribed during nitrogenase induction (Rice *et al.* 1982). Each fragment is named in the lowest part of Fig. 7.2. Fragments without a numerical designation are not detectably transcribed. All of the fragments known to contain *nif*-

```
                10                  20                  30                  40                  50
M T D E N I  R Q E  A F Y G K G G I  G K S  T T S  Q N T L  A A M A G M G Q R I  M I  V G C D P K A D S T R

                60                  70                  80                  90                 100
L  M L H S K A Q T T  V L  H L  A A E R G A V E D L E L H E V  M L T  G F R G V K C V E S  G G P E P G V G

                110                 120                 130                 140                 150
C A G R G I  I T A I  N F L E E N G A Y Q D L D F V S  Y D V L G D V V  C G G F A M P I  R E G K A Q E I

                160                 170                 180                 190                 200
Y I  V T S  G E M M A M Y A A N N I  A R G I  L K Y A H S  G G V R L G G L I  C N S  R K V D R E D E L I  M

                210                 220                 230                 240                 250
N L  A E R L N T Q M I  H F V P R D N I  V Q H A E L R R M T V N E Y A P D S  N Q G Q E Y R A L A K K I

                260                 270                 280                 290
N N D K L T I  P T P M E M D E L E A L K I  E Y G L L D D D T K H S E I  I  G K P A E A T N R S C R N
```

(a)

nifD

```
                10                  20                  30                  40                  50
M T P P E N K N L  V D E N K E L I  Q E V L K A Y P E K S  R K K R E K H L N V  H E E N K S  D C G V K S

                60                  70                  80                  90                 100
N I  K S  V P G V M T A R G C A Y A G S  K G V V W G P I  K D M I  H I  S H G P V G C G Y W S  W S  G R R N

                110                 120                 130                 140                 150
Y Y V G V T G I  N S  F G T M H F T S  D F Q E R D I  V F G G D K K L  T K L I  E E L D V L  F P L N R G V

                160                 170                 180                 190                 200
S I  Q S E C P I  G S I  G D D I  E A V A K K T S  K Q I  G K P V V P L R C F G F R G V S  Q S L G H H I  A

                210                 220                 230                 240                 250
N D A I  R D W I  F P E Y D K L K K E T R L D F E P S  P Y D V A L I  G D Y N I  G G D A W A S  R M L L E

                260                 270                 280                 290                 300
E M G L R V V A Q W S  G D G T L N E L I  Q G P A A K L V L I  H C Y R S  M N Y I  C R S  L E E Q Y G M P

                310                 320                 330                 340                 350
W M E F N F F G P T K I  A A S L R E I  A A K F D S K I  Q E N A E K V I  A K Y T P V M N A V L D K Y R

                360                 370                 380                 390                 400
P R L E G N T V M L Y V G G L R P R H V V P A F E D L G I  K V V G T G Y E F A H N D D Y K R T T H Y

                410                 420                 430                 440                 450
I  D N A T I  I  Y D D V T A Y E F E E F V K A K K P D L I  A S G I  K E K Y V F Q K M G L  P F R Q M H S

                460                 470                 480                 490
W D Y S  G P Y H G Y D G F A I  F A R D M D L S  L N S  P T W S L I  G A P W K K A A A K A K A A S
```

(b)

(Fig. 7.5 continued overleaf)

nifK

```
           10              20              30              40              50
M P Q N P E R T D H V D L F K Q P E Y T E L F E N K R K N F E G A H P P E E V E R V S E W T K S W D

           60              70              80              90             100
Y R E K N F A R E A L T V N P A K G C Q P V G A M F A A L G F E G T L P F V Q G S Q G C V A Y F R T

          110             120             130             140             150
H L S R H Y K E P C S A V S S S M T S D A A V F G G L N N M I E G M Q V S Y Q L Y K P K M I A V C T

          160             170             180             190             200
T C M A E V I G D D L G A F I T N S K N A G S I P Q D F P V P F A H T P S F V G S H I T G Y D N M M

          210             220             230             240             250
K G I L S N L T E G K K K A T S N G K I N F I P G F D T Y V G N N R E L K R M M G V M G V D Y T I L

          260             270             280             290             300
S D S S D Y F D S P N M G E Y E M Y P S G T K L E D A A D S I N A K A T V A L Q A Y T T P K T R E Y

          310             320             330             340             350
I K T Q W K Q E T Q V L R P F G V K G T D E F L T A V S E L T G K A I P E E L E I E R G R L V D A I

          360             370             380             390             400
T D S V A W I H G K K F A I Y G D P D L I I S I T S F L L E M G A E P V H I L C N N G D D T F K K E

          410             420             430             440             450
M E A I L A A S P F G K E A K V W I Q K D L W H F R S L L F T E P V D F F I G N S Y G K Y L W R D T

          460             470             480             490             500
S I P M V R I G Y P L F D R H H L H R Y S T L G Y Q G G L N I L N W V V N T L L D E M D R S T N I T

          510
G K T D I S F D L I R
```

(c)

FIG. 7.5. Complete amino acid sequences for the three proteins of the nitrogenase complex of *Anabaena* 7120, determined by translation of the gene sequences. Data for *nifH* are from Mevarech *et al.* (1980); for *nifD* from Lammers and Haselkorn (1983) modified at the C-terminal region as a consequence of the gene rearrangement described in the text; for *nifK* from Mazur and Chui (1982). Sequences are shown in the one-letter amino acid code. Underlined sequences are conserved in at least one other bacterial species. Cysteine residues likely to be important in ligand binding are circled.

genes are transcribed; that is comforting. In addition, the fragment to the left of *nifD* and the entire region to the left of *nifK* are transcribed. These may contain more *nif*-genes.

The qualitative transcription studies have been extended in several ways. One is to examine the size and stability of the messenger RNA corresponding to the cloned genes. So far this examination has been done only for the *nifH* gene, using pAn154.3 as the probe, with the result shown in Fig. 7.6 (S. J. Robinson, unpublished results). Here, three RNA preparations are compared. One is from cells grown on

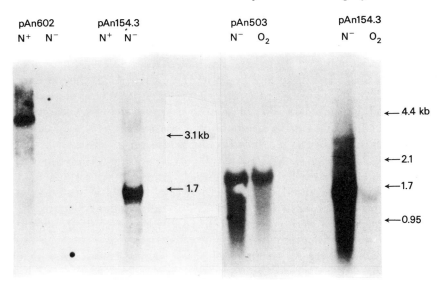

FIG. 7.6. The *nifH* gene is regulated by transcription in *Anabaena*. RNA was prepared from cells grown on ammonia, or induced anaerobically for nitrogenase (N-) or induced and then gassed with O$_2$ for 45 min. Each preparation was fractionated by electrophoresis, transferred to nitrocellulose paper, and probed with radioactive pAn154.3, which contains only the *nifH*-gene. The ammonia-grown cells have no *nif* messenger RNA; the induced cells have several sizes of *nifH* mRNA; the O$_2$-treated cells have lost their *nif* mRNA. Data kindly provided by Steven J. Robinson.

ammonia, the second from cells induced anaerobically for nitrogenase, and the third is from induced cells to which O$_2$ has been added for 45 min. First, it can be seen that nitrogenase induction results in the synthesis of several size classes of *nifH* messenger RNA. The largest of these is sufficiently long to code for both *nifH* and *nifD* proteins, while the more abundant 1.1 kb message is too small for both (Recently, a still larger RNA, 4.7 kb, has been detected with *nifH*, *nifD*, and *nifK* probes. When O$_2$ is admitted, both species of messenger RNA disappear. The result is observed because the induction was done anaerobically (argon plus DCMU), under which conditions heterocyst differentiation is incomplete and nitrogenase is induced in oxygen-sensitive cells. This result is of interest because it provides *one* of the reasons that nitrogenase proteins appear late in heterocyst differentiation: *nif*-gene transcription does not occur until anaerobiosis is established. Oxygen repression of *nif*-gene transcription is also a feature of *nif*-gene regulation in *Klebsiella*, where the responsible gene is *nifL* (Hill *et al.* 1981). Perhaps *Anabaena* has an analogue of *nifL*.

The cloned *Anabaena nif*-gene fragments and the knowledge of their sequences has made possible a deeper analysis of transcription based on determination of the promoter sequences for several genes. This analysis proceeds in two ways. In the first, a suitable DNA fragment which contains the suspected promoter is labelled at the 5'-end, annealed with messenger RNA from cells induced for nitrogenase, and then digested with nuclease S-1. The precise size of the DNA fragment protected from nuclease digestion by the messenger RNA is determined by comparison with a sequencing ladder on a urea gel. In the alternate procedure, a DNA fragment corresponding to a sequence within the gene is labelled at the 5'-end, annealed with messenger RNA, and then extended using avian myeloblastosis virus (AMV) reverse transcriptase and deoxynucleoside triphosphates. Again the product size is measured accurately. Ideally the two methods should agree and should indicate, within one or two nucleotides, the start site of transcription. Data for the determination of the start of *Anabaena nifH* messenger RNA, generously provided by Steven J. Robinson, are shown in Fig. 7.7. The promoter region of the *nifH*-gene sequence is shown in Fig. 7.8. The underlined residues in the -35 and -10 region are of considerable interest: they correspond to very bad promoter sequences in *E. coli*. This result is perfectly consistent with a requirement for specific transcriptional activation of the *nif*-genes, as described in an earlier section for *Klebsiella*.

The promoter identified by Robinson is not transcribed in *E. coli*. This is known because no *nifH* product is seen in *E. coli* cells containing the cloned *Anabaena nifH* gene unless there is a strong lambda promoter or plasmid promoter suitably upstream of the *Anabaena nifH* gene. Moreover, the *nifH* DNA fragment is not transcribed *in vitro*, either by *Anabaena* or *E. coli* RNA polymerase (Catherine Richaud, personal communication), while both polymerases recognize conventional *E. coli*-like promoters on other cloned *Anabaena* DNA fragments (Nilgun Tumer, personal communication). All of these observations suggest that *Anabaena nif*-genes must be activated for transcription or are transcribed by a modified RNA polymerase.

Further support for the modification of RNA polymerase during *Anabaena* heterocyst differentiation was provided by studies of the transcription of the *glnA* gene encoding glutamine synthetase. This enzyme is required for ammonia assimilation both in vegetative cells growing on ammonia and in heterocysts fixing nitrogen. The *glnA* gene was completely sequenced and the start sites for transcription were determined by S-1 nuclease protection and primer extension, using RNA prepared either from cells growing on ammonia or from cells induced anaerobically for nitrogenase (Tumer *et al.* 1983). In the former, there were several start sites, each of which contained -35 and -10

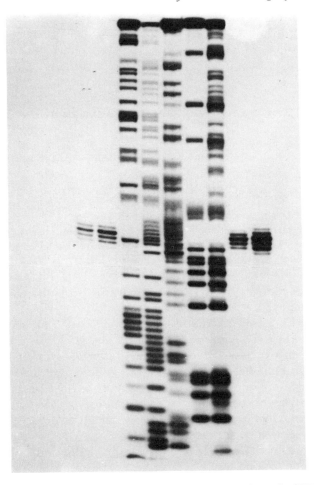

FIG. 7.7. Determination of the start site for transcription of *nifH* messenger RNA in *Anabaena*. The outside lanes contain end-labelled DNA fragments that had been annealed with RNA from cells induced for nitrogenase and then digested with nuclease S-1. The interior lanes contain the same end-labelled DNA fragment subjected to the Gilbert–Maxam sequencing reactions. Comparison of the S-1 lanes with the sequence ladder permits exact assignment of the start of the message.

sequences that correspond to consensus bacterial promoters. In cells induced for nitrogenase, none of these promoters is used. Instead, a new start site was observed, whose −35 and −10 sequences are nearly identical to those of the *nifH* gene. Thus, when the cell shifts to the nitrogen-fixation mode, the *glnA* gene shifts its transcriptional start from an *E. coli*-like promoter to a *nif* gene promoter. This striking

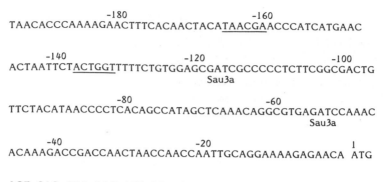

-180   -160
TAACACCCAAAAGAACTTTCACAACTACATAACGAACCCATCATGAAC

-140   -120   -100
ACTAATTCTACTGGTTTTTCTGTGGAGCGATCGCCCCCTCTTCGGCGACTG
            Sau3a

-80   -60
TTCTACATAACCCCTCACAGCCATAGCTCAAACAGGCGTGAGATCCAAAC
                                                      Sau3a

-40   -20   1
ACAAAGACCGACCAACTAACCAACCAATTGCAGGAAAAGAGAACA ATG

ACT GAC GAA AAC ATT AGA CAG ATA GCT TTC etc.

FIG. 7.8. Nucleotide sequence prior to the start of *nifH* messenger RNA in *Anabaena*. The nucleotide labelled + 1 is the first in the message, transcribed to the right. The − 10 and − 35 regions are seen *not* to resemble good bacterial promoters, which should be TTGACA at − 35 and TATAAT at − 10.

result implies that the transcription apparatus is modified to recognize the *nif* promoters.

It should be recalled that the repression of *nif*-gene expression in *Klebsiella* and in *Rhodopseudomonas* (Scolnik *et al.* 1983) by ammonia has been shown clearly to require functional glutamine synthetase. In other words, the crucial effector molecule in *nif*-gene regulation is glutamine or a metabolite of glutamine. Recall also that, in *Anabaena*, glutamine is the carrier of newly-fixed nitrogen from heterocyst to vegetative cell. It seems reasonable to expect, therefore, that glutamine will be a key effector molecule in *Anabaena* as well.

Experiments with the analogue methionine sulfoximine (MSX) confirm this expectation. MSX is an irreversible inhibitor of glutamine synthetase, with no other known reactivity in *Anabaena*. Addition of MSX to a culture otherwise repressed by ammonia results in differentiation of heterocysts and induction of nitrogenase (Stewart and Rowell 1975). Such MSX-treated cells actually excrete ammonia. The addition of glutamine blocks differentiation and represses nitrogenase (Haselkorn *et al.* 1980). Thus, there may be some shared elements of the mechanism for *nif* gene regulation in *Anabaena* and in *Klebsiella*.

It must be pointed out, however, that there are some facts that are difficult to accommodate in this picture. The marine *Anabaena* CA strain is repressed for both heterocyst differentiation and nitrogenase synthesis by nitrate but not by ammonia (Bottomley *et al.* 1979). When the reverse is true, as it often is, low nitrate reductase activity is blamed for a low intracellular concentration of ammonia, and therefore of glutamine. But no satisfactory explanation has yet been offered for this case of apparent refractoriness to external ammonia.

7.3 OXYGEN-SENSITIVE NITROGEN FIXATION IN CYANOBACTERIA

Oxygen-sensitive acetylene reduction was first observed in *Plectonema* under microaerophilic conditions; the activity could be observed merely by gassing a culture with argon (Stewart and Lex 1970). The more elaborate procedure for removing O_2 externally and internally described in Section 7.1.3, developed by Rippka and Waterbury, allowed the activity to be expressed in many unicellular strains as well (Rippka and Waterbury 1977).

Nitrogen fixation (acetylene reduction) requires, in addition to synthesis of the *nif*-gene products, sources of reductant, ATP, molybdenum, and iron. The latter two requirements are met in usual laboratory media but could be limiting in natural environments. ATP is produced by cyclic photophosphorylation but can also be produced by respiration of carbon reserves or, for some strains, external carbon sources such as fructose. Reductant means, ultimately, reduced pyridine nucleotide whose major source is probably the hexose monophosphate pathway. An alternate source of reductant is pyruvate, since a pyruvate: ferredoxin oxidoreductase has been found in heterocysts (Neuer and Bothe 1982). This enzyme activity, like nitrogenase, is extremely oxygen-sensitive. Another possible source is isocitrate, which yields NADPH when it is decarboxylated by isocitrate dehydrogenase to yield α-ketoglutarate (Bothe *et al.* 1980).

In every case examined, acetylene reduction activity was repressed by ammonia and induced by withdrawal of combined nitrogen. When O_2 is readmitted to an anaerobically-induced culture of *Plectonema* (or of *Anabaena*, for that matter) acetylene reduction activity is lost rapidly, within several hours. The extent of loss depends on the partial pressure of O_2. With low O_2, activity can be recovered without new protein synthesis. With 20 per cent O_2, the loss of activity is irreversible unless the *nif* genes are again induced and new proteins made (Rippka and Stanier 1978).

The inactivation by O_2 of the *nif*-proteins induced anaerobically in *Plectonema* leads to the degradation of those proteins within several hours. This was shown by monitoring the fate of polypeptide bands, identified as nitrogenase components, by polyacrylamide gel electrophoresis (Haselkorn *et al.* 1980). The disappearance of these bands, following the admission of O_2, is prevented by the addition of chloramphenicol at the time of O_2 exposure (R. Rippka, personal communication). This result implies that the degradation of O_2-inactivated nitrogenase components requires a protease that is synthesized subsequent to the admission of O_2.

Thus, it is now possible to list four reasons why O_2 is inimical to nitrogen fixation: in addition to competition for reductant and the irreversible inactivation of the nitrogenase complex, O_2 also provokes the proteolytic degradation of nitrogenase polypeptides and represses the transcription of nitrogenase messenger RNA.

7.4 AEROBIC NITROGEN FIXATION WITHOUT HETEROCYSTS

Having so thoroughly established the consequences of O_2 on nitrogen fixation, the discovery of aerobic nitrogen-fixing cyanobacteria was quite unexpected. At present, four marine species have been characterized to varying degrees, and many more isolates in the collection of Prof. A. Mitsui, both unicellular and filamentous, await further description. Three of the strains already described by others, *Gloeothece* (Wyatt and Silvey 1969), *Oscillatoria* (Stal and Krumbein 1981), and *Microcoleus* (Pearson *et al.* 1979), grow very slowly either in complete or nitrogen-free media. The fourth, called *Synechococcus*, grows much faster, with a doubling time of 17 h in nitrogen-free medium (Duerr 1981).

There is no clue as to how these cells, which generate O_2 photosynthetically, protect their nitrogenase from inactivation and induce *nif*-gene transcription. It might be imagined that, in the filamentous strains, there is some kind of functional differentiation that is the equivalent of heterocyst formation, but there is no morphological or biochemical evidence to favour that suggestion. The unicellular *Gloeothece*, by virtue of their sheath, form clumps in which there might also be suspected functional differentiation. However, in that case there are sheath-less mutants that fix nitrogen perfectly well under aerobic conditions (Kallas *et al.* 1983), requiring that an intracellular protective mechanism applicable to all cells be considered. Thus far, electron microscopic examination of *Gloeothece* has shown no differences between cells grown on ammonia or on nitrogen (Kallas *et al.* 1983).

The fast-growing marine *Synechococcus* has an extremely interesting growth curve in air. Nitrogenase is produced only during early log phase; enough nitrogen is fixed then for the cell to coast on its fixed nitrogen reserve until stationary phase (Duerr 1981). So in this case there is temporal rather than spatial differentiation. There remains the problem of protection, however, because O_2 evolution is continuous.

One important question is whether, in these species, the nitrogenase proteins are different from those of *Klebsiella*, *Anabaena*, etc. This question has been answered definitively for *Gloeothece* (Kallas *et al.* 1983) by hybridization of specific *nif* gene probes to restriction enzyme digests of total *Gloeothece* DNA. Specific probes for *nifH*, *nifD*, and *nifK* of *Ana-*

baena each hybridized to *Gloeothece* DNA, in such a way that the gene order must be H, D, K as it is in *Klebsiella*. These results mean that the nitrogenase proteins of *Gloeothece* are similar to those of *Anabaena* and *Klebsiella*, i.e. oxygen sensitive. The protective mechanism must be sought elsewhere.

7.5 NITROGEN FIXATION IN CYANOBACTERIAL SYMBIOTIC ASSOCIATIONS

The contribution of cyanobacteria to global nitrogen fixation is not restricted to the free-living systems just described. Cyanobacteria live in symbiotic association with ferns, mosses, liverworts, and some angiosperms. In all of these cases, the endosymbiont is capable of supplying fixed nitrogen to the host. Analysis of these symbiotic associations has been hampered, not only by the absence of conventional genetic systems, but also by difficulty in maintaining the host plants in the laboratory, in isolating the cyanobacterial partner in pure culture, and in reconstituting the association.

In spite of these difficulties, it has been possible to measure a number of biochemical parameters of the associations and to formulate models for the regulatory interactions between the partners. Considerable information has been obtained from the following systems: the lichens *Peltigera aphthosa* and *Peltigera canina*, which contain a symbiotic *Nostoc* (Stewart *et al.* 1980); the hornwort *Anthoceros*, which contains a *Nostoc* (Stewart and Rodgers 1977) that has been obtained in pure culture and reconstituted (J. C. Meeks, personal communication); the liverwort *Blasia*, which contains a *Nostoc* (Stewart and Rodgers 1977); the fern *Azolla*, which contains an *Anabaena* (Peters *et al.* 1981); and the angiosperm *Gunnera* (Silvester 1976), which contains a *Nostoc* recently obtained in pure culture and reconstituted (Bonnet and Silvester 1981).

These associations have certain features in common. In every case studied, the cyanobacterial partner fixes atmospheric N_2 and excretes a substantial portion of the fixed product as NH_4^+, which is taken up by the host tissue. Usually, photosynthesis is impaired in the cyanobacterial partner and the heterocyst frequency is high. In some cases, transport of reduced carbon from the host to the cyanobacteria has been demonstrated.

The mechanism by which the host persuades its partner to give up fixed nitrogen has been studied in several systems. Usually, the activity of glutamine synthetase in the symbiotic cyanobacteria is very low. Where such comparisons can be made quantitatively, the activity of GS in a symbiont is 5–20 per cent of that in the same strain when free-living. Unable to make glutamine, the symbiont excretes ammonia. In some associations the host uses glutamate dehydrogenase in others, glutamine synthetase, to assimilate the ammonia.

How does the host regulate GS of the symbiont? Based on the rapid and irreversible inhibition of GS by methionine sulfoximine, it was suggested that the host might elaborate a compound with similar properties. That may occur, but it is not necessary if instead the host regulates synthesis of the enzyme in the symbiont. Regulation at the level of synthesis has been shown in the *Azolla-Anabaena* association, in which the amount of glutamine synthetase antigen was determined by a quantitative radioimmune assay and found to be 5–10 per cent of that in free-living *Anabaena* (Orr and Haselkorn 1982).

Modern recombinant DNA methods are immediately applicable to several of the current questions concerning these associations. For example, identity of an isolated free-living organism with a symbiont can be established (or ruled out) unequivocally by comparing the pattern of restriction fragments in total DNA, prepared from the free-living organism and from the association, when probed with several cloned genes. Cloned genes can also be used to measure transcription directly, by DNA–RNA hybridization. The result for the structural gene for glutamine synthetase will be of particular interest.

7.6 PROSPECTS

Three areas appear to be particularly ripe for intensive study in the near future: the development of a system of genetic analysis in a nitrogen-fixing cyanobacterium; determination of the mechanism of oxygen protection in the aerobic non-heterocystous azotrophs; determination of regulatory mechanisms in the symbiotic associations.

Although much has already been done with *Anabaena* 7120, using recombinant DNA methods, it will not be possible to analyse any of the genes for which external probes do not exist unless a method is found for the re-introduction of DNA into *Anabaena*. Based on the successful attempts to transform unicellular cyanobacteria with DNA, it appears that restriction is the major governing variable. Since *Anabaena* contains so many activities (*Ava*I, *Ava*II, *Ava*III, at least) (cf. Reaston *et al.* 1982), it will be difficult to find non-restricting mutants or to purify all of the modifying activities in order to protect DNA *in vitro*.

Subsequent to the preparation of this review, a system for transfer of DNA into *Anabaena* by conjugation from *E. coli* was reported (Wolk, Vonshak, Kehoe, and Elhai 1984). The system is based on a shuttle vector that consists of three elements: pBR322 cut down to remove most of the *Ava*I and *Ava*II sites, antibiotic-resistance genes that function as selectable markers in *Anabaena*, and an *Anabaena* replicon derived from a cryptic endogenous plasmid. This vector replicates in both *E.*

coli and *Anabaena*. It is mobilized in *trans* from *E. coli* by a *col*K plasmid and the broad host range pilus for attachment to *Anabaena* is provided by RP4. In principle, this system will make it possible to complement Nif⁻ mutations and to create new Nif⁻ mutants by site-directed mutagenesis of cloned DNA fragments.

Discovery of a genetic system will be keenly felt in the study of the oxygen-protection mechanism too. However, it should be possible to start by isolating mutants that are Nif⁻ in air but Nif⁺ under anaerobic conditions and then to characterize the mutants biochemically. Of course, it would be better to know, by complementation tests, how many genes are involved in protection, but it is not necessary to wait for the genetic system to start looking.

The *Nostoc/Anthoceros* association seems to be the most promising, at present, for analysis of symbiotic systems. That is due to the ability, again, of mutant isolation among the free-living *Nostoc* in order to determine the genes involved in colonization, regulation of differentiation, *nif* gene expression, and so on. Here again, the ultimate limitation will be the ability to re-introduce DNA into the free-living *Nostoc*. Of course, it will also be necessary to do genetics of *Anthoceros* in order to determine fully the host's contribution to the association, but that subject is beyond the scope of this article.

REFERENCES

ADAMS, D. G. and CARR, N. G. (1981). *CRC Crit. Rev. Microbiol.* **9**, 45–100.

AUSUBEL, F. M. and CANNON, F. C. (1980). *Cold Spring Harbor Symp. Quant. Biol.* **45**, 487–92.

BONNETT, H. T. and SILVESTER, W. B. (1981). *New Phytol.* **89**, 121–8.

BOTHE, H., NEUER, G., KALBE, I., and EISBRENNER, G. (1980). In *Nitrogen fixation* (ed. W. D. P. Stewart and J. R. Gallon), pp. 83–112. Academic Press, New York.

BOTTOMLEY, P. J., GRILLO, J. F., VANBAALEN, C., and TABITA, F. R. (1979). *J. Bact.* **140**, 938–43.

CASSE, F., BOUCHER, C., JULLIOT, J. S., MICHEL, M., and DÉNARIÉ, J. (1979). *J. gen. Microbiol.* **113**, 229–42.

CURRIER, T. C., HAURY, J., and WOLK, C. P. (1977). *J. Bact.* **129**, 1556–62.

CURTIS, S. E. and HASELKORN, R. (1983). *Proc. natn. Acad. Sci. U.S.A.* **80**, 1835–9.

——(1984). *Plant Mol. Biol.* **3**, 249–58.

DRUMMOND, M., CLEMENTS, J., MERRICK, M., and DIXON, R. (1983). *Nature, Lond.* **301**, 302–7.

DUERR, E. O. (1981). Ph.D. dissertation, Univ. of Miami, Florida.

FISHER, R., TULI, R., and HASELKORN, R. (1981). *Proc. natn. Acad. Sci. U.S.A.* **78**, 3393–7.

FLEMING, H. and HASELKORN, R. (1973). *Proc. natn. Acad. Sci. U.S.A.* **70**, 2727–31.

—— (1974). *Cell* **3**, 159–70.

GIDDINGS, T., WOLK, C. P., and SHAMER-ILAN   (1981). *J. Bact.* **146**, 1067–74.

HALLENBECK, P. C., KOSTEL, P. J., and BENEMANN, J. R. (1979). *Eur. J. Biochem.* **98**, 275–84.

HASELKORN, R. (1978). *A. Rev. Pl. Physiol.* **29**, 319–44.

— MAZUR, B., ORR, J., RICE, D., WOOD, N., and RIPPKA, R. (1980). In *Nitrogen fixation*, Vol. 2 (ed. W. E. Newton and W. H. Orme-Johnson), pp. 259–78. University Park Press, Baltimore.

HILL, S., KENNEDY, C., KAVANAGH, E., GOLDBERG, P. B., and HANAU, R. (1981). *Nature, Lond.* **290**, 424–6.

KALLAS, T., RIPPKA, R., COURSIN, T., REBIERE, M.-C., TANDEAU DE MARSAC, N., and COHEN-BAZIRE, G. (1983). In *Photosynthetic prokaryotes: cell differentiation and function* (ed. G. C. Papageorgiou and L. Packer), pp. 281–302. Elsevier Biomedical, New York.

LAMMERS, P. J. and HASELKORN, R. (1983). *Proc. natn. Acad. Sci. U.S.A.* **80**, 4723–7.

LOCKAU, W., PETERSON, R. B., WOLK, C. P., and BURRIS, R. H. (1978). *Biochim. biophys. Acta* **502**, 298–308.

MAZUR, B. J. and CHUI, C.-F. (1982). *Proc. natn. Acad. Sci. U.S.A.* **79**, 6782–6.

— RICE, D., and HASELKORN, R. (1980). *Proc. natn. Acad. Sci. U.S.A.* **77**, 186–90.

MCLEAN, P. and DIXON, R. (1981). *Nature, Lond.* **292**, 655–6.

MEVARECH, M., RICE, D., and HASELKORN, R. (1980). *Proc. natn. Acad. Sci. U.S.A.* **77**, 6476–80.

MERRICK, M. (1982). *Nature, Lond.* **297**, 362–3.

MORTENSON, L. E. and THORNELEY, R. N. F. (1979). *A. Rev. Biochem.* **48**, 387–418.

NAGATANI, H. H. and HASELKORN, R. (1978). *J. Bact.* **134**, 597–605.

NEUER, G. and BOTHE, H. (1982). *Biochim. biophys. Acta.* **716**, 358–65.

— PAPEN, H., and BOTHE, H. (1983). In *Photosynthetic prokaryotes: cell differentiation and function* (ed. G. C. Papgeorgiou and L. Packer) pp. 219–42. Elsevier Biomedical, New York.

NIERZWICKI-BAUER, S. A., CURTIS, S. E., and HASELKORN, R. (1984). *Proc. natn. Acad. Sci. U.S.A.* **81**, 5961–75.

ORR, J. and HASELKORN, R. (1982) *J. Bact.* **152**, 626–35.

OW, D. W. and AUSUBEL, F. M. (1983). *Nature, Lond.* **301**, 307–13.

PEARSON, H. W., HOWSLEY, R., KJELDSEN, C. K., and WALSBY, A. E. (1979). *FEMS Microbiol. Lett.* **5**, 163–8.

PETERS, G. A., ITO, O., TYAGI, V. V. S., and KAPLAN, D. (1981). In *Genetic engineering of symbiotic nitrogen fixation and conservation of fixed nitrogen* (ed. J. M. Lyons *et al.*), pp 343–62. Plenum Press, New York.

REASTON, J., DUYVESTEYN, M. G. C., and DEWAARD, A. (1982). *Gene* **20**, 000–000.

RICE, D., MAZUR, B. J., and HASELKORN, R. (1982). *J. biol. Chem.* **257**, 13157–63.

RIPPKA, R., DERUELLES, J., WATERBURY, J., HERDMAN, M., and STANIER, R. Y. (1979). *J. gen. Microbiol.* **111**, 1–61.

— and STANIER, R. Y. (1978). *J. gen. Microbiol.* **105**, 83–94.

—and WATERBURY, J. (1977). *FEMS Microbiol. Lett.* **2**, 83–6.

ROBERTS, G. P. and BRILL, W. J. (1981). *A. Rev. Microbiol.* **35**, 207–35.

SCOLNIK, P. A., VIROSCO, J., and HASELKORN, R. (1983). *J. Bact.* **155**, 180–5.

SIBOLD, L., MELCK, D., and ELMERICH, E. (1981). *FEMS Microbiol. Lett.* **10**, 37–41.

SILVESTER, W. B. (1976). In *Symbiotic nitrogen fixation in plants* (ed. P. S. Nutman), pp. 521–38. Cambridge University Press.

STAL, L. J. and KRUMBEIN, W. E. (1981). *FEMS Microbiol. Lett.* **11**, 295–9.

STEWART, W. D. P. and LEX, M. (1970). *Arch. Mikrobiol.* **73**, 250–60.

— and RODGERS, G. A. (1977). *New Phytol.* **78**, 459–71.

— and ROWELL, P. (1975). *Biochem. biophys. Res. Commun.* **65**, 846–56.

—— and RAI, A. N. (1980). In *Nitrogen fixation* (ed. W. D. P. Stewart and J. R. Gallon), pp. 239–77. Academic Press, New York.

THOMAS, J., MEEKS, J. C., WOLK, C. P., SHAFFER, P. W., AUSTIN, S. M., and CHIEN, W.-S. (1977). *J. Bact.* **129**, 1545–55.

TUMER, N. E. ROBINSON, S. J., and HASELKORN, R. (1983). *Nature* **306**. 337–42.

WILCOX, M., MITCHISON, G. J., and SMITH, R. J. (1975). *Arch. Mikrobiol.* **130**, 219–23.

WOLK, C. P. (1978). *J. Bact.* **96**, 2138–43.

— THOMAS, J., SHAFFER, P. W., AUSTIN, S. M., and GALONSKY, A. (1976). *J. biol. Chem.* **251**, 5027–34.

— VONSHAK, A., KEHOE, P., and ELHAI, J. (1984). *Proc. natn. Acad. Sci. U.S.A.* **81**, 1561–5.

WYATT, J. T. and SILVEY, J. K. G. (1969). *Science* **165**, 908–9.

8 Genetics and molecular biology of symbiotic nitrogen fixation by *Rhizobium* sp. and *R. japonicum*

C. E. Pankhurst

8.1 INTRODUCTION

Fast-growing rhizobia, other than *Rhizobium trifolii*, *R. leguminosarum*, *R. phaseoli,* and *R. meliloti*, and slow-growing rhizobia, other than *R. japonicum* and *R. lupini*, have for several decades been referred to simply as *Rhizobium* sp. or in the case of rhizobia nodulating tropical legumes as 'cowpea rhizobia' or as members of the 'cowpea miscellany'. Although an unsatisfactory means of classification, this situation developed to accommodate the extreme diversity in nodulation capacity found amongst rhizobia isolated from the nodules of tropical, sub-tropical, and many temperate legumes. Some of these rhizobia appeared to be promiscuous and capable of nodulating a wide range of different legume species, while others were host specific (Vincent 1974; Dart 1977). The diverse nodulating properties of these rhizobia meant that it was not feasible to assign them species names on the basis of nodulation.

The recent impact of molecular biology and the awareness of the importance of symbiotic nitrogen fixation to the improved production of many tropical and sub-tropical legumes, has led to increased interest in the *Rhizobium* sp. as tools for nitrogen fixation research. For example, the genetic and biochemical factors leading to promiscuity of nodulation shown by some *Rhizobium* sp., including the ability to nodulate the non-legume *Parasponia andersonii* (Trinick 1973, 1982; Trinick and Galbraith 1981) may be significant for long term endeavours to extend the host range of rhizobia. The fact that several strains of *Rhizobium* sp. and *R. japonicum* develop nitrogenase activity under defined culture conditions is a desirable feature for the study of nitrogen fixation (*nif*)-gene expression and regulation in *Rhizobium* (Gibson *et al.* 1977;

Gresshoff *et al.* 1981; Kaneshiro *et al.* 1983; Agarwal and Keister 1983). The many examples where both fast- and slow-growing *Rhizobium* sp. effectively nodulate the same legume (Broughton and Dilworth 1971; Pankhurst 1977; Dreyfus and Dommergues 1981*a*) provide ways for defining host and *Rhizobium* contributions to the symbiosis and for characterizing 'infectivity genes'. Associated with these attractive prospects, however, one must be reminded of the problems of applying molecular biology to the study of these bacteria, particularly to the study of the slow-growing rhizobia. The extremely slow growth of some of these bacteria, coupled with their often poor growth on defined media (Chakrabarti *et al.* 1981) and their often high level of intrinsic resistance to antibiotics (Cole and Elkan 1979) make them less amenable to genetic analysis than their fast-growing counterparts. The bacterial cells are often difficult to lyse (Schwinghamer 1980) creating problems in isolating plasmids and other cellular components and they show great diversity in many biochemical properties (Vincent 1977). Despite this, however, significant progress has been made to our knowledge of both the classical and molecular genetics of these micro-organisms.

8.2 TAXONOMIC STATUS OF *Rhizobium* sp

The taxonomy of the genus *Rhizobium* has been recently reviewed (Graham 1976; Vincent 1977; Elkan 1981; Trinick 1982) and only certain aspects relevant to the status of *Rhizobium* sp. and *R. japonicum* will be discussed here. As described in *Bergy's manual for determinative bacteriology* (8th edn), the genus *Rhizobium* is divided into two groups; the fast-growing rhizobia, containing *R. trifolii*, *R. leguminosarum*, *R. phaseoli*, and *R. meliloti*, and the slow-growing rhizobia, containing *R. japonicum* and *R. lupini*. In addition to this is a miscellaneous grouping 'the cowpea miscellany' containing all *Rhizobium* sp. that cannot be distinguished as one of the recognized species.

The recognition of six *Rhizobium* species is historically based on the work of Fred *et al.* (1932) who evolved the idea of cross-inoculation groups. However, the assumption that each species of *Rhizobium* nodulates only plants within a specified cross-inoculation group, e.g. *R. japonicum* nodulates *Glycine* species, has lost credibility. Useful from a practical point of view, the shortcomings of the cross-inoculation (or plant affinity) concept is exemplified by the many examples of cross-infection shown by rhizobia for hosts within different affinity groups (Vincent 1974; Graham 1976; Trinick 1982). In view of the evidence that genes for nodulation and host specificity are plasmid borne in some rhizobia (e.g. Brewin *et al.* 1981), and the evidence that these plasmids can be transferred between rhizobia (Johnston *et al.* 1978) it is clear that plant affinity can no longer provide a sound basis for identifying natural relationships among rhizobia.

The major division of the genus *Rhizobium* into fast- and slow-growing groups is supported by studies of numerical taxonomy, DNA base ratio determinations, nucleic acid hybridization, RNA cistron similarities, immunology, composition of extracellular polysaccharides, carbohydrate utilization and metabolism, bacteriophage and antibiotic susceptibilities, protein composition, and types of intracellular inclusion bodies in bacteroids (Vincent 1977; Elkan 1981; Trinick 1982; Jordan 1982). In fact, the distinction between the two groups is so great that it is proposed to transfer the slow-growing rhizobia into a new genus, *Bradyrhizobium*, with at present one recognized species *Bradyrhizobium japonicum* (Buchanan 1980) comb. nov. (Jordan 1982). The slow-growing *Rhizobium* species *R. lupini* will not be specified in the genus *Bradyrhizobium* as recent DNA : DNA hybridization studies have failed to differentiate between *R. japonicum* and *R. lupini* (Hollis *et al.* 1981). It is proposed that slow-growing *Rhizobium* sp. (other than *R. japonicum*) be referred to as *Bradyrhizobium* sp. with the name of a host plant effectively nodulated by the bacterium in parentheses immediately following (Jordan 1982).

While these proposals simplify the taxonomic status of slow-growing rhizobia, the taxonomic position of fast-growing rhizobia not included in one of the named species, remains uncertain. Many of these rhizobia nodulate plant hosts normally nodulated by slow-growing strains. A recent example of this concerns fast-growing *Rhizobium* strains obtained from China that form effective nodules on wild soyabean and on an unbred soyabean cultivar from China (Keyser *et al.* 1982). These strains which nodulate commercial soyabean cultivars ineffectively have been verified as possessing the physiological attributes of fast-growing rhizobia (Sadowsky *et al.* 1983; Stowers and Eaglesham 1984). They also show low DNA : DNA homology (between 5 to 43 per cent) with type strains of *R. japonicum* and other *Rhizobium* species (Scholla, Moorefield, and Elkan 1984). Based on these data, it has been proposed that these rhizobia be grouped into a new *Rhizobium* species, *R. fredii* (Scholla and Elkan 1984). Similarly, fast-growing rhizobia that nodulate *Lotus* species show very low DNA : DNA homology (between 10 to 15 per cent) with other *Rhizobium* species and with slow-growing strains able to nodulate *Lotus* (Crow *et al.* 1981). These fast-growing rhizobia have also been grouped into a new *Rhizobium* species, *R. loti* (Jarvis *et al.* 1982). Perhaps when similar studies are performed with other defined groups of fast- and slow-growing rhizobia it will be possible to delineate further species.

8.3 MUTANT ISOLATION AND CHARACTERIZATION

An important requirement for the genetic and molecular analysis of any bacterium is the availability of an efficient system for introducing mutations into its genome. In this section, I shall consider the various methods that have been used to obtain symbiotic mutants of *R. japonicum* and *Rhizobium* sp. and discuss the properties of some of the mutants isolated. Some early studies of mutations that effect the symbiotic capacity of *Rhizobium* (mostly concerning *R. trifolii*, *R. leguminosarum*, and *R. meliloti*) have been reviewed (Dénarié *et al.* 1976; Schwinghamer 1977; Beringer *et al.* 1980; Kuykendall 1981).

8.3.1 Spontaneous mutants

One of the most interesting collections of spontaneous symbiotically defective mutants of *R. japonicum* was obtained by Kuykendall and Elkan (1976) as a result of purifying a presumptive pure culture of *R. japonicum* strain USDA110 into component clones. These clones (or sub-strains) differed as much as 20-fold in symbiotic nitrogen fixing ability. Kuykendall and Elkan (1977) examined these sub-strains for the presence of an inducible D-mannitol dehydrogenase (EC 1.1.1.67) and showed that the symbiotically competent derivatives lacked this enzyme. However, no general correlation between D-mannitol utilization and symbiotic inefficiency exists in *R. japonicum* as many symbiotically competent strains utilize D-mannitol (Kuykendall and Elkan 1977). Further studies with these sub-strains have shown that the inefficient derivatives have a greater capacity to assimilate ammonia, either in pure culture or as bacteroids (Upchurch and Elkan 1978) and that the efficient and inefficient types differ in their sensitivity to various antibiotics (Meyer and Pueppke 1980). However, no significant differences could be detected between strain USDA110 or its clonal derivatives in the *Eco*RI restriction pattern of total DNA isolated from these bacteria (Mielenz *et al.* 1979) or in the two-dimensional (PAGE) electrophoresis of proteins (Noel and Brill 1980). These strains are thus closely related and have probably arisen from spontaneous divergence from a common mother strain as a result of maintenance in rich laboratory media. Similar clonal derivatives showing significant differences in symbiotic capacity have been isolated from the slow-growing cowpea strain CB756 (Roughley 1976) and from several *R. trifolii* and *R. meliloti* strains (Brockwell *et al.* 1981). Such observations pose important questions concerning the stability of *Rhizobium* strains in both the field and laboratory.

Spontaneous mutants with altered capsule synthesis were isolated from a genetically marked strain of *R. japonicum* (Law *et al.* 1982).

These mutants were detected by their altered colony morphology and their altered ability to bind soyabean seed lectin. Three of the mutants failed to form a detectable capsule but were capable of nodulating and attaching to soyabean roots, indicating that the presence of a capsule physically surrounding the bacterium is not required for attachment, infection, or nodulation (Law *et al.* 1982).

8.3.2 Chemically-induced mutants

Mutants obtained as a result of chemical mutagenesis have the advantage over spontaneous mutants that are symbiotically defective in that there is more certainty that the phenotypic changes observed result from a single mutation. Maier and Brill (1976) mutagenized *R. japonicuum* strain 61A76 with nitrosoguanidine (NTG) and found five symbiotically defective mutants after screening 2 500 surviving clones for acetylene reduction with soyabean plants. Two of these mutants formed no nodules (Nod⁻) and three formed nodules that did not fix nitrogen (Nod⁺ Fix⁻). One of the Nod⁺ Fix⁻ mutants formed nodules similar to those formed by the wild-type parent strain, and was shown to produce an altered component II of nitrogenase (Maier and Brill 1976; Noel *et al.* 1982). The Nod⁻ mutants were subsequently shown by Maier and Brill (1978*a*) to lack a surface antigen characteristic of the parent strain 61A76. Within the same population of mutagenized cells of 61A76 examined by Maier and Brill (1976), two further mutants which had significantly greater symbiotic nitrogen-fixing ability that 61A76 were obtained (Maier and Brill 1978*b*). Both mutants were more effective than 61A76 in stimulating early growth of soyabeans (Thomas *et al.* 1983). In an analogous study, Williams (1981) isolated three symbiotically altered mutants from the slow-growing cowpea strain 101 following NTG treatment. Two of the mutants were ineffective but the third mutant produced more nodules and expressed greater acetylene-reducing activity than the wild-type strain.

Rhizobium strains that possess an active uptake hydrogenase (Hup⁺) are believed to be able to re-utilize some of the hydrogen evolved from nitrogenase during the catalytic reduction of dinitrogen to ammonia (Schubert and Evans 1976), and thus regenerate ATP (Emerich *et al.* 1979). The possession of an active hydrogenase should therefore lead to more efficient energy usage within the nodule and enhance nitrogen fixation (Evans *et al.* 1980), although this may occur only with certain *Rhizobium* strains and host plants (Gibson *et al.* 1981). Mutants of *R. japonicum* which are deficient in hydrogen uptake capacity (Hup⁻) and which spontaneously revert to the parent (Hup⁺) phenotype at a frequency consistent with that of a single point mutation, were isolated following mutagenesis with nitrous acid (Lepo *et al.* 1981). Soyabean plants inoculated with the H*up*⁻ strains had lower dry weights and contained less total nitrogen than plants inoculated with the parent H*up*⁺ strain, providing support for a beneficial role of the H*up*⁺ phenotype of *R. japonicum*—soyabean symbioses (Lepo *et al.* 1981). Similarly, H*up*⁻ mutants of a *Vigna radiata* (mungbean) *Rhizobium* strain fixed less nitrogen than the parental H*up*⁺ strain (Pahwa and Dorga 1981). Other H*up*⁻ mutants that also lack nitrogenase activity have been isolated (Moshiri *et al.* 1983). Of two mutants studied one appeared to be defective in haem synthesis while the other was defective in expression of nitrogenase components I and II. Revertants of the latter mutant were able to oxidize H_2 and also express nitrogenase activity suggesting that the original lesion was due to a single mutation, possibly in a regulatory gene (Moshiri *et al.* 1983).

The development of the appropriate conditions for the induction of nitrogenase in free-living cultures of some slow-growing *Rhizobium* strains (Pagan *et al.* 1975; Kurz and LaRue 1975; McComb *et al.* 1975) has provided a system suitable for screening the nitrogen-fixing capacity of different *Rhizobium* mutants, without involvement of the host plant. Using this approach Hua *et al.* (1981) screened 100 chlorate-resistant mutants of *R. japonicum* strain USDA110 (obtained after NTG mutagenesis) for nitrogenase activity in free-living culture. Five of the mutants had elevated levels of whole cell nitrogenase activity. Studies with one mutant (strain C33) showed several protein differences between the mutant and the wild-type strain on two-dimensional polyacrylamide gels. Strain C33 also lacked a nitrate reductase. However, this property did not affect nitrogenase activity as other nitrate reductase negative mutants showed no evidence of increased nitrogenase activity (Hua *et al.* 1981). The elevated nitrogenase activity of strain C33 is probably a consequence of an unknown NTG-induced mutation. Lack of an active nitrate reductase was also shown to have no effect on the symbiotic or asymbiotic nitrogenase activity of the slow-growing cowpea *Rhizobium* sp. strain 32H1 (Pagan *et al.* 1977). Similarly, chlorate resistant mutants of *R. japonicum* strain 61A76 lacking both assimilatory and dissimilatory nitrate reductase showed equal or better symbiotic nitrogen fixation than the parent strain (De Vasconcelos *et al.* 1980).

Glutamine auxotrophs of *Rhizobium* sp. strain 32H1 have been isolated following NTG treatment and penicillin enrichment for glutamine dependent mutants (Ludwig and Signer 1977). These mutants had low glutamine synthetase (EC 6.3.1.2) (GS) activity, developed little or no asymbiotic nitrogenase activity, and formed Nod⁺ Fix⁻ nodules on *Macroptilium atropurpureum*. Some prototrophic revertants regained the ability to express both asymbiotic and symbiotic nitrogenase activity while others did not. Revertants of the former type acquired a constitutive unadenylated GSI but remained GSII-negative while revertants of the latter type remained defective for GSI but synthesized GSII (Ludwig 1980). These results suggested that GSI or associated adenylation proteins are involved in *nif*-gene expression in *Rhizobium* sp. strain 32H1 while GSII is not so involved.

Apart from the glutamine auxotrophs of *Rhizobium* sp. strain 32H1 there have been few studies of auxotrophs of *R. japonicum* or *Rhizobium* sp. In two recent reports, Pilacinski and Schmidt (1981) used a carbenicillin-lysozyme enrichment technique to isolate methionine and pantothenic acid auxotrophs of *R. japonicum* strain USDA31 and Wells and Kuykendall (1983) used nitrous acid mutagenesis followed by tryptophan supplementation to isolate trypotophan auxotrophs of *R. japonicum* strain I-110 ARS. The latter mutants were further characterized with respect of specific lesions in the metabolic pathway leading to tryptophan biosynthesis.

8.3.3 Drug-resistant mutants

An association between drug resistance and alteration of symbiotic properties in both fast- and slow-growing rhizobia has been reported (Schwinghamer 1967; Pankhurst 1977; Pain 1979; Skotnicki *et al.* 1979). Significant differences between the fast- and the slow-growing rhizobia with respect to acquisition of resistance to certain antibiotics and associative loss of symbiotic ability was often apparent. For example, Pankhurst (1977) found that mutants of fast-growing *Lotus* rhizobia resistant to viomycin, neomycin, or kanamycin frequently formed Nod+ Fix- nodules on *Lotus* spp., whereas mutants of slow-growing *Lotus* strains resistant to the same antibiotics remained Nod+Fix+. In contrast, mutants of both groups of rhizobia resistant to D-cycloserine (an inhibitor of peptidoglycan synthesis) or rifampicin (an inhibitor of RNA polymerase) showed loss of symbiotic effectiveness (Pankhurst 1977). Resistance to D-cycloserine was also associated with loss of the ability of *Rhizobium* sp. strain 32H1 to develop both asymbiotic and symbiotic nitrogenase activity (Pankhurst and Craig 1979). In these mutants, resistance to D-cycloserine was associated with impairment to the L-alanine, D-alanine transport system (Pankhurst and Craig 1979).

The occurrence of multiple drug resistance amongst strains of slow-growing *Rhizobium* sp. and *R. japonicum*, appears to be a common phenomenon (Patterson and Skinner 1974; Levin and Montgomery 1974; Cole and Elkan 1979; Josey *et al.* 1979; Kremer and Peterson 1982) and maybe plasmid encoded (Cole and Elkan 1973, 1979). Such intrinsic drug-resistance, which may have survival value for these rhizobia in the natural environment, may prove useful for strain identification in ecological experiments (Kremer and Peterson 1982) and in genetic experiments for counter selection of donor strains after transfer of plasmids to these bacteria.

8.3.4 Transposon-induced mutants

The isolation of symbiotic mutants of *Rhizobium* using the genetic system of transposon mutagenesis has been used successfully in *R. trifolii, R. leguminosarum*, and *R. meliloti* (Beringer *et al.* 1978; Buchanan-Wollaston *et al.* 1980); Rolfe *et al.* 1980); Meade *et al.* 1982). A great advantage of transposon insertion, especially into a symbiotic gene, is that it 'marks' the gene both genetically and physically so that it can be subsequently mapped genetically or cloned. The successful use of transposon mutagenesis in *Rhizobium* sp. and *R. japonicum* has been reported (Cen *et al.* 1982); Rostas *et al.* 1984); Chua *et al.* 1985). Using three strains of *Rhizobium* sp. able to nodulate *Macroptilium atropurpureum* (Siratro) and the non-legume *Parasponia*, auxotrophic, Nod⁻ and Nod⁺ Fix⁻ mutants were isolated by selecting for transposition of the transposon Tn5 carrying kanamycin resistance (Cen *et al.* 1982). Symbiotic mutants were obtained at a frequency of about 0.9 per cent for each strain. Total DNA isolated from a number of presumptive Tn5-induced mutants was restricted with *Eco*RI and shown to contain a single copy of Tn5 on different fragments by specific hybridization with cloned Tn5 DNA. This result demonstrated the random nature of the Tn5 insertions. With strain CP283 which has the capacity to develop asymbiotic nitrogenase activity, Tn5-induced mutants that were Fix⁻ *in vitro* and Nod⁺Fix⁻ on *M. atropurpureum* plants were obtained (Cen *et al.* 1982). In similar experiments Rostas *et al.* (1984) and Chua *et al.* (1985) obtained symbiotically defective mutants of *R. japonicum* and a fast-growing *Lotus Rhizobium* strain respectively. Individual Tn5-induced Nod⁺Fix⁻ mutants of the *Lotus* strain NZP2037 were shown to be blocked at several different stages of nodule development (Chua *et al.* 1985).

As an alternative to random Tn5 mutagenesis, Ruvkun and Ausubel (1981) developed a method in *R. meliloti* where it is possible to replace wild-type genomic genes with cloned homologous sequences altered by Tn5 mutagenesis. This method of site-directed Tn5 mutagenesis has been used successfully by Hahn and Hennecke (1984) and Jagadish and Szalay (1984) to introduce Tn5 into the *nif* gene region of *R. japonicum* strain USDA110 and the slow-growing broad host range *Rhizobium* sp. strain IRc78, respectively. With the *R. japonicum* strain, USDA110 Tn5 insertions within *nifD*, and *nifH* caused a Nod⁺Fix⁻ phenotype, whereas mutants with insertions between or in the immediate vicinity of the *nifDK* or *nifH* genes were Nod⁺Fix⁺. This suggests the absence of essential nitrogen fixation genes between or adjacent to *nifDK* and *nifH* in this *R. japonicum* strain (Hahn *et al.* 1984).

However with *Rhizobium* sp. strain IRc78, transposon insertions in a *nifK* hybridizing region, or in regions spanning 10 kb downstream of *nifK*, revealed the presence of functional genes required for nitrogen fixation (Jagadish and Szalay 1984).

8.4 GENE TRANSFER

Three methods of gene transfer have been demonstrated for *Rhizobium*: conjugation, transduction, and transformation (see reviews by Schwinghamer 1977; Beringer *et al.* 1980; Kondorosi and Johnston 1981). Although much progress has been made with all three methods in *R. trifolii, R. leguminosarum,* and *R. meliloti* and gene linkage maps are available for each of these species (Kondorosi *et al.* 1980), studies with *R. japonicum* and *Rhizobium* sp. are not so advanced.

8.4.1 Conjugation

A natural conjugation system in a star-forming, pigmented derivative of *R. lupini* was described by Heuman (1968). Auxotrophic, antibiotic resistance and pigmentation markers were located on a circular chromosome, and the nature of the fertility system examined in some detail (Heuman *et al.* 1974; Heuman and Springer 1977). However, because the strain of *R. lupini* used failed to nodulate lupins, these studies have not led to the mapping of the genes concerned with symbiotic functions. In *R. japonicum*, an R-plasmid coding for resistance to penicillin, neomycin, and chloramphenicol was detected and could be transferred by conjugation to *Agrobacterium tumefaciens* (Cole and Elkan 1973). This is the only known report of natural conjugative gene transfer in *R. japonicum*.

The introduction of conjugative plasmids [especially those of the P-1 incompatability class (see review by Holloway 1979)] into *R. japonicum* and *Rhizobium* sp. in order to promote chromosomal gene transfer, has been studied. Kuykendall (1979) demonstrated that transfer of the P-1 group plasmids R1822 and pRD1 into *R. japonicum* strain USDA110, occurred at a frequency of about 10^{-3}. Transfer of these plasmids between genetically marked sublines of USDA110 also occurred at a frequency of about 10^{-3}. Pilacinski and Schmidt (1981) obtained similar results with the transfer of the P-1 group plasmids RP4 and R68.45 into three serologically distinct *R. japonicum* strains. Transfer to a fourth strain, USDA38, was unsuccessful. Despite this, the transfer of these plasmids to three very different *R. japonicum* strains suggests that it is likely that these plasmids will be transferable to most *R. japonicum* cultures. Attempts by Pilacinski and Schmidt (1981) to show transfer of chromosomal markers using R68.45 in USDA31 were unsuccessful.

Transfer of pRD1 and R68.45 to seven slow-growing strains of cowpea *Rhizobium* sp. was studied by Kennedy *et al.* 1981). Transfer at a frequency of 10^{-4} to 10^{-6} occurred for five strains but was not detected for the other two. With strain CB756, transconjugants containing pRD1 or R68.45 had a lowered growth rate, showed significantly delayed and lowered asymbiotic nitrogenase activity, and formed nodules on *M. atropurpureum* plants that were significantly less effective than those formed by the plasmid-free parent strain. The drug-resistance and *tra* genes of both plasmids were stably maintained in the CB756 transconjugants (Kennedy *et al.* 1981).

The high frequency of transfer of P-1 group plasmids into some *R. japonicum* and slow-growing *Rhizobium* sp. strains, the expression of the antibiotic-resistance markers they contain and their stability is similar to what is observed in *R. trifolii, R. leguminosarum,* and *R. meliloti* (Beringer 1974; Kondorosi *et al.* 1977; Scott and Ronson 1982). However, the reduction in growth rate of strain CB756 transconjugants containing P-1 group plasmids has not been reported for any other *Rhizobium* strain, and is an interesting result. Similarly, the finding by both Pilacinski and Schmidt (1981) and Kennedy *et al.* (1981) of *Rhizobium* strains into which the P-1 group plasmids tried were not transferable is of interest. Such strains may have a DNA restriction system which degrades unmodified DNA.

8.4.2 Transformation

Transformation is the most widely reported method of gene transfer in *Rhizobium* (see reviews by Balassa 1963; Gabor-Hotchkiss 1972; Dénarié *et al.* 1976; Schwinghamer 1977; Kondorosi and Johnston 1981). Despite this, however, no transformation system has been developed to a stage suitable for fine-scale chromosomal mapping, and only recently have reports appeared concerning the transformation of *Rhizobium* with plasmid DNA (Bullerjahn and Berzinger 1982; Kiss and Kolman 1982).

For *R. japonicum*, Raina and Modi (1972) and Doctor and Modi (1976*a,b*) have demonstrated a low frequency of transformation (0.02 to 0.08 per cent) for several unlinked chromosomal markers. Competence of the *R. japonicum* cells occurred during late log-phase and was enhanced by the presence of casamino acids and Mg^{2+} ions (Raina and Modi 1972; Doctor and Modi 1976*b*). A competence factor with a molecular weight of 82 000 has been isolated and partially purified from these cells (Dandekar *et al.* 1978).

Intergeneric transformation between *R. japonicum* and *Azotobacter vinelandii* has been reported (Bishop *et al.* 1977; Maier *et al,* 1978). In both reports, *Azotobacter nif⁻* mutants were transformed to *nif⁺* by *R. japonicum* 61A76 DNA. Of 50 *nif⁺* transformants tested by Maier *et al.*

(1978) three were found to contain an O-antigen related polysaccharide that is believed to be specifically required for nodulation by *R. japonicum*. Such transfer of *R. japonicum nif* genes as well as genes required for nodulation to *Azotobacter* could provide a useful system for identifying genes involved in symbiosis.

8.4.3 Transduction

Transduction, a useful technique for fine-linkage mapping of bacterial genes, has been studied in only a few *Rhizobium* strains. Generalized transduction in *R. meliloti* was reported by Kowalski (1967) and in *R. leguminosarum* by Buchanan-Wollaston (1979). Transduction in *R. japonicum* mediated by a temperate phage was briefly reported by Doctor and Modi (1976*b*). In the *R. japonicum* strain tested co-transduction of *ura* and *arg* alleles was observed at a high frequency (3×10^{-3}) but transductants for two other markers tested were not obtained suggesting that this was a case of specialized transduction (Doctor and Modi 1976*b*). Transduction was also demonstrated for the non-nodulating *R. lupini* strain studied by Heuman *et al.* (1974). More recently generalized transduction in *R. japonicum* strain D211 by phage M-1 has been observed (Shah *et al.* 1983). This enabled a linkage map or D211 for 10 different genetic markers to be constructed by co-transduction analysis. Both ultraviolet irradiation of the phage lysate (Shah *et al.* 1983) and addition of D-glucosamine to the transduction mixture (Shah *et al.* 1981) increased the recovery of transductants.

Studies of transduction with other fast- and slow-growing *Rhizobium* sp. have not been reported. However, the morphology and host range of virulent phages of fast- and slow-growing rhizobia that nodulate *Lotus* species has been examined (Patel 1976). Cross-infection between phage isolated for fast-growing *Lotus* rhizobia and slow-growing *Lotus* rhizobia was not observed.

8.5 PLASMIDS

Perhaps the single most important development in the genetic analysis of *Rhizobium* was the discovery that genes required for nitrogen fixation, nodulation, and host-specificity are plasmid borne in some strains (Higashi 1967; Dunican *et al.* 1976; Johnston *et al.* 1978; Zurkowski and Lorkiewicz 1979; Nuti *et al.* 1979; Buchanan-Wollaston *et al.* 1980; Banfalvi *et al.* 1981). Genes affecting symbiotic effectiveness (Brewin *et al.* 1980) and ability to induce polygalacturonase in legume roots (Palomares *et al.* 1978) are also plasmid encoded. These findings together with the development of systems for transferring these symbiotic (*sym*) plasmids between *Rhizobium* strains (Johnston *et al.* 1978; Hirsch 1979; Brewin *et al.* 1980; Beynon *et al.* 1980; Kondorosi *et al.*

1982) have resulted in a temporary switch from the study of chromosomal to plasmid genes. Recent studies of *Rhizobium* plasmids have been reviewed (Dénarié *et al.* 1981) and only data relevant to *R. japonicum* and *Rhizobium* sp. will be considered here.

8.5.1 Identification

Plasmids were first detected in *R. japonicum* and *R. lupini* using ethidium bromide-caesium chloride or sucrose density gradients (Klein *et al.* 1975; Sutton 1975; Nuti *et al.* 1977). More recently, Gross *et al.* (1979) used the alkaline denaturation procedure developed by Currier and Nester (1976) followed by agarose gel-electrophoresis, to detect one to four plasmids of molecular weight 48 to 138 Mdal in eight non-mucoid strains of *R. japonicum*. Using similar procedures, Scott and Tait (1981), Cantrell *et al.* (1982), and Sadowsky and Bohlool (1983) have also detected plasmids in *R. japonicum* strains. However, in all of these studies strains which appeared to lack plasmids were found. Cantrell *et al.* (1982) found a correlation between the absence of plasmids and the presence of an active hydrogenase (Hup^+ phenotype) in the strains they studied. The absence of plasmids in the Hup^+ strains was thoroughly investigated using different isolation procedures including the 'in gel' lysis method developed by Eckhardt (1978). The Eckhardt procedure has been used successfully to detect plasmids with molecular weights of 300 to 400 Mdal in *R. trifolii, R. leguminosarum,* and *R. meliloti* (Dénarié *et al.* 1981; Banfalvi *et al.* 1981; Rosenberg *et al.* 1982). No evidence for the presence of a DNAase that could interfere with the detection of plasmid DNA could be found in the plasmid-less strains and no plasmids were seen in soyabean nodule bacteroids (Cantrell *et al.* 1982). Thus it seems that some *R. japonicum* strains either contain no plasmids or have plasmids with molecular weights in excess of 400 Mdal which cannot be detected by presently available techniques.

Large plasmids (35 to >300 Mdal) have been demonstrated in fast-growing rhizobia that nodulate soyabeans (Masterson *et al.* 1982; Sadowsky and Bohlool 1983; Heron and Pueppke 1984) and in fast-growing *Rhizobium* sp. that nodulate tropical and sub-tropical legumes (Broughton *et al.* 1984*a*). In the latter rhizobia, the number of plasmids present in individual strains had no relationship to the number of different host legumes they could nodulate (Broughton *et al.* 1984*a*). A single large plasmid has been reported in two fast-growing *Lotus Rhizobium* strains (Pankhurst 1983*a*).

8.5.2 Transfer

Apart from an R-plasmid coding for penicillin, neomycin, and chloramphenicol resistance in *R. japonicum* (Cole and Elkan 1973) naturally-occurring genetic markers on plasmids in *R. japonicum* or other

Rhizobium sp. have not been reported. In contrast, self-transmissible plasmids specifying bacteriocin production have been found in *R. leguminosarum* (Hirsch 1979) and a plasmid specifying melanin production in plate culture has been found in *R. phaseoli* (Beynon *et al.* 1980). The bacteriocinogenic plasmids described by Hirsch (1979) were all transferred between strains of *R. leguminosarum* at high frequencies (10^{-2} per recipient) and one of them (pRL1JI) also carried genes for the nodulation of *P. sativum* (Johnston *et al.* 1978).

Insertion of a transposon into a *Rhizobium* plasmid to facilitate its transfer to other bacteria was first reported by Johnston *et al.* (1978). Since then, *sym*-plasmids in *R. trifolii*, *R. leguminosarum*, and *R. meliloti* have been labelled with the transposon Tn5 and transferred to other rhizobia and *Agrobacterium* (Hooykaas *et al.* 1981; Buchanan-Wollaston *et al.* 1980; Kondorosi *et al.* 1982). Using a similar approach, the single large plasmid of the fast-growing *Lotus Rhizobium* strain NZP2037 (Pankhurst *et al.* 1983*a*) and the large plasmid of the broad host range *Rhizobium* sp. strain NGR234 (Broughton *et al.* 1984*a*; Morrison *et al.* 1984) have been transferred to other rhizobia.

8.6 SYMBIOSIS GENES

8.6.1 Location of *nif*- and *nod*-genes

A direct physical approach for demonstrating the presence of nitrogen fixation (*nif*)-genes in rhizobia was made possible by the finding that the DNA sequences of the structural *nif*-genes were highly conserved amongst nitrogen fixing bacteria (Nuti *et al.* 1979; Ruvkun and Ausubel 1980). The procedure utilized recombinant plasmids carrying *nif*-genes from *Klebsiella pneumoniae* (Cannon *et al.* 1977). One particular plasmid, pSA30, which carries the nitrogenase structural genes H, D, K, and part of E (Cannon *et al.* 1979) has been used successfully to identify (via DNA : DNA hybridization) DNA fragments carrying similar genes in *Rhizobium*. These genes have been located on a plasmid in *R. leguminosarum* and *R. meliloti* strains (Nuti *et al.* 1979; Prakash *et al.* 1981; Banfalvi *et al.* 1981; Rosenberg *et al.* 1981; Hombrecher *et al.* 1981).

Similar studies using pSA30 or cloned *R. meliloti nifHDK* genes as hybridization probes, have located *nif*-genes on large plasmids in several fast-growing *Rhizobium* sp. (Masterson *et al.* 1982; Pankhurst *et al.* 1983*b*; Broughton *et al.* 1984*a*). However, with two fast-growing strains of *Lotus* rhizobia, the single large plasmid present did not hybridize with these probes (Pankhurst *et al.* 1983*a*). When total DNA and plasmid DNA was isolated from one of these strains, strong hybridization between *nifHDK* and the total DNA, but not the plasmid DNA, was found (Pankhurst *et al.* 1983*a*). Similar results have been obtained for several strains of *R. japonicum* (Haugland and Verma 1981; Masterson *et al.* 1982). Thus while the *nif*-genes are clearly present on plasmids in *R. leguminosarum*, *R. phaseoli*, *R. trifolii*, *R. meliloti*, and many fast-growing *Rhizobium* sp., they appear to be located elsewhere on the genome (on the chromosome or on a very large undetected megaplasmid) in the strains of *R. japonicum* and fast-growing *Lotus* rhizobia so far studied.

Two lines of evidence have clearly established that genes for nodulation (*nod*) are also plasmid-borne in *R. leguminosarum*, *R. phaseoli*, *R. trifolii*, and *R. meliloti*. Firstly, complete or partial curing of a plasmid in these strains is correlated with loss of nodule forming ability (Zurkowski and Lorkiewicz 1979; Beynon *et al.* 1980; Banfalvi *et al.* 1981; Zurkowski 1982). Secondly, transfer of a plasmid from one *Rhizobium* strain to another may be correlated with transfer of nodule-forming ability (Johnston *et al.* 1978; Beynon *et al.* 1980; Scott and Ronson 1982; Kondorosi *et al.* 1982). Evidence that the plasmid containing *nod* functions in these rhizobia also contains the *nif*-genes is shown by co-transfer of the *nif*-genes (demonstrated by hybridization with a *nif* probe) in the transfer experiments and by simultaneous loss of nodule-forming ability and *nif* hybridization in the curing experiments. More recently, direct evidence that the *nif*- and the *nod*-genes arc closely linked in *R. leguminosarum*, *R. trifolii*, and *R. meliloti* has been obtained (Downie *et al.* 1983; Scott *et al.* 1984; Kondorosi *et al.* 1984).

A plasmid location of *nod* as well as *nif*-genes has been confirmed by plasmid curing experiments in a fast-growing broad host range *Rhizobium* sp. (Morrison *et al.* 1983) and a fast-growing soyabean *Rhizobium* strain (Sadowsky and Bohlool 1983). Direct physical linkage of *nod*- and *nif*- genes to a *sym* plasmid using cloned *nod*- and *nif*-genes as hybridization probes has also been demonstrated for several fast-growing *Rhizobium* sp. that nodulate tropical and sub-tropical legumes (Broughton *et al.* 1984*a*). In contrast, *nod*-genes were not located on the single large plasmid present in two fast-growing *Lotus Rhizobium* strains as plasmid cured derivatives of these strains were still able to nodulate *Lotus* spp. (Chua *et al.* 1985). The location of *nod*-genes in slow-growing *Rhizobium* sp. and *R. japonicum* has not yet been established.

8.6.2 Physical characterization of *nif*-genes

Physical analysis of the *nifHDK*-genes from *R. meliloti* is well advanced and includes a complete DNA sequence of the *nifH*-gene (Torok and Kondorosi 1981) and part of the *nifD*-gene (Ruvkun *et al.* 1982). Identification of *nifH*, *nifD*, and *nifK* which code for the single subunit of the nitrogenase Fe protein (component II) and for the α- and β-subunits of the MoFe protein (component I) of nitrogenase, respectively, was determined by DNA hybridization to cloned *K. pneumoniae nif*-genes and by comparison of *nif*-gene sequences for the two bacteria (Ruvkun *et al*. 1982). As in *K. pneumoniae* the *nifHDK* genes in *R. meliloti* are transcribed in the direction *nifHDK* and are part of a single operon (Ruvkun *et al*. 1982; Corbin *et al*. 1983). In addition, conservation of promoter sequences amongst the *R. meliloti nif*-gene promoters and those of *K. pneumoniae* has been demonstrated (Better *et al*. 1983).

In contrast, the physical organization of the *nifHDK* genes in the *Rhizobium* sp. and *R. japonicum* strains so far examined appears to be quite different. In the slow-growing *Rhizobium* sp. strains ANU289 (Scott *et al*. 1983) and IRc78 (Yun and Szalay 1984) and in *R. japonicum* strain USDA110 (Kaluza *et al*. 1983; Adams *et al*. 1984) the *nifH* and *nifDK* genes are located on separate operons. In ANU289 these two operons are separated by at least 13 kb (Scott *et al*. 1983) while in USDA110 *nifH* is located on the 3′ side (downstream) at a distance of 17 kb from the *nifDK* operon (Fischer and Hennecke 1984). A number of similarities have been observed between the nucleotide sequences of the *nifH* and *nifDK* promoters of USDA110 and those of *nif* specific operons in *K. pneumoniae* and *R. meliloti* (Adams and Chelm 1984). This suggests that despite organizational differences a similar mechanism for transcriptional control of *nif*-gene expression may function in all three organisms. The identification of a DNA region homologous with the *K. pneumoniae nifA* gene between the *nifH* and *nifDK* operons in USDA110 (Adams *et al*. 1984) supports this suggestion. The *nifA* homologous region is in an identical position with respect to the *nifH* promoter (6 kb upstream) as is the *nifA*-like locus described in *R. meliloti* (Szeto *et al*. 1984).

Transcription of *nif*-genes in the endosymbiont or bacteroid form of *Rhizobium* sp. or *R. japonicum* has not been studied. However, the transcriptional start points and the complete nucleotide sequences of the *nifH* and the *nifD* genes in *R. japonicum* strain USDA110 are known (Fuhrmann and Hennecke 1984; Kaluza and Hennecke 1984). The *nifH* gene is 882 nucleotides long, coding for 294 amino acids which give a polypeptide with a molecular weight of 31 525, while the *nifD* gene is 1 545 nucleotides long, coding for 515 amino acids which make up a dinitrogenase α-subunit with a molecular weight of 57 918. Extensive sequence homology between the *nifH* and *nifD* genes and their respective counterparts from other nitrogen-fixing bacteria is apparent (Fuhrmann and Hennecke 1984; Kaluza and Hennecke 1984).

Two other fast-growing *Rhizobium* sp. strains have been subjected to preliminary analysis of their *nif*-gene regions. The sequence of the *nifH* promoter region of the *Rhizobium* sp. strain BTAil (which forms stem as well as root nodules on *Aeschynomene indica*) has been determined (Legocki *et al*. 1984). The 200-bp nucleotide sequence upstream to the *nifH* structural gene showed substantial homology to the corresponding *nifH* regions of *Rhizobium* sp. strains ANU289 (Scott *et al.* 1983), IRc78 (Yun and Szalay 1983), and *R. japonicum* strain USDA110 (Adams and Chelm 1984). By fusing the *nifH* promoter region of BTAil to the *lacZ* gene of *Escherichia coli* and inserting it into a 'non-essential' region of the BTAil chromosome, expression of β-Galactosidase in stem nodules formed by the engineered strain on *A. indica* was achieved (Legocki *et al.* 1984). The second *Rhizobium* sp. strain (ORS571) forms stem and root nodules on *Sesbania rostrata* (Dreyfus and Dommergues 1981*b*). This strain also has the unusual property of growing on dinitrogen as a sole nitrogen source (Elmerich *et al.* 1982). A 13 kb *Bam*H1 fragment carrying *nifHDK* genes has been cloned from ORS571 and shown to complement on ORS571 mutant defective in both asymbiotic and symbiotic nitrogenase activity. With this strain therefore it should be possible to study the biochemistry and genetics of nitrogen fixation with bacteria grown under non-symbiotic conditions.

8.7 CONTRIBUTION OF *Rhizobium* sp. AND *R. japonicum* TO THE MOLECULAR ANALYSIS OF ROOT NODULE DEVELOPMENT

Concomitant with our expanding knowledge concerning the location, isolation, and physical characterization of symbiosis genes in *Rhizobium*, new avenues for exploring the physiological functions of these genes in nodule development and function are appearing. A good example of this is the transfer of *sym*-genes from *Rhizobium* to *Agrobacterium*. In the majority of cases reported so far (Hooykaas *et al.* 1981; Hooykaas *et al.* 1982; Kondorosi *et al.* 1982; Hirsch *et al.* 1984) the *Agrobacterium* transconjugants develop Nod⁺Fix⁻ nodules on the host legume normally nodulated by the *Rhizobium* donor. This means that genes necessary for binding (recognition), infection, and nodule initiation are being expressed in *Agrobacterium*. The products of these genes may be recognized by examining the *Agrobacterium* transconjugants for altered properties, e.g. changes in the composition of exopolysaccharides or lipopolysaccharides. The extent to which nodule development occurs with the transconjugants will also define the limit to which the *sym*-genes can function in their new host bacterium.

There are several examples where *Rhizobium* sp. and *R. japonicum* can be specifically useful for analysis of *sym*-genes. Some of these, e.g. nodule development by both fast- and slow-growing strains on the same host, broad-host range, asymbiotic nitrogenase activity, have been mentioned, but it is considered useful to examine them in more detail in order to emphasize their research possibilities.

8.7.1 Nodule development by fast- and slow-growing *Rhizobium* strains on the same host

There are numerous examples where an individual host legume forms Nod⁺Fix⁺ nodules with fast- and slow-growing *Rhizobium* strains, e.g. *Lotus pedunculatus* (Pankhurst 1977); *Vigna sinensis* (Broughton and Dilworth 1971); *Glycine max* (Keyser *et al.* 1982). As there is generally very little DNA : DNA homology between fast- and slow-growing rhizobia (Jarvis *et al.* 1977; Jordan 1982) these rhizobia must contain as a minimum, a similar complement of *sym*-genes plus other ancillary genes required for nodulation of the particular host. Cross-hybridization studies between such strains may assist in identifying and characterizing such genes. In addition, analysis of components both types of rhizobia have on their surface may detect a common factor required for nodulation. Comparative studies of nodule enzymes (Boland *et al.* 1978) or nodule specific proteins (nodulins) (Verma 1981) may also reveal to what extent the host plant response is modified when in symbiosis with constrasting *Rhizobium* genotypes.

8.7.2 Broad-host range

The broad-host range displayed by many slow-growing and a few fast-growing *Rhizobium* sp. (Trinick 1980, 1982) including for some strains the capacity to nodulate the non-legume *Parasponia andersonii* (Trinick and Galbraith 1981) is another property important for the analysis of nodule development. For example, the capacity of *R. trifolii, R. leguminosarum*, and *R. meliloti* transconjugants carrying the *sym*-plasmid from the broad-host range *Rhizobium* sp. strain NGR234 to nodulate Siratro (Morrison *et al.* 1984) indicates that genes for broad-host range are linked to other *sym*-genes and that ability to nodulate a range of legumes is not entirely dependent on other strain characteristics.

Aspects of nodule development directly under the control of the host plant genotype are readily seen in experiments in which different species are nodulated by the same *Rhizobium* strain. Thus it is known that the plant determines the structure of the globin chains of leghaemoglobin (Dilworth 1969; Broughton and Dilworth 1971), the size and shape of the bacteroids (Dart 1977), and bacteroid viability (Sutton and Patterson 1980). Nodule hydrogenase activity may also be host controlled (Gibson *et al.* 1981).

The number of plant genes involved in nodule development is unknown (see Chapter 11, this volume for discussion) and they may not necessarily all be chromosomal genes as recent evidence from studies of *Lotus* spp. hybrids suggests a degree of maternal inheritance of nodule forming ability (R. M. Greenwood, personal communication). Clearly further comparisons of nodule development by the same *Rhizobium* strain on different legumes will widen knowledge of the hosts contribution to the symbiosis. As many of the tropical legumes that could be used in such studies form large nodules, detailed comparisons of, for example, the 'nodulins' (Verma 1981) they contain should prove very useful.

8.7.3 Asymbiotic nitrogen fixation

The report by Holsten *et al.* (1971) of nitrogen fixation *in vitro* by *R. japonicum* grown in co-culture with soyabean cells provided the primary impetus to efforts in a number of laboratories that culminated in the demonstration of asymbiotic nitrogen fixation by free-living rhizobia (Pagan *et al.* 1975; Kurz and LaRue 1975; McComb *et al.* 1975; Keister 1976). Some of the initial interest in the *in vitro* system was directed toward its potential as a system to study the infection process. However, it soon became apparent that the plant cell-bacterium association was not entirely comparable to infection *in vivo* (Child and LaRue 1976). Moreover, the behaviour of the plant cells in the *in vivo* systems did not always correspond to the genetic constitution of the host plant from which they were derived. Nevertheless, the ability of rhizobia to fix nitrogen when grown under defined conditions resulted in a useful tool for genetical analysis of the bacterium.

Not all rhizobia appear to be capable of asymbiotic nitrogenase activity (Gibson *et al.* 1977; Kaneshiro *et al.* 1978). In fact, only a small number of slow-growing *Rhizobium* sp. and *R. japonicum* strains have been found to develop significant levels of nitrogenase activity under defined media conditions. Other slow-growing strains and wild-type strains of fast-growing *Rhizobium* species show little or no activity under the same conditions (Gibson *et al.* 1977; Kaneshiro *et al.* 1978; Lorkiewicz *et al.* 1978; Pankhurst 1981). The recent report of a fast-growing *Rhizobium* sp. that exhibits asymbiotic nitrogenase activity and can grow at the expense of dinitrogen (Elmerich *et al.* 1982) is currently an exception.

Methods used to detect nitrogenase activity in *Rhizobium* strains under defined conditions include agar culture (Gibson *et al.* 1976; Pankhurst and Craig 1978; Wilcockson and Werner 1978), static liquid culture (Evans and Keister 1976), shaken liquid culture (Keister and Evans 1976; Scott *et al.* 1979; Werner and Stripf 1978), and chemostat culture (Bergersen *et al.* 1976). Essential requirements for induction include a low concentration of O_2 near the cells (about 1 μM) (Keister and Evans 1976; Bergersen *et al.* 1976), provision of a carboxylic acid, e.g. succinate, and a source of combined nitrogen, e.g. glutamine (Gibson *et al.* 1976; Bergersen *et al.* 1976; Wilcockson and Werner 1978; Pankhurst 1981; Kaneshiro and Kurtzman 1982). Using chemostat cultures of *Rhizobium* sp. strain 32H1, Bergersen and Turner

(1978) have shown that derepression of nitrogenase synthesis occurs when the concentration of dissolved O_2 in the medium becomes limiting. The fixed nitrogen is exported into the medium as ammonia where it is partly used by the bacteria for growth. In the presence of excess ammonia, nitrogenase activity will continue provided the O_2 concentration is sufficiently low to prevent growth of the bacteria. Under these conditions, the activity of glutamine synthetase, an enzyme which appears to act as a positive control element for nitrogenase synthesis in *K. pneumoniae* (Tubb 1974) is high. However, when growth of the bacteria is promoted by increasing the supply of O_2, the glutamine synthetase becomes adenylylated and nitrogenase activity declines (Bergersen and Turner 1978).

Significant analogies between the properties of O_2 limited cultures of *Rhizobium* sp. strain 32H1 and bacteroids of this strain isolated from cowpea nodules are apparent. Bacteroids and cultures have similar nitrogenase activities, both are produced in O_2 limited systems, and both types of cell have best nitrogenase activity at the low free O_2 concentrations which prevail during deoxygenation of oxyleghaemoglobin. Newly-fixed nitrogen appears in the surrounding medium as ammonium, being assimilated from there by host-cell glutamine synthetase in nodules (Robertson *et al.* 1975) and by bacterial glutamine synthetase in cultures (Bergersen and Turner 1978). In both systems it seems to be an advantage that the O_2-limited conditions tend to prevent adenylylation of the glutamine synthetase in the bacteria, thus preventing repression of nitrogenase in the presence of products of nitrogen fixation. Nitrogen-fixing agar cultures of *Rhizobium* sp. strain 32H1 also contain a significant population of swollen pleiomorphic cells which resemble the bacteroids of 32H1 in cowpea nodules (Pankhurst and Craig 1978; Van Brussel *et al.* 1979; Kaneshiro *et al.* 1983). Similar pleiomorphic cells have been observed in nitrogen-fixing liquid cultures of this strain (C. E. Pankhurst, unpublished observations). The development of these cells is probably due to the reduced O_2 tension in these cultures and thus a consequence rather than a prerequisite for nitrogenase synthesis.

The question why only certain strains of *R. japonicum* and slow-growing *Rhizobium* sp. appear capable of significant asymbiotic nitrogenase activity is an important one. While the optimum medium constituents for asymbiotic nitrogenase activity by strains possessing this capability (Nase$^+$) varies (Gibson *et al.* 1976; Bergersen and Gibson 1978; Pankhurst 1981; Kaneshiro and Kurtzman 1982), exhaustive testing of alternate media conditions has not induced nitrogenase in Nase$^-$ strains (Pankhurst 1981; Kaneshiro and Kurtzman 1982). The exoploysaccharides produced by five Nase$^+$ strains all contained significant amounts of a 6-deoxyhexose, but a similar sugar was found in the exopolysaccharide of one of four Nase⁻ strains examined (Kennedy and Pankhurst 1979). This appeared to rule out the possible role of exopoly-saccharide composition influencing oxygen transfer into the rhizobial cell as an important factor associated with nitrogenase induction. However, Agarwal and Keister (1983) reported that *R. japonicum* strains that show little or no nitrogenase activity in liquid culture had a shorter generation time, consumed more O_2, and produced more exopolysac-charide under aerobic or microaerophilic conditions than Nase⁺ strains. Eight strains that were Nase⁺ synthesized an acidic exopolysaccharide containing rhamnose and 4-0-methylglucuronic acid while 16 strains that were Nase⁻ produced an exopolysaccharide containing glucose, mannose, galacturonic acid, and galactose (Huber *et al.* 1984).

A highly significant correlation between the level of natural resistance to the RNA polymerase inhibitor rifampicin and ability of slow-growing *Rhizobium* strains to express nitrogenase activity in agar culture has been reported (Pankhurst *et al.* 1982). Nase⁺ strains were found to be 5 to 100 times more resistant to rifampicin than Nase⁻ strains. The reason for this is unknown at present, but it is interesting to note that a role for RNA polymerase in the differentiation of nitrogen-fixing bacteroids has been inferred from the observation that several rifampicin-resistant mutants of both fast- and slow-growing *Rhizobium* strains form Nod⁺Fix⁻ nodules (Pankhurst 1977; Pain 1979). However, Regensburger and Hennecke (1983) could detect no differences in the electrophoretic properties of RNA polymerase isolated from aerobically grown and nitrogen-fixing cells of *R. japonicum* strain USDA110.

The possibility that the host plant provides factors necessary for derepression of nitrogenase in free-living cultures of *Rhizobium* should not be overlooked. Indeed, the importance of plant cell factors has been emphasized by the experiments of Reporter and co-workers (Reporter 1976; Bednarski and Reporter 1978; Storey *et al.* 1979) who describe fac-tors produced in legume plant cell - *Rhizobium* co-culture that cause in-duction to nitrogenase activity in a wide range of rhizobia under micro-aerophilic conditions. The factors contained in this plant-conditioned medium (PCM) seem to affect oxidative phosphorylation in rhizobia and also cause loss of exopolysaccharides from the *Rhizobium* cell surface (Reporter 1981). The active fractions of PCM responsible for derepres-sion contain a Cu-peptidoglucan and a Cu-metallothionen (Storey and Reporter 1980; Reporter 1981). Metallothionens appear to sequester tran-sition metals preventing harm to plants and presumably animals (Reporter 1981). Little is known of their occurrence or induction in plant cells.

8.7.4 Opines

There has been much speculation that at some stage during the nodula-tion process, DNA sequences are transferred from the invading

Rhizobium cells and become incorporated into the legume cell nucleus in a way analogous to the transfer of the T-DNA from the Ti-plasmid of *Agrobacterium* during crown gall induction (Lepidi *et al.* 1976; Broughton *et al.* 1984*b*; Vance 1983). Evidence for such a DNA transfer from *Rhizobium* to the legume cell nucleus is lacking although Lepidi *et al.* (1976) detected uptake of *Rhizobium* DNA by *Vicia faba* root cells after exposure of root apexes to viable *Rhizobium* cells. Using bacteria-free callus tissue grown from 'pseudo-nodules' incited on *Medicago sativa* mutants by *R. meliloti* (Vance and Johnson 1983) as a source of potentially transformed plant DNA, Broughton *et al.* (1984*b*) showed the presence of a hybridizing band with this DNA not present with control root DNA when the *R. meliloti sym*-plasmid was used as the probe. Unfortunately the hydridizing region on the *R. meliloti sym*-plasmid could not be localized.

A major feature of crown gall cells is their capacity to synthesize opines, unusual α-N-substituted amino-acid derivatives (see review by Dellaporta and Pesano 1981). Opine synthesis in tumour cells is *Agrobacterium* strain specific and is coded for by specific Ti-plasmid sequences that are transferred as part of the T-DNA to the plant cell (Hernalsteens *et al.* 1980). The occurrence of unusual amino acids in legume root nodules has also been reported (Greenwood and Bathurst 1968; 1978; Freney and Gibson 1975). In a comprehensive study Greenwood and Bathurst (1978) examined the amino-acid composition of nodules formed on species of 16 legume genera in eight different cross-inoculation groups (*Lotus-lupinus, Carmichaelia, Trifolium, Pisum, Vicia, Medicago, Phaseolus*, goat's rue, and *Vigna* groups). In all cases differences in the amino-acid patterns observed were determined by the *Rhizobium* strain. In cases where nodules were formed by the same *Rhizobium* strain on different legumes, the amino-acid patterns were closely similar. With some legumes, e.g. *Carmichaelia* or *Lotus*, the nodules formed by different *Rhizobium* strains showed many diverse amino-acid patterns and were useful in distinguishing between *Rhizobium* strains. With other legumes fewer amino-acid patterns were found, e.g. *Trifolium* species have only two distinct patterns characterized by high or low levels of bound γ-aminobutyric acid. Many unknown compounds were detected in the nodules examined by Greenwood and Bathurst (1978). These compounds were characteristic of different *Rhizobium* strains, occurred in a bound form in the nodule, and were not detected in root tissue. One of these compounds from *Lotus tenuis* nodules has been identified as 2,4-diamino-3-methyl-butanoic acid (Shaw *et al.* 1981).

Although the role of these unusual amino acids in root nodules is unknown, it is tempting to speculate that they fulfil a role for *Rhizobium* that opines fulfil for *Agrobacterium* in crown gall tumours. Clearly, more information is required concerning the synthesis of these compounds in nodules, whether the genes for their synthesis are present on *Rhizobium* plasmids and whether the *Rhizobium* specifying them can utilize them for growth. Undoubtedly, *Rhizobium* sp. that produce distinctly different compounds or are able to nodulate a range of different legumes will be valuable in these analyses.

8.8 CONCLUDING REMARKS

Genetic and molecular biology studies of *Rhizobium* sp. and *R. japonicum* have now reached an interesting stage. Many of the techniques that have yielded valuable information in identifying and characterizing symbiotic genes in *R. trifolii*, *R. leguminosarum*, *R. phaseoli*, and *R. meliloti*, e.g. transposon mutagenesis and plasmid mobilization, are now being applied to similar analyses in *Rhizobium* sp. and *R. japonicum*. It is to be hoped that inherent difficulties in working with the slow-growing rhizobia will not deter investigators from pursuing these studies. The rewards are obvious. Understanding the molecular basis of broad-host-range, asymbiotic nitrogen fixation, or opine synthesis in nodules are but three challenging areas of research where *Rhizobium* sp. will be invaluable. Coupled with this is the fact that many of these rhizobia fix nitrogen with the world's most important grain legumes. Increased study of these bacteria will thus stimulate the development of molecular biology technologies in those countries where these legumes are of paramount importance.

REFERENCES

ADAMS, T. H. and CHELM, B. K. (1984). *J. molec. & Appl. Genet.* **2**, 392.
— McCLUNG, C. R., and CHELM, B. K. (1984). *J. Bact.* **159**, 857.
AGARWAL, A. K. and KEISTER, D. L. (1983). *Appl. & Environ. Microbiol.* **45**, 1592.
BALASSA, G. (1963). *Bact. Rev.* **27**, 228.
BANFALVI, Z., SAKANYAN, V., KONCZ, C., KISS, A., DUSHA, I., and KONDOROSI, A. (1981). *Molec. & gen. Genet.* **184**, 318.
BEDNARSKI, M. and REPORTER, M. (1978). *Appl. & environ. Microbiol.* **36**, 115.
BERGERSEN, F. J. and TURNER, G. L. (1978). *Biochim. biophys. Acta.* **538**, 406.
— Gibson, A. H. (1978). In *Limitations and potentials for biological nitrogen fixation in the tropics* (ed. J. Döbereiner, R. H. Burris, and A. Hollaender), p. 263. Plenum Press, New York.
BERINGER, J. E. (1974). *J. gen. Microbiol.* **84**, 188.
— Beynon, J. L., BUCHANAN-WOLLASTON, A. V., and JOHNSTON, A. W. B. (1978). *Nature, Lond.* **276**, 633.
— BREWIN, N. J., and JOHNSTON, A. W. B. (1980). *Heredity* **45**, 161.
BETTER, M., LEWIS, B., CORBIN, D., DITTA, G., and HELINSKI, D. R. (1983). *Cell* **12**, 721.

BEYNON, J. L., BERINGER, J. E., and JOHNSTON, A. W. B. (1980). *J. gen. Microbiol.* **120**, 421.

BISHOP, P. E., DAZZO, F. B., APPLEBAUM, E. R., MAIER, R. J., and BRILL, W. J. (1977). *Science* **198**, 938.

BOLAND, M. J., FORDYCE, A. M., and GREENWOOD, R. M. (1978). *Aust. J. Pl. Physiol.* **5**, 553.

BREWIN, N. J., BERINGER, J. E., BUCHANAN-WOLLASTON, A. V., JOHNSTON, A. W. B., and HIRSCH, P. R. (1980). *J. gen. Microbiol.* **116**, 261.

— BEYNON, J. L., and JOHNSTON, A. W. B. (1981). In *Genetic engineering of symbiotic nitrogen fixation and conservation of fixed nitrogen.* (ed. J. M. Lyons, R. C. Valentine, D. A. Phillips, D. W. Rains, and R. C. Huffaker), p. 65. Plenum Press, New York.

BROCKWELL, J., GAULT, R. R., MYTTON, L. R., SCHWINGHAMER, E. A., and GIBSON, A. H. (1981). In *Current perspectives in nitrogen fixation.* (ed. A. H. Gibson and W. E. Newton), p. 433. Australian Academy of Science, Canberra.

BROUGHTON, W. J. and DILWORTH, M. J. (1971). *Biochem. J.* **125**, 1075

— HEYCKE, N., MEYER, A. H., and PANKHURST, C. E. (1984*a*). *Proc. natn. Acad. Sci. U.S.A.* **81**, 3093.

— SAMREY, U., PANKHURST, C. E., SCHNEIDER, G. M., and VANCE, C. P. (1984*b*). In *Advances in nitrogen fixation research* (ed. C. Veeger and W. E. Newton), p. 279. Martinus Nijhoff, The Hague.

BUCHANAN-WOLLASTON, A. V. (1979). *J. gen. Microbiol.* **112**, 135.

— BERINGER, J. E., BREWIN, N. J., HIRSCH, P. R., and JOHNSTON, A. W. B. (1980). *Molec. & gen. Genet.* **178**, 185.

BULLERJAHN, G. S. and BENZINGER, R. H. (1982). *J. Bact.* **150**, 421.

CANNON, F. C., RIEDEL, G. E., and AUSUBEL, F. M. (1977). *Proc. natn. Acad. Sci. U.S.A.* **74**, 2963.

——— (1979). *Molec. & gen. Genet.* **174**, 59.

CANTRELL, M. A., HICKOK, R. E., and EVANS, H. J. (1982). *Arch. Microbiol.* **131**, 102.

CEN, Y., BENDER, G. L., TRINICK, M. J., MORRISON, N. A., SCOTT, K. F., GRESSHOFF, P. M., SHINE, J., and ROLFE, B. G. (1982). *Appl. & environ. Microbiol.* **43**, 233.

CHAKRABARTI, S., LEE, M. S., and GIBSON, A. H. (1981). *Soil biol. Biochem.* **13**, 349.

—CHILD, J. J. and LA RUE, T. A. (1976). In *Proc. 1st Int. Symp. Nitrogen Fixation.* (ed. W. E. Newton and C. J. Nyman), p. 447. Washington State University Press.

CHUA, K. Y., PANKHURST, C. E., MACDONALD, P. E., HOPCROFT, D. H., JARVIS, B. D. W., and SCOTT, D. B. (1985). *J. Bact* **162**, 335.

COLE, M. A. and ELKAN, G. H. (1973). *Antimicrobiol. chemother. Agents* **4**, 248.

—— (1979). *Appl. & environ. Microbiol.* **37**, 867.

CORBIN, D., BARRAN, L., and DITTA, G. (1983). *Proc. natn. Acad. Sci. U.S.A.* **80**, 3005.

CROW, V. L., JARVIS, B. D. W., and GREENWOOD, R. M. (1981). *Int. J. syst. Bact.* **31**, 152.

CURRIER, T. C. and NESTER, E. W. (1976). *Anal. Biochem.* **66**, 431.

DART, P. J. (1977). In *A treatise of dinitrogen fixation*, Section III (ed. R. W.F. Hardy and W. S. Silver), p. 367. John Wiley, New York.

DELLAPORTA, S. L. and PESANO, R. L. (1981). In *Int. Rev. Cytol.*, Suppl. **13**, (ed. K. L. Giles and A. G. Atherly), p. 83. Academic Press, London.

DÉNARIÉ, J., TRUCHET, G., and BERGERON, B. (1976). In *Symbiotic nitrogen fixation in plants* (ed. P. S. Nutman), p. 47. Cambridge University Press.

— BOISTARD, P. and CASSE-DELBART, F. (1981). In *Int. Rev. Cytol., Suppl.* **13** (ed. K. L. Giles and A. G. Atherly), p. 225. Academic Press, London.

DILWORTH, M. J. (1969). *Biochim. biophys. Acta.* **184**, 432.

DOCTOR, F. and MODI, V. V. (1976a). *J. Bact.* **126**, 997.

— — (1976b). In *Symbiotic nitrogen fixation in plants* (ed. P. S. Nutman), p. 69. Cambridge University Press.

DOWNIE, J. A., MA, Q-S, KNIGHT, C. D., HOMBRECHER, G., and JOHNSTON, A. W. B. (1983). *Eur. Molec. Biol. Orgn. J.* **2**, 947.

DREYFUS, B. L. and DOMMERGUES, Y. R. (1981a). *Appl. & environ. Microbiol.* **41**, 97.

— — (1981b). *FEMS Microbiol. Lett.* **10**, 313.

ECKHARDT, T. (1978). *Plasmid* **1**, 584.

ELKAN, G. H. (1981). In *Int. Rev. Cytol.*, Suppl. **13** (ed. K. L. Giles and A. G. Atherly), p. 1. Academic Press, London.

ELMERICH, C., DREYFUS, B. L., REYSSET, G., and AUBERT, J. P. (1982). *Eur. Molec. Biol. Orgn J.* **1**, 499.

EMERICH, D. W., RUIZ-ARGUESO, T., CHING, T., and EVANS, H. J. (1979). *J. Bact.* **137**, 153.

EVANS, H. J., EMERICH, D. W., RUIZ-ARGUESO, T., MAIER, R. J., and ALBRECHT, S. L. (1980). In *Nitrogen fixation*, Vol. 2 (ed. W. E. Newton and W. H. Orme-Johnson), p. 69. University Park Press, Baltimore.

EVANS, W. R. and KEISTER, D. L. (1976). *Can. J. Microbiol.* **22**, 949.

FISCHER, H.-M. and HENNECKE, H. (1984). *Molec. & gen. Genet.* **196**, 537.

FRED, E. B., BALDWIN, I. L., and McCOY, R. (1932). *Root nodule bacteria and leguminous plants.* University of Wisconsin Press, Madison.

FRENEY, J. R. and GIBSON, A. H. (1979). *Aust. J. Pl. Physiol.* **2**, 663.

FUHRMANN, M. and HENNECKE, H. (1984). *J. Bact.* **158**, 1005.

GABOR-HOTCHKISS, M. (1972). In *Uptake of informative molecules by living cells* (ed. L. Ledoux), p. 222. North-Holland, Amsterdam.

GIBSON, A.H., SCOWCROFT, W. R., CHILD, J. J., and PAGAN, J. D. (1976). *Arch. Microbiol.* **108**, 45.

— — and PAGAN, J. D. (1977). In *Recent developments in nitrogen fixation* (ed. W. E. Newton, J. R. Postgate, and C. Rodringuez-Barruew), p. 387. Academic Press, London.

— — DREYFUS, B. L., LAWN, R. J., SPRENT, J. I., and TURNER, G. L. (1981). In *Current perspectives in nitrogen fixation* (ed. A. H. Gibson and W. E. Newton), p. 373. Australian Academy of Science, Canberra.

GRAHAM, P. H. (1976). In *Symbiotic nitrogen fixation in plants* (ed. P. S. Nutman), p. 99. Cambridge University Press,.

GREENWOOD, R. M. and BATHURST, N. O. (1968). *N. Z. J. Sci.* **11**, 280.

— — (1978). *N. Z. J. Sci.* **21**, 107.

GRESSHOFF, P. M., CARROLL, B., MOHAPATRA, S. S., REPORTER, M., SHINE, J., and ROLFE, B. G. (1981). In *Current perspectives in nitrogen fixation* (ed. A. H. Gibson and W. E. Newton), p. 209. Australian Academy of Science, Canberra.

GROSS, D. C., VIDAVER, A. K., and KLUCAS, R. V. (1979). *J. gen. Microbiol.* **114**, 257.

HAHN, M. and HENNECKE, H. (1984). *Molec. & gen. Genet.* **193**, 46.

— MEYER, L., STUDER, D., REGENSBURGER, B., and HENNECKE, H. (1984). *Plant molec. Biol.* **3**, 159.

HAUGLAND, R. and VERMA, D. P. S. (1981). *J. molec. appl. Genet.* **1**, 205.

HERNALSTEENS, J. P., VAN VLIET, F., DE BEUCKELEER, M., DEPICKER, A., ENGLER, G., LEMMERS, M., HOLSTERS, M.,VAN MONTAGU, M., and SCHELL, J. (1980). *Nature, Lond.* **287**, 654.

HERON, D. S. and PUEPPKE, S. G. (1984). *J. Bact.* **160**, 1061.

HEUMANN, W. (1968). *Molec. & gen. Genet.* **102**, 132.

— and Springer, R. (1977). *Molec. & gen. Genet.* **150**, 73.

— KAMBERGER, W., PÜHLER, A., RORSCH, A., SPRINGER, R., and BURKARDT, H. J. (1974). In *Proc. 1st int. Symp. Nitrogen Fixation* (ed. W. E. Newton and C. J. Nyman), p. 383. Washington State University Press, Pullman.

HIGASHI, S. (1967). *J. gen. appl. Microbiol.* **13**, 391.

HIRSCH, A. M., WILSON, K. J., JONES, J. D. G., BANG, M., WALKER, V. V., and AUSUBEL, F. M. (1984). *J. Bact.* **158**, 1133.

HIRSCH, P. R. (1979). *J. gen. Microbiol.* **113**, 219.

HOLLIS, A. B., KLOOS, W. E., and ELKAN, G. H. (1981). *J. gen. Microbiol.* **123**, 215.

HOLLOWAY, B. W. (1979). *Plasmid* **2**, 1.

HOLSTEN, R. D., BURNS, R. C., HARDY, R. W. F., and HERBERT, R. R. (1971). *Nature, Lond.* **232**, 173.

HOOYKAAS, P. J. J., VAN BRUSSELL, A. A. N., DEN DULK-RAS, H., VAN SLOGTEREN, G. M. G., and SCHILPEROORT, R. A. (1981). *Nature, Lond.* **291**, 351.

— SNIJDEWINT, F. G. M., and SCHILPEROORT, R. A. (1982). *Plasmid* **8**, 73.

HOMBRECHER, G., BREWIN, N. J., and JOHNSTON, A. W. B. (1981). *Molec. & gen. Genet.* **182**, 133.

HUA, S. T., SCOTT, D. B., and LIM, S. T. (1981). In *Genetic engineering of symbiotic nitrogen fixation and conservation of fixed nitrogen* (ed. J. M. Lyons, R. C. Valentine, D. A. Phillips, D. W. Rains, and R. C. Huffaker), p. 95. Plenum Press, New York.

HUBER, T. A., AGARWAL, A. K., and KEISTER, D. (1984). *J. Bact.* **158**, 1168.

JAGADISH, M. N. and SZALAY, A. A. (1984). *Molec. & gen. Genet.* **196**, 290.

JARVIS, B. D. W., MACLEAN, T. S., ROBERTSON, I. G. C., and FANNING, G. R. (1977). *N. Z. J. Agric. Res.* **20**, 235.

— PANKHURST, C. E., and PATEL, J. J. (1982). *Int. J. syst. Bact.* **32**, 378.

JOHNSTON, A. W. B., BEYNON, J. L., BUCHANAN-WOLLASTON, A. V., SETCHELL, S. M., HIRSCH, P. R., and BERINGER, J. E. (1978). *Nature, Lond.* **276**, 136.

JORDAN, D. C. (1982). *Int. J. syst. Bact.* **32**, 136.

JOSEY, D. P., BEYNON, J. L., JOHNSTON, A. W. B., and BERINGER, J. E. (1979). *J. appl. Bact.* **46**, 343.

KALUZA, K., FUHRMANN, M., HAHN, M., REGENSBURGER, B., and HENNECKE, H. (1983). *J. Bact.* **155**, 915.

— and HENNECKE, H. (1984). *Molec. & gen. Genet.* **196**, 35.

KANESHIRO, T., CROWELL, C. D., and HANRAHAN, R. F. (1978). *Int. J. syst. Bact.* **28**, 27.

— and KURTZMAN, M. A. (1982). *J. appl. Bact.* **52**, 201.

— BAKER, F. L., and JOHNSON, D. E. (1983). *J. Bact.* **153**, 1045.

KEISTER, D. L. (1976). *J. Bact.* **123**, 1265.

— and EVANS, W. R. (1976). *J. Bact.* **129**, 149.

KENNEDY, C., DREYFUS, B., and BROCKWELL, J. (1981). *J. gen. Microbiol.* **125**, 233.

KENNEDY, L. D. and PANKHURST, C. E. (1981). *Microbios* **23**, 167.

KEYSER, H. H., BOHLOOL, B. B., HU, T. S., and WEBER, D. F. (1982). *Science* **215**, 1631.

KISS, G. B. and KALMAN, Z. (1982). *J. Bact.* **150**, 465.

KLEIN, G. E., JEMISON, P., HAAK, R. A., and MATTYSSE, A. G. (1975). *Experientia* **31**, 532.

KONDOROSI, A., KISS, G. B., FORRAI, T., VINCZE, E., and BANFALVI, Z. (1977). *Nature, Lond.* **268**, 525.

— VINCZE, E., JONHSTON, A. W. B., and BERINGER, J. E. (1980). *Molec. & gen. Genet.* **178**, 403.

— and JOHNSTON, A. W. B. (1981). In *Int. Rev. Cytol.*, suppl. 13 (ed. K. L. Giles and A. G. Atherly), p. 191. Academic Press, London.

— KONDOROSI, E., PANKHURST, C. E., BROUGHTON, W. J., and BANFALVI, Z. (1982). *Molec. & gen. Genet.*, **188**, 433.

— BANFALVI, Z., and KONDOROSI, A. (1984). *Molec. & gen. Genet.* **193**, 445.

KOWALSKI, M. (1967). *Acta microbiol. pol.* ser. A. **16**, 7.

KREMER, R. J. and PETERSON, H. (1982). *Appl. & environ. Microbiol.* **43**, 636.

KURZ, W. G. W. and LaRUE, T. A. (1975). *Nature, Lond.* **256**, 407.

KUYKENDALL, L. D. (1979). *Appl. & environ. Microbiol.* **37**, 862.

— (1981). In *Int. Rev. Cytol.*, suppl. 13 (ed. K. L. Giles and A. G. Atherly), p. 299. Academic Press, London.

— and ELKAN, G. H. (1976). *Appl. & environ. Microbiol.* **32**, 511.

— — (1977). *J. gen. Microbiol.* **98**, 291.

LAW, I. J., YAMAMOTO, Y., MORT, A. J., and BAUER, W. D. (1982). *Planta* **154**, 100.

LEGOCKI, R. P., YUN, A. C., and SZALAY, A. A. (1984). *Proc. natn. Acad. Sci. U.S.A.* **81**, 5806.

LEPIDI, A. A., NUTI, M. P., BERNACCHI, G., and NEGLIA, R. (1976). *Pl. Soil* **45**, 555.

LEPO, J. E., HICKOK, R. E., CANTRELL, M. A., RUSSELL, S. A., and EVANS, H. J. (1981). *J. Bact.* **146**, 614.

LEVIN, R. A. and MONTGOMERY, M. P. (1974). *Pl. Soil.* **41**, 669.

LORKIEWICZ, Z., RUSSA, R., and URBANIK, T. (1978). *Acta microbiol. pol.* **27**, 5.

LUDWIG, R. A. (1980). *Proc. natn. Acad. Sci. U.S.A.* **77**, 5817.

— and SIGNER, E. R. (1977). *Nature, Lond.* **267**, 245.

MAIER, R. J. and BRILL, W. J. (1976). *J. Bact.* **127**, 763.
—— (1978*a*). *J. Bact.* **133**, 1295.
—— (1978*b*). *Science* **201**, 448.
— BISHOP, P. E., and BRILL, W. J. (1978). *J. Bact.* **134**, 1199.
MASTERSON, R. V., RUSSELL, P. R., and ATHERLY, A. G. (1982). *J. Bact.* **152**, 928.
McCOMB, J. A., ELLIOTT, J., and DILWORTH, M. J. (1975). *Nature, Lond.* **256**, 409.
MEADE, H. M., LONG, S. R., RUVKUN, G. B., BROWN, S. E., and AUSUBEL, F. M. (1982). *J. Bact.* **149**, 114.
MEYER, M. C. and PUEPPKE, S. G. (1980). *Can. J. Microbiol.* **26**, 606.
MIELENZ, J. R., JACKSON, L E., O'GARA, F. and SHANMUGAM, K. T. (1979). *Can. J. Microbiol.* **25**, 803.
MORRISON, N. A., HAU, C. Y., TRINICK, M. J., SHINE, J., and ROLFE, B. G. (1983). *J. Bact.* **153**, 527.
— CEN, Y. H., CHEN:, H. C., PLAZINSKI, J., RIDGE:, R., and ROLFE, B. G. (1984). *J. Bact.* **160**, 483.
MOSHIRI, F., STULTS, L., NOVAK, P., and MAIER, R. J. (1983). *J. Bact.* **155**, 926.
NOEL, K. D. and BRILL, W. J. (1980). *Appl. & environ. Microbiol.* **40**, 931.
— STACEY, G., TANDON, S. R., SILVER, L. E., and BRILL, W. J. (1982). *J. Bact.* **152**, 485.
NUTI, M. P., LEDEBOER, A. M., LEPIDI, A. A., and SCHILPEROORT, R. A. (1977). *J. gen. Microbiol.* **100**, 241.
— LEPIDI, A. A., PRAKASH, R. K., SCHILPEROORT, R. A., and CANNON, F. C. (1979). *Nature, Lond.* **282**, 533.
PAGAN, J. D., CHILD, J. J., SCOWCROFT, W. R., and GIBSON, A. H. (1975). *Nature, Lond.* **256**, 406.
— SCOWCROFT, W. R., DUDMAN, W. F., and GIBSON, A. H. (1977). *J. Bact.* **129**, 718.
PAIN, A. N. (1979). *J. appl. Bact.* **47**, 53.
PAKWA, K. and DOGRA, R. C. (1981). *Arch. Microbiol.* **129**, 380.
PALOMARES, A., MONTOYA, E., and OLIVARES, J. (1978). *Microbios* **21**, 33.
PANKHURST, C. E. (1977). *Can. J. Microbiol.* **23**, 1026.
— (1981). *J. appl. Bact.* **50**, 45.
— and Craig, A. S. (1978). *J. gen. Microbiol.* **106**, 207.
—— (1979). *J. gen. Microbiol.* **110**, 117.
— SCOTT, D. B., and RONSON, C. W. (1982). *FEMS Microbiol. Lett.* **15**, 137.
— BROUGHTON, W. J., and WIENEKE, U. (1983*a*). *J. gen. Microbiol.* **129**. 2535.
— BACHEM, C., KONDOROSI, E., and KONDOROSI, A. (1983*b*). In *Molecular genetics of the bacteria–plant interaction* (ed. A. Pübler), p. 169. Springer-Verlag, Berlin, Heidelberg, New York.
PATEL, J. J. (1976). *Can. J. Microbiol.* **22**, 204.
PATTERSON, A. C. and SKINNER, F. A. (1974). *J. appl. Bact.* **37**, 239.
PILACINSKI, W. P. and SCHMIDT, E. L. (1981). *J. Bact:* **145**, 1025.
PRAKASH, R. K., SCHILPEROORT, R. A., and NUTI, M. P. (1981). *J. Bact.* **145**, 1129.

RAINA, J. L. and MODI, V. V. (1972). *J. Bact.* **111**, 356.

REGENSBURGER, B. and HENNECKE, H. (1983). *Arch. Microbiol.* **135**, 103.

REPORTER, M. (1976). *Pl. Physiol.* **57**, 651.

— (1981). In *Current perspectives in nitrogen fixation* (ed. A. H. Gibson and W. E. Newton), p. 214. Australian Academy of Science, Canberra.

ROBERTSON, J. G., FARDEN, K. J. F., WARBURTON, M. P., and BANKS, J. (1975). *Aust. J. Pl. Physiol.* **2**, 265.

ROLFE, B. G., DJORDJEVIC, M., SCOTT, K. F., HUGHES, J. E., BADENOCH-JONES, J., GRESSHOFF, P. M., CEN, Y., DUDMAN, W. F., ZURKOWSKI, W., and SHINE, J. (1981). In *Current perspectives in nitrogen fixation* (ed. A. H. Gibson and W. E. Newton), p. 142. Australian Academy of Science, Canberra.

ROSENBERG, C., BOISTARD, P., DÉNARIÉ, J., and CASSE-DELBART, F. (1981). *Molec. & gen. Genet.* **184**, 326.

— CASSE-DELBART, F., DUSHA, I., DAVID, M., and BOUCHER, C. (1982). *J. Bact.* **150**, 402.

ROSTAS, K., SISTA, P. R., STANLEY, J., and VERMA, D. P. S. (1984). *Molec. & gen. Genet.* **197**, 230.

ROUGHLEY, R. J. (1976). In *Symbiotic nitrogen fixation in plants* (ed. P. S. Nutman). p. 125. Cambridge University Press.

RUVKUN, G. B. and AUSUBEL, F. M. (1980). *Proc. natn. Acad. Sci. U.S.A.* **77**, 191.

—— (1981). *Nature, Lond.* **289**, 85.

— SUNDARESAN, V., and AUSUBEL, F. M. (1982). *Cell.* **29**, 551.

SADOWSKY, M. J. and BOHLOOL, B. B. (1983). *Appl. & environ. Microbiol.* **46**, 906.

— KEYSER, H. H., and BOHLOOL, B. B. (1983). *Int. J. syst. Bact.* **33**, 716.

SCHOLLA, M. H. and ELKAN, G. (1984). *Int. J. syst. Bact.* **34**, 484.

— MOOREFIELD, J. H., and ELKAN, G. (1984). *Int. J. syst. Bact.* **34**, 283.

SCHUBERT, K. R. and EVANS, H. J. (1976). *Proc. natn. Acad. Sci. U.S.A.* **73**, 1207.

SCHWINGHAMER, E. A. (1967). *Antonie van Leeuwenhoek* **33**, 121.

— (1977). In *A treatise of dinitrogen fixation* Sect. III (ed. R. W. F. Hardy and W. S. Silver), p. 577. John Wiley, New York.

— (1980). *FEMS Microbiol. Lett.* **7**, 157.

SCOTT, D. B., HENNECKE, H., and LIM, S. T. (1979). *Biochim. biophys. Acta.* **565**, 365.

— and TAIT, R. C. (1981). In *Genetic engineering of symbiotic nitrogen fixation and conservation of fixed nitrogen* (ed. J. M. Lyons, R. C. Valentine, D. A. Phillips, D. W. Rains, and R. C. Huffarker), p. 137. Plenum Press, New York.

— and Ronson, C. W. (1982). *J. Bact.* **151**, 36.

—COURT, C. B., RONSON, C. W., SCOTT, K. F., WATSON, J. M., SCHOFIELD, P. R., and SHINE, J. (1984). *Arch. Microbiol.* **139**, 151.

SCOTT, K. F., ROLFE, B. G., and SHINE, J. (1983). *DNA* **2**, 147.

SHAH, K., PATEL, C., and MODI, V. V. (1983). *Can. J. Microbiol.* **29**, 33.

SHAH, R., SOUSA, S., and MODI, V. V. (1981). *Arch. Microbiol.* **130**, 262.

SHAW, G. T., ELLINGHAM, P. J., and NIXON, L. N. (1981). *Phytochem.* **20**, 1853.

SKOTNICKI, M. L., ROLFE, B. G., and REPORTER, M. (1979). *Biochem. biophys. Res. Commun.* **86**, 968.

STOREY, R., RAINEY, K., POPE, L., and REPORTER, M. (1979). *Pl. Sci. Lett.* **14**, 253.

— and REPORTER, M. (1980). *Aust. J. Pl. Physiol.* **7**, 251.

STOWERS, M. D. and EAGLESHAM, A. R. J. (1984). *Pl. Soil* **77**, 3.

SUTTON, W. D. (1974). *Biochim. biophys. Acta.* **366**, 1.

— and PATTERSON, A. D. (1980). *Planta* **148**, 287.

SZETO, W. W., ZIMMERMAN, J. L., SUNDARESAN, V., and AUSUBEL, F. M. (1984). *Cell* **36**, 1035.

THOMAS, R. J., JOKINEN, K., and SCHRADER, L. E. (1983). *Crop Sci.* **23**, 453.

TÖROK, I, and KONDOROSI, A. (1981). *Nucl. Acids Res.* **9**, 5711.

TRINICK, M. J. (1973). *Nature, Lond.* 244, 459.

— (1980). *J. appl. Bacteriol.* **49**, 39.

— (1982). In *Nitrogen fixation* Vol. 2 (ed. W. J. Broughton), Chapter 3. Oxford University Press, Oxford.

— and GALBRAITH, J. (1980). *New Phytol.* **86**, 17.

TUBB, R. S. (1974). *Nature, Lond.* **251**, 481.

UPCHURCH, R. G. and ELKAN, G. H. (1978). *J. gen. Microbiol.* **104**, 219.

VAN BRUSSEL, COSTERTON, J. W., and CHILD, J. J. (1979). *Can. J. Microbiol.* **25**, 352.

VANCE, C. P. (1983). *A. Rev. Microbiol.,* **37**, 399.

— and JOHNSON, L. E. B. (1983). *Can. J. Bot.* **61**, 93.

VERMA, D. P. S. (1981). In *Current perspectives in nitrogen fixation* (ed. A. H. Gibson and W. E. Newton), p. 205. Australian Academy of Science, Canberra.

VINCENT, J. M. (1974). In *The biology of nitrogen fixation* (ed. A. Quispel), p. 265. North-Holland, Amsterdam.

— (1977). In *A treatise of dinitrogen fixation* Sect. III (ed. R. W. F. Hardy and W. S. Silver), p. 277. John Wiley, New York.

WELLS, S. E. and KUYKENDALL, L. D. (1983). *J. Bact.* **156**, 1356.

WERNER, D. and STRIPF, R. (1978). *Z. Naturf.* **33**, 859.

WILCOCKSON, J. and WERNER, D. (1978). *J. gen. Microbiol.* **108**, 151.

WILLIAMS, P. M. (1981). *Pl. Soil* **60**, 349.

YUN, A. C. and SZALAY, A. A. (1984). *Proc. natn. Acad. Sci. U.S.A.* **81**, 7358.

ZURKOWSKI, W. (1982). *J. Bact.* **150**, 999.

— and LORKIEWICZ, Z. (1976). *J. Bact.* **128**, 481.

—— (1979). *Genet. Res.* **32**, 311.

9 Plasmid molecular genetics of *Rhizobium leguminosarum*, *Rhizobium trifolii*, and *Rhizobium phaseoli*

R. J. M. van Veen, R. J. H. Okker,
P. J. J. Hooykaas, and R. A. Schilperoort

9.1 INTRODUCTION

The bacterial family Rhizobiaceae comprises the genus *Agrobacterium*, which has the capacity to induce the plant diseases crown gall and hairy root on many dicotyledonous plants (Riker 1930; Smith and Townsend 1907) and the genus *Rhizobium* which is able to induce the formation of nitrogen-fixing nodules on the roots of leguminous plants (Beijerinck 1888; Hellriegel and Wilforth 1889). In all cases, the interaction between bacterium and plant leads to cell proliferations but the mechanism of infection is quite different for both genera. Rhizobia have a limited host-range and have been classified according to their host plants (see Vincent 1974). Numerical taxonomic analyses, as well as DNA homology studies have resulted in the division of rhizobia into two groups (DeLey *et al.* 1966; DeLey 1968; Heberlein *et al.* 1967; Vincent 1974). The first class contains bacteria that are fast-growing, i.e. they have a generation time of about three hours and consist of several species. *Rhizobium leguminosarum*, *R. trifolii,* and *R. phaseoli,* which are closely related and probably form one bacterial species (see below), belong to this group. *R. meliloti*, *R. loti* and the more recently-discovered fast-growing *Rhizobium japonicum* strains also belong to this group but are more distantly related to *R. leguminosarum* and to each other (Sadowsky *et al.* 1983; Scholla *et al.* 1984).

Slow-growing bacteria such as *R. lupini* and most *R. japonicum* strains, having a generation time of 6–8 h, are very different from bacteria of the fast-growing group and recently were placed in a separate genus called *Bradyrhizobium* (Jordan 1982). Fast- and slow-growing rhizobia not only show differences in their growth characteristics, but also differ in their metabolic activities, the produc-

224

tion of acid or alkali and their DNA composition (Gibbins and Gregory 1972; Moffet and Colwell 1968; Martinez-Drets *et al.* 1977; Stowers and Eaglesham 1983).

Plasmids have been detected in all bacterial species belonging to the family of the *Rhizobiaceae*. The involvement of plasmids in the interaction between the bacterium and the plant leading to tumours, hairy roots, or root-nodules has long been a matter of speculation. In the case of *A. tumefaciens* it is very well documented that in virulent strains a large plasmid, called the *tumour inducing*, or Ti-plasmid is responsible for the interaction between the bacterium and the plant leading to crown gall formation (for recent reviews see *Molecular biology of plant tumours*, Academic Press 1982; and Hooykaas and Schilperoort 1984). In this chapter, recent developments of the work with plasmids present in *R. leguminosarum*, *R. trifolii*, and *R. phaseoli* are presented. Special emphasis will be given to those plasmids that are involved in the process of symbiotic nitrogen fixation—the so-called Sym-plasmids. Studies on symbiotic nitrogen fixation are scientifically of interest not only to understand the process of nitrogen fixation but also with regard to the problem of plant cell differentiation.

9.2 DETECTION AND LARGE-SCALE ISOLATION OF RHIZOBIAL PLASMIDS

Based on the observation that the ability to induce nodules is an unstable characteristic in rhizobia, there have been speculations that plasmids are involved in nodulation by rhizobia (Dunican and Tierney 1974; Higashi 1967; Zurkowski *et al.* 1973). Using methods developed for plasmid isolation in *E. coli*, small plasmids have been detected in some *Rhizobium* strains (Dunican *et al.* 1976; Sutton 1974). Nuti *et al.* (1977) were the first to demonstrate the presence of very large plasmids in different *Rhizobium* species by sedimentation analysis of lysates on alkaline sucrose gradients and by dye buoyant density centrifugation. Casse *et al.* (1979) devised a procedure for the rapid detection of large plasmids in fast-growing *Rhizobium* strains. Using this procedure a large number of wild-type isolates of *Rhizobium* were screened (Casse *et al.* 1979; Prakash *et al.* 1980). Large plasmids were observed in all these strains. Most of the strains examined were found to contain more than one plasmid and the number of plasmids could be as many as seven. Recently, a new method for plasmid detection was presented (Eckhardt 1978; modified by Rosenberg *et al.* 1982). This method involves lysis of the bacteria in the slot of an agarose gel immediately followed by electrophoresis of the DNA. By this method megaplasmids of a molecular weight of about 1000 (Burkhardt and Burkhardt 1984) were detected in all *R. meliloti* strains studied.

For the physical mapping of *Rhizobium* plasmids with restriction enzymes and for the localization on physical maps of functions determined by these plasmids, a large amount of pure plasmid DNA is required. Various methods for the 'large-scale' isolation of rhizobial plasmids have been developed (e.g. Nuti *et al.* 1977; Prakash *et al.* 1981), most of which are adapted from procedures described for the isolation of *Agrobacterium* Ti-plasmids (Currier and Nester 1976; Koekman *et al.* 1979). The procedure we have used is the following (see Prakash *et al.* 1981).

1. Bacteria are lysed at an osnotic concentration identical to that in the original culture.
2. Lysis occurs in TE buffer (50 mM TRIS, 20 mM EDTA pH = 8.0), in the presence of 0.25 mg ml^{-1} pronase E and 1 per cent (w/v) Sodium Dodecyle Sulphate (SDS).
3. Alkali treatment and neutralization of the lysate is performed as described for Ti-plasmid isolation.
4. Precipitation of chromosome-membrane complexes is performed for, at maximum, four hours on ice.
5. After salt precipitation the supernatant is brought to 10 per cent (w/v) polyethylene glycol (PEG) and the plasmid DNA is precipitated overnight (4 °C).
6. After PEG precipitation, caesium chloride-ethidium bromide gradients are prepared and run as described for Ti-plasmid isolation.
7. After centrifugation the plasmid band is recovered under yellow light, extensively washed with isoamylalcohol to remove ethidium bromide, dialysed against 3 per cent (w/v) NaCl to remove caesium chloride, and the DNA is precipitated with 96 per cent (w/v) ethanol.

By this method we routinely isolate 20–50 μg of plasmid DNA from one litre cultures.

9.3 GENETICS OF FAST-GROWING *Rhizobium* SPECIES

9.3.1 Mapping of the chromosome

Geneticists had been studying *Rhizobium* long before the discovery of plasmids. Auxotrophic mutants and mutants resistant against antibiotics, bacteriophages, D-amino acids, and amino acid analogs have been obtained. The influence of such mutations of the nodulating- and nitrogen-fixing capacities of the strains have been examined (for reviews, see Beringer *et al.* 1980; Dénarié *et al.* 1976). *Rhizobium* strains do not, or at a very low frequency, promote the transfer of chromosomal DNA. Chromosomal markers could be transferred when an *incP*-1 R-plasmid was present in the donor strain. Using R68.45 (Beringer and Hopwood 1976) to promote chromosome transfer of auxotrophic mutants, circular genetic maps of the chromosomes of *R. leguminosarum* 300 (Beringer *et al.* 1978*b*), three *R. meliloti* strains 2011,41, and GR4 (Meade and Signer 1977; Kondorosi *et al.* 1977; Casadesus and Olivares 1979) and a *R. trifolii* strain MS55 (Megias *et al.* 1982) have been obtained. A chromosomal linkage map of *A. tumefaciens* was also constructed (Hooykaas *et al.* 1982*b*). Comparison of these maps showed remarkable similarities (Hooykaas *et al.* 1982*b*; Kondorosi *et al.* 1980). Recombination between chromosomal DNA of *R. meliloti* strains 2011 and 41 was found to occur at high frequencies in crosses between these strains. High frequencies of recombination were detected between different *R. leguminosarum* strains as well (Johnston and Beringer 1977). In crosses between a *R. meliloti* strain and a *R. leguminosarum* strain virtually no chromosomal recombination was detected (Johnston *et al.* 1978*c*). *Rhizobium meliloti* and *R. leguminosarum* belong to two different species taxonomically (see Section 9.1), which could explain the lack of recombination between their chromosomes. However, the arrangement of the chromosomal markers in these strains, and also in an *A. tumefaciens* strain, is similar (see Fig. 9.1) suggesting a close genetic relationship between these bacterial species, in spite of taxonomic differences and lack of recombination. Complementation studies using R-prime plasmids, consisting of R68.45 and pieces of *R. meliloti* 2011 chromosomal DNA, were successful with certain mutants of *R. meliloti* 41, *R. leguminosarum* 300 and *A. tumefaciens* C58. These R-prime plasmids did not complement *E. coli* markers, although in one case a mutated plasmid could be isolated which complemented an *E. coli* marker and at the same time remained able to complement *Rhizobium* markers (Johnston *et al.* 1978*a*). The nature of the mutation is unknown. In chromosomal mapping experiments, DNA involved in the determination of the host range of nodulation was never

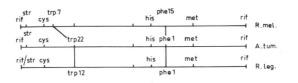

FIG. 9.1. Comparison of the linkage maps of *A. tumefaciens* C58 with those of *R. leguminosarum* 300 and *R. meliloti* 41. The data for the latter two organisms are re-arranged from Konderosi *et al.* (1980). Loci at corresponding positions have been indicated. The solid lines connecting *trp* and *phe* genes, respectively, show genes for which an identical function has been demonstrated (Taken from *Molec. & gen. Genet.* **188**, 12–17).

transferred and consequently could not be mapped on the chromosome (Banfalvi *et al.* 1981). Some auxotrophic mutants have been found which fail to fix nitrogen in the root nodule; also those *Agrobacterium* strains that have acquired the capacity to nodulate after introduction of a Sym-plasmid do not fix nitrogen (see below). Whether this is due to the energetic and metabolic state of the bacteroids in the nodule or whether the chromosome contains genes involved in the process of nitrogen fixation is not clear. Over the last years much genetic research has turned to the plasmids of fast-growing *Rhizobium* species. For the identification of the genetic information located on these plasmids, methods for plasmid gene transfer amongst *Rhizobia* and *Agrobacteria* have been developed. In the following sections we will describe: methods for plasmid transfer, markers located on *Rhizobium* plasmids and the construction of mutants, the expression of symbiotic plasmids in *R. meliloti* and *A. tumefaciens*, interactions between two (symbiotic) plasmids, and DNA transformation and bacteriophage-mediated transduction in *Rhizobium*.

9.3.2 Identification of Sym-plasmids

In order to be able to detect markers present on plasmids it is necessary to cure and/or transfer these plasmids to new hosts. A pre-requisite for curing- and transfer-experiments is that a selectable marker be present on these plasmids. As *Rhizobium* plasmids are generally cryptic, a selectable marker has to be introduced into them first. Transposons have been widely used for this purpose. Beringer *et al.* (1978a) devised a method for the introduction of transposon Tn5 (Km^r) at random sites in the genome of *Rhizobium* strains. The method is based on the suicide character of R : : Mu plasmids in *Rhizobium* and *Agrobacterium* (Boucher *et al.* 1977; Van Vliet *et al.* 1978; Klapwijk *et al.* 1980).

Transposon Tn5-mutated *R. leguminosarum* strains were the first to be used in conjugation experiments. From some strains, transfer of Km^r occurred at a high frequency, indicating that Tn5 had inserted into a conjugative plasmid. The plasmids transferred appeared to determine the production of a medium-sized bacteriocin (see below and Johnston *et al.* 1978b). Most importantly, one of the labelled plasmids (pRL1JI) conferred the ability to nodulate peas to *R. trifolii* and *R. phaseoli* strains, indicating that the plasmid at least carries a determinant for host range.

Hooykaas *et al.* (1981) labelled a plasmid of *R. trifolii* stain RCC5 with Tn5. Conjugation experiments showed that pSym5, like pRL1JI, carried determinants for host specificity. When transferred to *Rhizobium leguminosarum* the transconjugants had the ability to nodulate both clover plants and pea plants. In addition, the plasmid was cured by heat treatment (37 °C) and the resultant kanamycin sensitive colonies were unable to nodulate clover plants (Hooykaas *et al*. 1981). These bacteria lacked the Tn5-labelled plasmid. By direct screening of *Rhizobium trifolii* strains incubated at high temperature (Zurkowski and Lorkiewicz 1979; Zurkowski 1982) efficient segregation of non-nodulating derivatives was found. Upon re-introduction of the plasmid, resultant strains again induced nitrogen-fixing nodules on their host. It was concluded that the labelled plasmid carried some if not most genes involved in symbiotic nitrogen fixation, and the plasmid was called a symbiotic (Sym)-plasmid (pSym5) (Hooykaas *et al*. 1981). Lamb *et al*. (1982) labelled a plasmid of *R. phaseoli* 8002 and found this plasmid to be a conjugative Sym-plasmid.

Besides Sym-plasmids, other conjugative plasmids have been detected in *R. leguminosarum*. Bacteriocinogenic plasmids were amongst the first conjugative plasmids that were detected (Hirsch 1979, 1980). Alternative ways to promote transfer in bacteria with non-conjugative plasmids involved in symbiosis had to be found. These plasmids can be mobilized by conjugative endogenous plasmids present in the *Rhizobium* strain (Brewin *et al*. 1980*a,b*) or by introduced broad-host-range conjugative R-plasmids (Hooykaas *et al*. 1980, 1982*a,c*; Scott and Ronson 1982; Zurkowski 1981). Mobilization of non-conjugative plasmids by *inc*P-1 R-plasmids has been used to promote transfer of *Agrobacterium* Ti and Ri (Root inducing) plasmids (Hooykaas *et al*. 1977; 1982*a*). The mechanism of plasmid mobilization, as described by Hooykaas *et al*. (1980) occurs in two steps. First, DNA homology is created between the conjugative R-plasmid and the non-conjugative plasmid via transposition of a transposon or insertion sequence present on either plasmid. During this transposition event, an unstable co-integrate between the two replicons is formed which is transferred into the recipient. In the recipient, the co-integrate dissociates into its components, leaving a copy of the transposon on both replicons. If Tn5 is introduced at random sites into a *Rhizobium* strain containing an R-plasmid next to the endogenous non-conjugative plasmid, the transposon can insert into the chromosome, the Tra⁻ plasmid of interest or the *inc*P-1 plasmid. With conjugation experiments the three possibilities can be distinguished. This procedure has successfully been used to identify non-conjugative Sym-plasmids in *R. leguminosarum* RCC1001 (Hooykaas *et al*. 1982*c*) and *R. phaseoli* RCC3622 as well as Ri-plasmid of *A. rhizogenes* NCPPB1855 (Hooykaas *et al*. 1982*a*).

R. leguminosarum RCC1001 was chosen for research in our laboratory because of the small number of plasmids present in this strain. Two plasmids are present. One has a size of 150 MDal while the other is about 400 MDal. After mutagenesis with transposon Tn5 no trans-ferable kanamycin resistance was found from any of the Km^r strains, indicating that neither of the two plasmids in RCC1001 is conjugative.

One out of 56 independent Tn5 insertions, however, could be mobilized by a conjugative *inc*P-1 plasmid. Transconjugants from matings between this isolate and either *R. trifolii* RCC5 or a Sym-plasmid-5 cured derivative of RCC5 as recipients, were able to induce nitrogen-fixing nodules on *Vicia* (Hooykaas *et al.* 1982*c*). The 150 MDal plasmid of *R. leguminosarum* RCC1001 was termed pSym-1, analogous to pSym-5, the symbiotic plasmid of *R. trifolii* RCC5. Zurkowski (1981) used RP4 for the transfer of nodulation determinants of *R. trifolii* 24. Transconjugants that were able to nodulate clover plants had, besides RP4, acquired a plasmid (pWZ2) of strain 24. Earlier it was reported that loss of this plasmid resulted in loss of nodulation properties (Zurkowski and Lorkiewicz 1978, 1979). Scott and Ronson (1982) transferred a Sym-plasmid from *R. trifolii* by means of R68.45. It can be concluded that, in strains of *R. leguminosarum, R. trifolii,* and *R. phaseoli,* genes involved in host specificity, nodulation, and nitrogen fixation are present on conjugative or non-conjugative plasmids which can be transferred amongst the above, mentioned species without loss of function (In Section 9.3.5 the interaction between different rhizobial plasmids present in one bacterium will be discussed).

9.3.3 Properties of different rhizobial plasmids

Rhizobium plasmids are cryptic in most cases. No selectable marker is present that is functional in the bacterium. To identify genes located on these plasmids, selectable marks had to be introduced (as described above). A transposon Tn5-labelled plasmid can be followed in conjugation experiments as well as curing experiments.

Besides the genes coding for nodulation (*nod*) nitrogen fixation (*nif, fix*), host range (*hsp*), and transfer (*tra*), other genetic traits are present on rhizobial plasmids. Some plasmids contain genes coding for hydrogen uptake (*hup*), other plasmids determine bacteriocin production (*mep*), a function which results in repression of chromosomal genes coding for small bacteriocin production (*rps*) and melanin-production (*mel*).

The rhizobial nitrogenase complex evolves H_2 in an ATP-dependent side reaction, which means a loss of energy. *Rhizobium* strains with a hydrogenase uptake system (*hup*) are able to partially recycle the ATP that is consumed in the side reaction (Nelson and Salminen 1982; Schubert and Evans 1976). As a consequence, such strains are probably more efficient in nitrogen fixation.

The *hup* system is located at least in one case on a non-conjugative Sym-plasmid, pRL6JI of *R. leguminosarum* (Brewin *et al.* 1980*a*). After transfer of this plasmid to a nodulation deficient, Hup⁻ *R. leguminosarum* strain by means of a mobilizing self-transmissible plasmid, the evolution of hydrogen was reduced ten-fold compared to nodulation proficient, Hup⁻ strains. The plasmid was tested in several *Rhizobium* strains that were Hup⁻. The effect of the introduction of the *hup* system appeared to be dependent on the genetic background of the strain. The Hup⁻ strain that was the most efficient N_2-fixing one, was stimulated 31 per cent in N_2 fixation by the introduction of the Hup⁺ Sym-plasmid (De Jong *et al.* 1982).

Several conjugative plasmids determine bacteriocin production. Hirsch isolated 97 *R. leguminosarum* strains from the field. Three of these contained a conjugative plasmid coding for medium-bacteriocin production. Bacteriocins are molecules, produced by bacteria, that kill or severely inhibit in growth closely related, sensitive strains. They can be considered as highly specific antibiotics. Three molecular-size classes of bacteriocins are discernible: (1) the size class *small*, giving large inhibition zones when applied in a well on a lawn of sensitive indicator bacteria; (2) the size class *medium*, giving small inhibition zones in this test; and (3) the size class *large* giving no inhibition zones. The size of the inhibition zone is inversely related to the molecular size of the bacteriocin.

The three conjugative plasmids mentioned earlier each coded for production of a different *medium* (*mep*) bacteriocin (Hirsch 1979), as was demonstrated by the absence of cross-resistance against the heterologous bacteriocins. The strains were all immune to their own *medium* bacteriocin. This makes it likely that the gene(s) encoding immunity to their own *medium* bacteriocin are present on the self-transmissible plasmid together with the genes for bacteriocin production.

Many field isolates of *R. leguminosarum* produce a *small* bacteriocin. Until now, only one type of *small* bacteriocin has been found, since all *small* producers are immune to the *small* bacteriocin of all other producers. The *small* bacteriocin gene(s) have never been ascribed to a rhizobial plasmid. However, after introduction of a conjugative rhizobial plasmid with *medium* bacteriocin production, many small-producing strains are converted to non-producing small senstive strains (Wyffelman *et al.* 1983). The bacteriocin-related genetic traits of rhizobial plasmids are abbreviated as, e.g. Rps (*r*epression *p*roduction *s*mall) and Ris (*r*epression *i*mmunity *s*mall). As an example, introduction of pJB5JI into small-producing *Rhizobium* strains results in strains that no longer produce small bacteriocin and have become sensitive to small bacterocin. Small-bacteriocin-resistant strains are easily isolated from such strains on plates containing small bacteriocin. The small-resistant strains were either cured of pJB5JI or contained deletions (Wyffelman *et al.* 1983; Priem and Wyffelman 1984). The Mep and Rps traits can be useful markers for rhizobial conjugative plasmids. There are no indications that these genes have any role in the development of the symbiosis. Sym-plasmids from *R. phaseoli*, determining nodulation and nitrogen fixation on *Phaseolus* were shown to code for melanin production, giving rise to dark brown colonies on tyrosine-containing media (Beynon *et al.* 1980). This trait is closely linked to the nodulation genes but appears not to be of any importance to symbiosis (Lamb *et al.* 1982; P.J.J. Hooykaas and H. Den Dulk-Ras, unpublished). Melanin production can be considered a useful marker in the genetics of *R. phaseoli* plasmids.

Detailed analysis of the molecular processes of nodule formation and nitrogen fixation needs rhizobial strains mutated at various stages in nodulation. Three conditions must be fulfilled before a fruitful search for such mutated strains can be initiated: (1) an efficient way to induce localizable mutations; (2) a nodulation and nitrogen-fixation test system that can handle large numbers of strains suspected to have mutations in nodulation or fixation genes; and (3) the possibility to divide mutants into several clearly recognizable phenotypes, e.g. mutants blocked in root hair curling, infection thread formation, etc. The *E. coli* suicide plasmid pJB4JI is commonly used for the introduction of random transposon Tn5 mutations in rhizobial Sym-plasmids. The Tn5-induced mutations are clearly tagged by the transposon-induced Km resistance and the transposon constitutes a unique sequence in the plasmid DNA which can be recognized in Southern hybridization experiments and restriction enzyme analyses. On the other hand, the polarity of transposon mutations is a drawback, and therefore the search for methods of inducing non-polar mutations must be continued. Mutation induction with a suicide plasmid containing transposon Tn7 or Tn1831 have not been successful (Beringer *et al.* 1978*a*; C. Wyffelman, personal communication). Several strategies to mutagenize rhizobial plasmids with transposons other than Tn5 are under study in this laboratory and elsewhere. Two rapid tests for screening of mutants have been described. Rolfe *et al.* (1980) used seedling of *T. repens* in petri dishes in rapid screening tests for *R. trifolii* and van Brussel *et al.* (1982) described the use of *Vicia sativa* for nodulation-tests with *R. leguminosarum*. In the latter test system, nodules are formed in 4–6 days and plant culture is on agar slants in 180 × 13 mm culture tubes.

Most problems in programmes of mutation screening are encountered in the screening of the different mutant phenotypes. Mutants in root-hair curling, root adsorption, and nitrogen fixation are easily scored, but mutants in other stages of symbiosis can only be screened by elaborative (electron) microscopical work. No easy solution seems at hand for this problem at the moment. Mutants with Tn5 leading to defects in nodulation and nitrogen fixation have been reported for both *R. trifolii* and *R. leguminosarum* (Buchanan-Wollaston *et al.* 1980; Ma *et al.* 1982; Scott, Hughes, and Gresshoff 1982). Buchanan-Wollaston *et al.* (1980) isolated 160 strains with a Tn5-insertion in the self-transmissible Sym-plasmid pRL1JI. After transfer of the plasmids with Tn5-insertions to a nodulation-deficient deletion mutant of *R. leguminosarum*, 9 per cent of the isolates did not form nodules whereas 9 per cent did form nodules but no nitrogen was fixed in these. Ma *et al.* (1982) isolated and analysed eight Tn5 mutants in the self-transmissible Sym-plasmid pRL1JI that failed to fix nitrogen any longer. Two phenotypes were discernible, one type producing leghaemoglobin but with a low content of bacteroids in the root cells, and another type which was probably disturbed in the release of bacteria from the infection thread. Using Tn5-encoded Km resistance as a marker, fragments of the Sym-plasmid containing the Tn5 insert were cloned. These clones were used in further analysis of the mutants. The inserts of Tn5 were present in two clusters, one cluster within 4 kb of the structural genes for nitrogenase, the other cluster at a distance of 30 kb from the first. Plasmids with a mutation present in one of the clusters could complement mutations in the other cluster.

Scott *et al.* (1982) mutagenized *R. trifolii* randomly by means of the pJB4JI plasmid. They reported that 0.3—0.5 per cent of the strains with Tn5 had a defect in symbiotic properties. DNA from these mutant strains was cloned. The clones were used as a probe to identify wild-type DNA containing symbiotic genes in a gene bank. After re-cloning the wild-type DNA a broad-host-range vector, restoration of symbiotic properties by the wild-type DNA in the mutant strains was studied. In cases where restoration was observed, it appeared to be due either to complementation (a high proportion of cells became symbiotically proficient) or recombination between the cloned wild-type DNA and the Tn5-containing DNA (a low proportion of cells became symbiotically proficient).

9.3.4 Expression of Sym-plasmids in *R. meliloti* and *A. tumefaciens*

As has been mentioned earlier, Sym-plasmids from either *R. leguminosarum* or *R. trifolii* and *R. phaseoli* are fully expressed within these closely related hosts. It was of interest to see whether the different Sym-plasmids are also stably maintained and expressed in host-bacteria less closely related to the original host. Therefore, conjugation experiments were performed with *Rhizobium meliloti* strains, which induce nodules on *M. sativa*, *A. tumefaciens* strains which give tumours on many dicotyledonous plants, and *E. coli*. Transconjugants of *R. meliloti*, carrying the Sym-plasmids responsible for nodulation on *Vicia* or *Trifolium*, plants did give rise to nodulation, but the nodules arose only rarely and were very small. No acetylene reduction was detected in these nodules. More information is available about the expression of

Sym-plasmids of *R. leguminosarum* (pSym-1 and pJB5JI), *R. trifolii* (pSym-5), and *R. phaseoli* (pSym-9) in *A. tumefaciens* (Hooykaas *et al.* 1981; 1982*c*; Van Brussel *et al.* 1982). When pSym-5 was conjugated into an *Agrobacterium* strain, cured of its Ti-plasmid, the transconjugants induced nodule formation on *Trifolium* roots. The nodules appeared after a longer period of incubation (4–5 weeks) than did the nodules induced by the *R. trifolii* parental strain. Moreover, no free nitrogen was fixed in these nodules and the *Trifolium* plants, grown on a nitrogen-free medium, in consequence deteriorated rapidly. Re-isolates from the nodules induced by *A. tumefaciens* confirmed that the reisolated bacteria were indeed *Agrobacterium* without contaminating *Rhizobium*. Microscopical examination of the roots inoculated with *Agrobacterium* pSym-5 showed marked root-hair curling. (*Agrobacterium* without Sym-plasmids had no effect on the root hairs.) Electron microscope studies showed plant cells filled with bacteria surrounded by membranes. No bacteroid-like structures were observed. The Sym-plasmid apparently is expressed within *Agrobacterium* but no functional nodules are formed. The lack of nitrogen fixation in nodules induced by *Agrobacterium* pSym-5 is in contrast with earlier results of Stanley and Dunican (1979), who reported the transfer and expression in *Agrobacterium* of *nif*-genes of *R. trifolii*. They mobilized genetic information, using an *inc*P-1 plasmid, and screened the transconjugants for *in vitro* nitrogen-fixation capacities by the acetylene reduction assay. The nodulation capacity of the Nif⁺ transconjugants was not tested.

The introduction of *R. leguminosarum* Sym-plasmids (pSym-1 and pJB5JI) into *Agrobacterium* resulted in the formation of delayed nodules on *Vicia* roots (Hooykaas *et al.* 1982*c*; Van Brussel *et al.* 1982). No nitrogen fixation was detected in these nodules either. Similar results also were obtained when a Sym-plasmid of *R. phaseoli* (pSym-9) was conjugated into *Agrobacterium*. This gave strains that were able to induce spherical nodules, which are typical for *R. phaseoli* on beans (Hooykaars *et al.* 1984) but no nitrogen was fixed in these nodules. There are indications that different *Agrobacterium* strains harbouring Sym-plasmids do not give identical responses on the plant. In addition, there is a difference in response when one strain of *Agrobacterium* harbouring a Sym-plasmid in response when one strain of *Agrobacterium* harbouring a Sym-plasmid is used on different host-plants (Van Brussel *et al.* 1982; Hooykaas *et al.* 1982*c*).

Stable maintenance of large rhizobial plasmids in *E. coli* might facilitate the mutation and manipulation of specific genes involved in nodulation and nitrogen fixation. Hernalsteens *et al.* (1978) found that Ti-plasmids of *A. tumefaciens* could be maintained in *E. coli* as unstable co-integrate replicons with a broad host range *inc*P-1 plasmid. In recombination deficient *E. coli* strains, these co-integrates were stable.

A co-integrate between pTi and R772 has been obtained which is completely stable in *Agrobacterium* and *E. coli* (*recA* and *recA*⁺) (Hooykaas *et al.* 1980; Hille *et al.* 1982). We also tested whether *Rhizobium* plasmids could be transferred to plasmid-less *E. coli* strains, either *rec*A or *recA*⁺. We found that conjugative plasmids of *Rhizobium* cannot replicate in *E. coli* and behave as suicide-plasmids (R. J. M van Veen and P. J. J. Hooykaas, unpublished). In order to test a non-conjugative Sym-plasmid (pSym-1) for maintenance in *E. coli*, a broad-host-range R-plasmid was introduced into a *Rhizobium* strain containing pSym-1. Transconjugants from conjugation experiments with this strain and *rec*A⁻ or *rec*A⁺ *E. coli* strains, arose at a frequency of 10^{-5} per recipient. In the *E. coli* transconjugants a very large plasmid, larger than pSym-1 and a fainter band of the R-factor could be detected in Eckhardt gels. Back-crosses between these *E. coli* strains and *Rhizobium* (cured of pSym) or *Agrobacterium* resulted in transconjugants able to induce nitrogen-fixing nodules on *Vicia*. In these *Rhizobium* transconjugants, plasmid bands of a molecular weight identical to pSym-1 and the R-plasmid were visible on agarose gels. These results suggest that an unstable co-integrate of pSym and R-plasmid was present in the *E. coli* strains and that after re-transfer to *Agrobacterium* these co-integrates dissociate into their component R- and Sym-plasmids. Possibly these unstable R : : Sym co-integrates can be used in site-directed mutagensis experiments in *E. coli*.

9.3.5 Interactions between different *Rhizobium* plasmids

Incompatibility is defined as the inability of two plasmids to be stably maintained in one-host bacterium in the absence of selection for their presence. Incompatible plasmids are placed in one incompatibility group (*inc* group). In *Rhizobium* strains, up to seven plasmids can be stably maintained. Sym-plasmids identified in *R. leguminosarum*, *R. trifolii*, and *R. phaseoli* have a considerable amount of DNA homology (see below) and are also functionally very much alike. With a few exceptions, however, no incompatibility, has been found between different Sym-plasmids (Hooykaas *et al.* 1982*c*). In most cases, Sym-plasmids can be introduced by means of conjugation into *Rhizobium* strains already harbouring a Sym-plasmid without the loss of the resident Sym-plasmid in the recipient strain. Even so, there are reports which indicate that two Sym-plasmids present in one *Rhizobium* strain can cause interference in nodulation or nitrogen fixation (Beynon 1980; Brewin *et al.* 1980*b,c*; 1982; Djordjevic *et al.* 1982). A *R. phaseoli* strain containing the *R. leguminosarum* Sym-plasmid pJB5JI, nodulated pea plants but nodules appeared later than after inoculation with the *R. leguminosarum* parental strain. Moreover, the nodules, remained smaller than normal. In contrast, *R. phaseoli* strains containing pJB5JI but lacking the ability to nodulate beans, showed normal nodulation on pea plants (these strains no longer possessed the *R. phaseoli* Sym-plasmid). This type of interaction between two Sym-plasmids was described as physiological incompatibility (Beringer 1980). Recent results on the behaviour of pJB5JI in *R. trifolii* showed similar inter-actions between Sym-plasmids (Djordjevic *et al.* 1982). *R. trifolii* transconjugants containing two Sym-plasmids (one from *R. leguminosarum* and one from its own *R. trifolii*) were variable in their symbiotic phenotypes. Some could nodulate both pea and clover plants and nitrogen fixation took place in the nodules. This is what can be expected when both Sym-plasmids are properly expressed without any interaction between them. The majority, however, fixed little or no nitrogen on either clover or pea plants. One of the transconjugants could not induce nodules on pea plants (this turned out to be the result of a deletion in the pJB5JI plasmid). Sub-culturing through nodule passage on peas, clover, or on agar plants resulted in instability of the recovered bacteria with regard to nitrogen fixation capacity. Surpris-ingly, transfer of plasmid pJB5JI from re-isolated Fix⁻ bacteria back to *R. leguminosarum* gave rise to transconjugants which had regained the capacity to nodulate and to fix nitrogen on pea plants. No detectable changes in the plasmid profiles of the Fix⁻ *R. trifolii* strains were detected when compared with the initial transconjugants. The authors suggested a regulatory 'switch' mechanism to explain this reversible Fix⁻ phenotype. The interactions between plasmids present in three closely-related *R. leguminosarum* strains have been studied intensively (Brewin *et al.* 1980*a,b,c*; 1982; Johnston *et al.* 1982). In each of these three strains (*R. leguminosarum* 248, 300 and 128C63) a Sym-plasmid has been identified (pRL1JI, pRL10JI, and pRL6JI respectively), sub-sequently tagged with transposon Tn5, and their interaction studied. Besides these Sym-plasmids, other plasmids have been identified in these strains, some of which have no effect on symbiosis by the bacteria. Remarkably, one of these cryptic plasmids was found to be incompatible with the Sym-plasmid of *R. phaseoli* 1233 (Johnston *et al.* 1982). The Sym-plasmids of strains 248 and 128C63 were found to be incompatible with some bacteriocinogenic plasmids from strains of *R. leguminosarum.* Recombinant plasmids arose at high frequencies when the Sym-plasmids were conjugated into the bacteriocinogenic strains (Brewin *et al.* 1982). Recombinant plasmids were also obtained between compatible plasmids but at lower frequencies (Brewin *et al.* 1980). DNA homology studies between these plasmids have not yet been performed, but studies with other Sym-plasmids (see below) in-dicate that large regions of DNA homology exist between different Sym-plasmids of *R. leguminosarum, R. trifolii,* and *R. phaseoli* which could facilitate inter-plasmid recombination. A generalized picture that can be drawn from these results implies that functions involved in nodulation and nitrogen fixation are linked to a large number of different extra-chromosomal replicons. Sym-plasmids in general are compatible with one another, but when two Sym-plasmids are present in one strain these strains sometimes behave abnormally and segregate into cells with recombinant plasmids or cells with abnormal symbiotic properties.

9.3.6 Transformation and transduction of *Rhizobium* strains of the *R. leguminosarum* cluster

The availability of efficient methods for transduction and transformation of *Rhizobium* strains is of great importance for the analysis of symbiotic functions. However, successful transformation with linear, double-stranded DNA has not been reported, which seems to indicate that *Rhizobium* is not competent for this type of transformation. Transformation with circular plasmid DNA has been reported recently for *R. leguminosarum* (Bulerjahn and Benzinger 1982), using a method combining elements of transformation procedures for *E. coli* (Mandel and Higa 1970) and *A. tumefaciens* (Holsters *et al.* 1978). Use of plasmid DNA isolated from *R. leguminosarum* and the low frequency of transformation (10^{-8}) makes the procedure of limited value. Similar results have been reported for transformation of *R. meliloti* (Kiss and Kalman 1982). Transformation of *R. trifolii* with plasmid DNA isolated from another strain of *R. trifolii* has been reported to be efficient and simple (Kowalczuk *et al.* 1981). Incubation of DNA for 90 min with late exponential cells yielded transformation frequencies of $1.4–2.6 \times 10^{-5}$. However, no details of control experiments were provided.

Transformation of our lab strains of *R. leguminosarum* and *R. trifolii* with homologous plasmid DNA and the methods mentioned above were unsuccessful.

Transduction of *R. leguminosarum* has been achieved using the rhizobial phage RL38 and is rather efficient, especially after ultraviolet irradiation of the phage preparation (Buchanan-Wollaston 1979). General transduction occurs, i.e. every marker has about the same frequency of transduction. The maximal efficiency obtained was 2×10^{-5} per recipient. The phage can also be used to transfer DNA by transduction from *R. leguminosarum* to *R. trifolii* and between *R. trifolii* strains, but not from *R. trifolii* to *R. leguminosarum*. A drawback of phage RL38 is its large size (120 Mdal; Beringer *et al.* 1980), making it difficult to obtain high resolution in transduction analysis (C. Wyffelman, personal communication). A more systematic search for smaller, efficiently-transducing *Rhizobium* phages seems worthwhile. Using transduction with phage RL38, transposition mutagenesis of the chromosome (Beringer *et al.* 1978*a*) and of a Sym-plasmid of *Rhizobium leguminosarum* (Buchanan-Wollaston *et al.* 1980) was verified.

9.4 PHYSICAL MAPPING OF FUNCTIONS AND CONSERVED REGIONS ON Sym-PLASMIDS

Today, much genetic evidence for the presence of genes involved in nodulation and nitrogen fixation on rhizobial plasmids is available. In the expectation that the structural *nif*-genes would be evolutionary conserved, the structural *nif* (*nifKDH*) genes from *Klebsiella pneumoniae* cloned in an *E. coli* cloning vector were used as probes on different DNA fractions from *Rhizobium* strains (Cannon *et al.* 1979) in order to find their cellular location in *Rhizobium*. In this way, biochemical evidence was obtained for the presence of *nif*-genes on *Rhizobium* plasmids (Nuti *et al.* 1979; Krol *et al.* 1980; Prakash *et al.* 1981; Hombrecher *et al.* 1981). The sequence of *R. meliloti* DNA, hybridizing to this probe, has been cloned, and was used as a probe on whole plasmid preparations of *R. leguminosarum* strains (Krol *et al.* 1982*a*). Results identical to those with the *nif* genes from *K. pneumoniae* were obtained. The homology between plasmid DNA of rhizobial strains and the structural *nif*-genes from *K. pneumoniae* is limited to *nifDH* (Krol *et al.* 1980).

Prakash *et al.* (1982*b*) constructed a restriction map of pSym-1, a 150 MDal. Sym-plasmid of *R. leguminosarum RCC 1001*. On this physical map of region of homology with the structural *nif*-genes from *K. pneumoniae*, and the reportedly large homologous region of pSym-1 with Sym plasmids of *R. trifolii* and *R. phaseoli* (Prakash *et al.* 1981) were localized. In Fig. 9.2(a) (p. 239), it can be seen that a region conserved in different Sym-plasmids (pSym-1, pSym-5, pSym-9) includes the structural *nif*-genes. The size and strong conservation of this region suggests the possibility that around the structural *nif*-genes other genes involved in nitrogen fixation and nodulation are located. Ma *et al.* (1982) mapped a number of Nif⁻ mutants to confirm this idea of clustering in another *R. leguminosarium* Sym-plasmid). Besides this region overlapping the structural *nif*-genes, two other conserved regions were identified. The function of these regions is unknown at present.

The expression of plasmid pSym-1, in free-living bacteria and in bacteroids has been studied by isolating RNA from late log phase bacteria and bacteroids. The polynucleotide kinase labelled RNA was used as a probe in Southern blot hybridization with restricted pSym-1 (Prakash *et al.* 1982*a*). In Fig. 9.2(b) (p.240) it can be seen that the expression that is detected, is within the region where the *nif*-genes are located. The genes coding for the nitrogenase enzyme are strongly expressed but the expressed area is larger than the fragments homologous

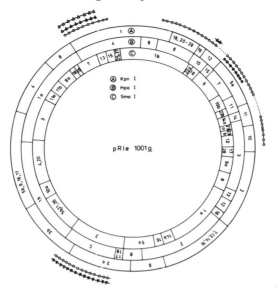

FIG. 9.2(a). A circular map order of *Hpa*I, *Sma*I, and *Kpn*I fragments of pRle1001a. The lines outside the map indicate the region of pRle1001a homologous with *K. pneumoniae* structural *nif* genes, ●; pSym-5,   ; pTiAch5, ∧; pTiC58, ▲. The weak homology with pSym-5 is represented by dotted lines (Taken from *Plasmid* **7**, 271–80).

to the *K. pneumoniae* structural *nif*-genes. Some other parts of pSym-1 are also expressed in the bacteroids. Genes located on pSym-1 which are expressed at low levels would not have been found using this kind of labelled RNA. More sensitive procedures, using radioactive cDNA prepared from the isolated RNA's are needed for detection of low level expression of genes. The region of plasmid DNA expressed in free-living bacteria of late log phase is not located in one of the regions conserved in different Sym-plasmids. In early log phase bacteria, no expression of plasmid genes was detected (Krol *et al.* 1980). Plasmids other than Sym-plasmids present in the *Rhizobium* strains were not found to hybridize to RNA isolated from bacteroids (Krol *et al.* 1982a). The RNA's present in bacteroids of a *R. leguminosarum* strain have been separated on agarose gels and hybridized to the structural *nif* genes from *K. pneumoniae* (Krol *et al.* 1982b). It was concluded that, as in *A. vinelandii* and *K. pneumoniae*, the genes for the subunits of the components of the enzyme nitrogenase are located in one operon.

The conservation of Sym-plasmid DNA sequences is not restricted to *Rhizobium* but extends to other members of the *Rhizobiaceae*. Prakash *et al.* (1981) found homology between Sym-plasmids of rhizobia and Ti-plasmids of *A. tumefaciens*. On the physical map of pSym-1, this

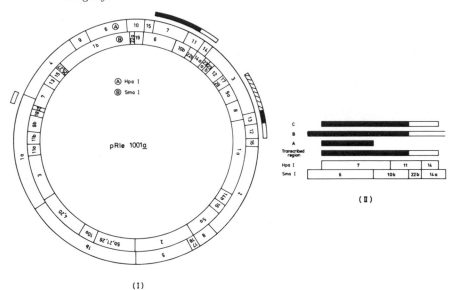

(I)

FIG. 9.2(b). (I) Organization of the transcribed regions on the restriction endonuclease map of plasmid pRle1001a. The homologous region of pRle1001a
with RNA from cultured *R. leguminosarum* 1001 is shown by ■ and with RNA
from bacteroids of *R. leguminosarum* 1001 as ■ (strong homology) and □
(weak homology). (II) Details of the transcribed region carrying *nif* genes.
(A) The region homologous with the *K. pneumoniae nif* DH; (B) and (C) the
region conserved on pSym-5 and pSym-9 respectively (Taken from *Plasmid* **7**,
281–6).

homology is located within two areas also conserved in rhizobial Sym-
plasmids, but not in the area where the structural *nif*-genes are located.
On the physical and genetic map of the Ti-plasmids, the homology
could be placed in regions involved in conjugative transfer of Ti-
plasmids, and in the virulence region (a region essential for the
transformation by the bacterium of plant cells). Large parts of the *vir*
region are homologous between different Ti-plasmids (Engler *et al.*
1981). The homology of pSym and pTi is not located in these parts,
but in other areas which may represent insertion sequences. No
homology with the T-region (that part of the Ti-plasmid which can be
found as T-DNA in plant tumour cells) could be detected. Both
Rhizobium and *Agrobacterium* interact with plant cells. Whether these
data concerning plasmid DNA homology, mean that common steps
exist in the interaction between bacteria and the plant has yet to be
investigated. Of course such common steps could equally well be encoded by chromosomal genes.

The presence of large plasmids in fast-growing *Rhizobium* species, on which most functions involved in the interaction between the bacterium and the host plant are located has opened a molecular genetic approach for identification and manipulation of symbiotic genes. Transposon mutagenesis *in vivo* and of the cloned fragments of the plasmid in *E. coli* cloning vectors *in vitro* will probably result in a detailed functional map of Sym-plasmids in the near future.

ACKNOWLEDGEMENTS

We gratefully acknowledge the help of our colleagues of the *Molbas-Research group*, and the Dutch Program Committee for Biotechnology for financial aid to R.v.V.

REFERENCES

BANFALVI, Z., SAKANJAN, V., KONCZ, C., KISS, A., DUSHA, I., and KONDO-ROSI, A. (1981). *Molec. & gen. Genet.* **184**, 318–25.

BEIJERINCK, M. W. (1888). *Bot. Ztg* **46**, 726, 741, 757, 781, 797.

BEYNON, J. L., BERINGER, J. E., and JOHNSTON, A. W. B. (1979). *J. gen. Microbiol.* **120**, 421–9.

BERINGER, J. E. and HOPWOOD, D. A. (1976). *Nature, Lond.* **246**, 291–3.

— BEYNON, J. L., BUCHANAN-WOLLASTON, A. V., and JOHNSTON, A. W. B. (1978a). *Nature, Lond.* **276**, 633–4.

— HOGGAN, S. A., and JOHNSTON, A. W. B. (1978b). *J. gen. Microbiol.* **104**, 201–7.

— BREWIN, N. J., and JOHNSTON, A. W. B. (1980). *Heredity* **45**, 161–86.

BOUCHER, C., BERGERON, B., BARATE DE BERTALMIO, M., and DÉNARIÉ, J. (1977). *J. gen. Microbiol.* **98**, 253–63.

BREWIN, N. J., DE JONG, T. M., PHILLIPS, D. A., and JOHNSTON, A. W. B. (1981a). *Nature, Lond.* **288**, 77–9.

— BERINGER, J. E., and JOHNSTON, A. W. B. (1980c). *J. gen. Microbiol.* **120**, 413–20.

— WOOD, E. A., JOHNSTON, A. W. B., DIBB, N. J., and HOMBRECHER, G. (1982). *J. gen. Microbiol.* **128**, 1817–27.

BUCHANAN-WOLLASTON, A. V. (1979). *J. gen. Microbiol.* **112**, 135–42.

— BERINGER, J. E., BREWIN, N. J., HIRSCH, P. R., and JOHNSTON, A. W. B. (1980). *Molec. & gen. Genet.* **178**, 185–90.

BULLERJAHN, G. G. and BENZINGER, R. H. (1982). *J. Bact.* **150**, 421–4.

BURKHARDT, B. and BURKHARDT, H. J. (1984). *J. molec. Biol.* **175**, 213–18.

CANNON, F. C., RIEDEL, G. E., and AUSUBEL, F. M. (1979). *Molec. & gen. Genet.* **174**, 59–66.

CASADESUS, J. and OLIVARES, J. (1979). *Molec. & gen. Genet.* **174**, 203–9.

CASSE, F., BOUCHER, C., JULLIOT, J. S., MICHEL, M., and DÉNARIÉ, J. (1979). *J. gen. Microbiol.* **113**, 229–42.

CURRIER, T. C. and NESTER, E. W. (1976). *Analyt. Biochem.* **66**, 431–41.

DE JONG, T. M., BREWIN, N. J., JOHNSTON, A. W.B., and PHILLIPS, D. A. (1982). *J. gen. Microbiol.* **128**, 1829–38.

DELEY, J., BERNAERTS, M. J., RASSEL, A., and GUILMOT, J. (1966). *J. gen. Microbiol.* **43**, 7–17.

— (1968). *A. Rev. Phytopathol.* **6**, 63–90.

DÉNARIÉ, J., TRUCHET, G. and BERGERON, B. (1976). In *Symbiotic nitrogen fixation in plants* (ed. P. S. Nutman), pp. 47–61. Cambridge University Press.

DJORDEJEVIC, M. A., ZURKOWSKI, W., and ROLFE, B. G. (1982). *J. Bact.* **151**, 560–8.

DUNICAN, L. K. and TIERNEY, A. B. (1974). *Biochem. biophys. Res. Commun.* **57**, 67–72.

— O'GARA, R., and TIERNEY, A. B. (1976). In *Symbiotic nitrogen fixation in plants* (ed. P. S. Nutman), pp. 77–90. Cambridge University Press.

EKHARDT, T. (1978). *Plasmid* **1**, 584–8.

ENGLER, G., DEPICKER, A., MAENHAUT, R., VILLAROEL, R., VAN MONTAGU, M., and SCHELL, J. (1981). *J. Molec. Biol.* **152**, 183–208.

GIBBINS, A. M. and GERGORY, K. F. (1972). *J. Bact.* **111**, 129–41.

GRAHAM, P. H. (1964). *J. gen. Microbiol.* **35**, 511–17.

HEBERLEIN, G. T., DELEY, J., and TYTGAT, R. (1967). *J. Bact.* **94**, 116–24.

HELLRIEGEL, H. and WILFARTH, H. (1888). *Beilageh. Z. Ver. Rubenzucker-Industrie Deutsch. Reichs*, p. 234.

HERNALSTEENS, J. P., DE GREVE, H., VAN MONTAGU, M., and SCHELL, J. (1978). *Plasmid* **1**, 218–25.

HIGASHI, S. (1967). *J. gen. appl. Microbiol.* **13**, 391–403.

HILLE, J., KLASEN, I., and SCHILPEROORT, R. A. (1982). *Plasmid* **7**, 107–18.

HIRSCH, P. R. (1979). *J. gen. Microbiol.* **113**, 219–28.

— VAN MONTAGU, M., JOHNSTON, A. W. B., BREWIN, N. J., and SCHELL, J. (1980). *J. gen. Microbiol.* **120**, 403–12.

HOLSTERS, M., DE WAELE, D., DEPICKER, A., MESSENS, E., VAN MONTAGU, M., and SCHELL, J. (1978). *Mol. & gen. Genet.* **163**, 181–7.

HOMBRECHER, G., BREWIN, N. J., and JOHNSTON, A. W. B. (1981). *Mol. & gen. Genet.* **182**, 133–6.

HOOYKAAS, P. J. J., KLAPWIJK, P. M., NUTI, M. P., SCHILPEROORT, R. A., and RÖRSCH, A. (1977). *J. gen. Microbiol.* **98**, 477–84.

— DEN DULK-RAS, H., and SCHILPEROORT, R. A. (1980). *Plasmid* **4**, 64–75.

— VAN BRUSSEL, A. A. N., DEN DULK-RAS, H., VAN SLOGTEREN, G. M. S., and SCHILPEROORT, R. A. (1981). *Nature, Lond.* **291**, 351–3.

—— (1984). In *Molecular genetics of plants* (Advances in genetics, Vol. 22) (J. G. Scandalios), pp. 209–83. Acadamic Press, New York.

— DENDULK-RAS, H., and SCHILPEROORT, R. A. (1982*a*). *Plasmid* **8**, 94–6.

— PEERBOLTE, R., REGENSBURG-TUÏNK, A. J. G., de VRIES, P., and SCHILPEROORT, R. A. (1982*b*). *Molec. & gen. Genet.* **188**, 12–17.

— SNIJDEWINT, F. G. M., and SCHILPEROORT, R. A. (1982*c*). *Plasmid* **8**, 73–82.

— VAN BRUSSEL, A. A. N., VAN VEEN, R. J. M., and WYFFELMAN, C. A. (1984). In *Advances in nitrogen fixation research* (ed. C. Veeger and W. E. Newton), pp. 661–6. Pudoc/Nijhoff.

JORDAN, D. C. (1982). *Int. J. syst. Bact.* **32**, 136–9.

JOHNSTON, A. W. B. and BERINGER, J. E. (1977). *Nature, Lond.* **267**, 611–13.
— BEYNON, J. L., BUCHANAN-WOLLASTON, A. V., SETCHELL, S. M., HIRSCH, P. R., and BERINGER, J. E. (1978*b*). *Nature, Lond.* **276**, 634–6.
— SETCHELL, S. M., and BERINGER, J. E. (1978*c*). *J. gen. Microbiol.* **104**, 209–18.
— HOMBRECHER, G., BREWIN, N. J., and COOPER, M. C. (1982). *J. gen. Microbiol.* **128**, 85–93.
KISS, G. B. and KALMAN, Z. (1982). *J. Bact.* **150**, 465–70.
KLAPWIJK, P. M., VAN BREUKELEN, J., KOREVAAR, K., OOMS, G., and SCHILPEROORT, R. A. (1980). *J. Bact.* **141**, 129–36.
KOEKMAN, B. P., OOMS, G., KLAPWIJK, M. P., and SCHILPEROORT, R. A. (1979). *Plasmid* **2**, 347–57.
KOWALCZUK, E., SKORUPSKA, A., and LORKIEWICZ, Z. (1981). *Molec. & gen. Genet.* **183**, 388–91.
KONDOROSI, A., KISS, G. B., FORRAI, T., VINZE, C. E., and BANFALVI, Z. (1977). *Nature, Lond.* **268**, 525–7.
— VINCZE, E., JOHNSTON, A. W. B., and BERINGER, J. E. (1980). *Molec. & gen. Genet.* **178**, 403–8.
KROL, A. J. M., HONTELEZ, J. G. J., VAN DEN BOS, R. C., and VAN KAMMEN, A. (1980). *Nucl. Acids Res.* **8**, 4337–47.
—— and VAN KAMMEN, A. (1982*a*). *J. gen. Microbiol.* **128**, 1839–47.
—— ROOZENDAAL, B., and VAN KAMMEN, A. (1982*b*). *Nucl. Acids Res.* **10**, 4147–57.
LAMB, J. W., HOMBRECHER, G., and JOHNSTON, A. W. B. (1982). *Mol. & gen. Genet.* **186**, 449–52.
MA, Q. S., JOHNSTON, A. W. B., and HOMBRECHER, G. (1982). *Molec. & gen Genet.* **187**, 166–71.
MANDEL, M. and HIGA, A. (1970). *J. molec. Biol.* **53**, 159–62.
MARTINEZ-DRETS, G., GARDIOL, A., and ARIAS, A. (1977). *J. Bact.* **130**, 1139–43.
MEADE, H. M. and SIGNER, E. R. (1977). *Proc. natn. Acad. Sci. U.S.A.* **74**, 2076–8.
MEGIAS, M., CAVIEDES, M. A., PALOMARES, A. J., and PEREZ-SILVA, J. (1982). *J. Bact.* **149**, 59–64.
MOFFETT, M. L. and COLWELL, R. R. (1968). *J. gen. Microbiol.* **51**, 245–55.
NELSON, L. M. and SALMINEN, S. O. (1982). *J. Bact.* **151**, 989–95.
NUTI, M. P., LEDEBOER, A. M., LEPIDI, A. A., and SCHILPEROORT, R. A. (1977). *J. gen. Microbiol.* **100**, 241–8.
— LEPIDI, A. A., PRAKASH, R. K., SCHILPEROORT, R. A., and CANNON, F. C. (1979). *Nature, Lond.* **282**, 533–5.
PRAKASH, R. K., HOOYKAAS, P. J. J., LEDEBOER, A. M., KIJNE, J. W., SCHILPEROORT, R. A., NUTI, M. P., LEPIDI, A. A., CASSE, F., BOUCHER, C., JULLIOT, J. S., and DÉNARIÉ, J. (1980). In *Proc. 3rd Int. Symp. Nitrogen Fixation* (ed. W. E. Newton and W. H. Orme-Johnston), pp. 139–64. University Park Press, Baltimore.
— SCHILPEROORT, R. A., and NUTI, M. P. (1981). *J. Bact.* **145**, 1129–36.
— VAN BRUSSEL, A. A. N., QUINT, A., MENNES, A. M., and SCHILPEROORT, R. A. (1982*a*). *Plasmid* **7**, 281–6.

— VAN VEEN, R. J. M., and SCHILPEROORT, R. A. (1982*b*). *Plasmid* **7**, 271–80.

— and SCHILPEROORT, R. A. (1982*c*). *J. Bact.* **149**, 1129–34.

PRIEM, W. J. E. and WYFFELMAN, C. A. (1984). In *Advances in nitrogen fixation research (C. Veeger and W. E. Newton)*, p. 717. Pudoc/Nijhoff.

RIKER, A. J. (1930). *J. agric. Res.* **41**, 507–40.

ROLFE, B. G., GRESSHOFF, P. M., and SHINE, J. (1980). *Pl. Sci. Lett.* **19**, 277–84.

ROSENBERG, C., CASSE-DELBART, F., DUSHA, I., DAVID, M., and BOUCHER, C. (1982). *J. Bact.* **150**, 402–6.

SADOWSKY, M. J., KEYZER, H. H., and BOHLOOL, B. B. (1983). *Int. J. syst. Bact.* **33**, 716–22.

SCHUBERT, K. R. and EVANS, H. J. (1976). *Proc. natn. Acad. Sci. U.S.A.* **73**, 1207–11.

SCOTT, K. F., HUGHES, J. E., and GRESSHOFF, P. M. (1982). *J. molec. appl. Genet.* **1**, 315–26.

SCOTT, D. B. and RONSON, C. W. (1982). *J. Bact.* **151**, 36–43.

SMITH, E. F. and TOWNSEND, C. O. (1907). *Science* **25**, 671–3.

STANLEY, J. and DUNICAN, L. K. (1979). *Molec. & gen. Genet.* **74**, 211–20.

STOWERS, M. D. and EAGLESHAM, A. R. J. (1983). *J. gen. Microbiol.* **130**, 3651–5.

SUTTON, W. D. (1974). *Biochim. biophys. Acta* **366**, 1–10.

VAN BRUSSEL, A. A. N., TAK, T., WETSELAAR, A., PEES, E., and WIJFFELMAN, C. A. (1982). *Pl. Sci. Lett.* **27**, 317–25.

VAN VLIET, F., SILVA, B., VAN MONTAGU, M., and SCHELL, J. (1978). *Plasmid* **1**, 446–55.

VINCENT, J. M. (1974). In *The biology of nitrogen fixation* (ed. A. Quispel), pp. 265–341. North-Holland, Amsterdam.

— (1980). In *Proc. 3rd Int. Symp. on Nitrogen Fixation* (ed. W. E. Newton and W. H. Orme-Johnson), pp. 103–29. University Park Press, Baltimore.

WYFFELMAN, C. A., PEES, E., VAN BRUSSEL, A. A. N., and HOOYKAAS, P. J. J. (1983). *Molec. Gen. Genet.* **192**, 171–6.

ZURKOWSKI, W., HOFFMAN, M., and LORKIEWICZ, L. (1973). *Acta microbiol. pol.* Ser A5, 55–60.

— and ZURKOWSKI, Z. (1978). *Genet. Res.* **32**, 311–114.

— and LORKIEWICZ, L. (1979). *Arch. Microbiol.* **123**, 195–201.

— (1981). *Molec. & gen. Genet.* **181**, 522–4.

— (1982). *J. Bact.* **150**, 999–1007.

10 Molecular biology of symbiotic nitrogen fixation by *Rhizobium meliloti*

Adam Kondorosi

10.1 INTRODUCTION

Rhizobium meliloti, a species of *Rhizobiaceae*, forms symbiotic associations with a limited number of leguminous plants, including *Melilotus*, *Medicago* or *Trigonella*. The agriculturally most important partner is *Medicago* (alfalfa, lucerne), a major forage crop in many countries. For this reason *R. meliloti* has attracted particular attention in comparison to other species of *Rhizobium*. *R. meliloti* is a typical fast-growing *Rhizobium*, forming a taxonomically distinct group within the *Rhizobiaceae* (Graham 1964; Kondorosi *et al.* 1980; Elkan 1981).

One of the most important aspects of the research reviewed below is the identification and molecular analysis of symbiotic genes in *R. meliloti*. In *R. meliloti* a large indigenous plasmid (megaplasmid) carries the majority of symbiotic genes. The organization of this gene cluster on the megaplasmid will be discussed in detail.

Recent advances in methods for bacterial genetics and in recombinant DNA technology have contributed considerably to the progress in the molecular studies of symbiotic nitrogen fixation. Conversely, due to the increased interest in biological nitrogen fixation, some of the techniques now available for studying various gram-negative bacteria were elucidated on rhizobia, some specifically on *R. meliloti*. These latter methods, together with the various gene transfer systems for *R. meliloti*, are discussed first.

Genetic studies have been carried out on several *R. meliloti* strains which has obvious disadvantages. For this reason, genetic or physical mapping, nucleotide or amino acid sequence determinations, etc., have to be carried out for several strains. On the other hand, our knowledge, of the characteristic properties of the species *R. meliloti* may be generalized. The studies reviewed below have been performed mainly on three strains of very diverse geographical origin. *R. meliloti* 41 (Rm41) was isolated in Hungary, *R. meliloti* 1021 (Rm1021) was isolated in Australia; with earlier designations such as SU47 or Rm2011 or RCR2011 (Rm1021 is a spontaneous streptomycin resistant derivative of RCR2011), *R. meliloti* 102F34 (Rm102F34) was found in the USA. Data on strains *R. meliloti* L5-30 (RmL5-30 from Poland) and *R. meliloti* GR4 (RmGR4 from Spain) will also be discussed.

10.2 GENE TRANSFER SYSTEMS

The three main types of bacterial genetic information transfer; conjugation, transduction, and transformation have all been reported for *R. meliloti*. Since this aspect of *Rhizobium* genetics was extensively reviewed (Schwinghamer 1977; Beringer *et al.* 1980; Kondorosi and Johnston 1981), this section will deal with these aspects only briefly.

10.2.1 Conjugation

All *R. meliloti* isolates tested so far contained indigenous plasmids including the megaplasmid (see Section 10.6). Self-transmissibility of these plasmids, however, was observed in only a few cases (Bedmar and Olivares 1980; J. Dénarié, personal communication; Banfalvi *et al.* 1985).

R-plasmids belonging to the P-1 incompatibility group, such as RP4, were shown to be transmissible between a wide range of gram-negative bacteria, including *R. meliloti* (Datta *at al.* 1971; Beringer 1974; Dixon *et al.* 1976). These plasmids are stably maintained in *R. meliloti* and their antibiotic resistance markers, with a few exceptions, are expressed (Kondorosi and Johnston 1981). P-1 plasmids with thermosensitive mutations affecting maintenance in *E. coli* are phenotypically thermoresistant in *R. meliloti* (Hooykaas *et al.* 1982*b*).

Mobilization of the *R. meliloti* chromosome was promoted by certain P-1 type plasmids, depending on the *R. meliloti* strain used. Plasmid RP4 mediated chromosomal gene transfer in Rm2011 (= Rm1021) at a fairly high frequency (Meade and Signer 1977) but it was inefficient in Rm41 (Kondorosi *et al.* 1977*a*). The plasmid R68.45, a variant of R68 with enhanced chromosome-mobilizing ability in many gram-negative bacteria (Haas and Holloway 1976) promoted high-frequency chromosome transfer in Rm41 (Kondorosi *et al.* 1977*a*) and in RmGR4 (Casadesús and Olivares 1979). A R68.45 derivative, suppressing an *E. coli dnaG* mutation, was even more efficient in chromosome mobilization (Ludwig and Johansen 1980).

Based on linkage analysis of auxotrophic and antibiotic resistance markers, linkage maps of three strains, Rm41 (Kondorosi *et al.* 1977*a*), Rm2011 (Meade and Signer 1977), and RmGR4 (Casadesús and Olivares 1979) were constructed. The linkage maps of Rm41 and Rm2011 were compared, and showed complete homology in chromosome recombination experiments (Kondorosi *et al.* 1980).

In crosses between *R. meliloti* and *R. leguminosarum*, however, recombinants were obtained at a very low frequency, suggesting that the chromosomes of the two species share little homology (Johnston *et al.* 1978*b*; Kondorosi *et al.* 1980). On the other hand, the relatedness of the two species is supported by the similarity in the arrangements of particular auxotrophic and resistance markers on the two chromosomal maps (Kondorosi *et al.* 1980). The chromosomal map of *Agrobacterium tumefaciens* also shows some similarity to these *Rhizobium* maps (Hooykaas *et al.* 1982).

Derivatives of P-type R-plasmids carrying sections of the *R. meliloti* chromosome have been constructed either *in vitro* (Julliot and Boistard 1979; Vincze *et al.* 1981) or *in vivo* (Johnston *et al.* 1978*b*; Kiss *et al.* 1980; Banfalvi *et al.* 1983). It was shown that the chromosome mobilizing ability of R68.45 is necessary for R-prime formation *in vivo*.

R68.45 also interacts with the indigenous plasmids of *R. meliloti*. On this basis, R-primes carrying sections of the megaplasmid were also isolated (Banfalvi *et al.* 1983). These R-primes were derived from strains carrying Tn5 on the megaplasmid. Since Tn5 is unable to transpose to new sites on *R. meliloti* DNA (Meade *et al.* 1982; Forrai *et al.* 1983), transfer of the kanamycin resistance marker into *E. coli* was followed, providing a simple way to select transconjugants containing R-primes.

The various R-primes were used to map new mutations (Kiss *et al.* 1980), to compare linkage maps (Kondorosi *et al.* 1980; Hooykaas *et al.* 1982), and to transfer *R. meliloti* genes into other bacterial genera (Johnston *et al.* 1978*b*; Banfalvi *et al.* 1983). Conversely, R-primes carrying genes of other bacterial species have been used to suppress *R. meliloti* mutations (Dixon *et al.* 1976).

P-1 type plasmids were also useful in the mobilization of indigenous *R. meliloti* plasmids. Stable co-integration of RP4 with a 140 Mdal plasmid of Rm41 (pRme41*a*) was detected and this co-integrate was transmissible to other bacterial species, such as *E. coli* (Vincze *et al.* 1981). Co-integration of RP4 with the megaplasmid was attempted by using RP4-primes carrying two fragments of the megaplasmid in inverted orientation. This type of homology between the R-prime and the megaplasmid allowed the co-integration of the two DNA molecules, but the co-integrates were unstable and resulted in the formation of R-primes (Dusha *et al.* 1983; Julliot *et al.* 1984). This R-prime isolation technique seems to be very useful for constructing sets of R-primes spanning the entire megaplasmid. Restriction fragments at the ends of inserts can be identified, cloned into RP4, and R-primes covering the adjacent megaplasmid regions can be isolated.

Mobilization of the *sym* megaplasmid was achieved when a small fragment of RP4 carrying the *mob* region was inserted into the megaplasmid (Kondorosi *et al.* 1982). First, the *mob* region together with a kanamycin resistance marker (Km^r) was inserted *in vitro* into a fragment of pRme41*b* cloned into pBR322. The recombinant plasmid formed was mobilized into *R. meliloti* by pJB3JI, a Km^s derivative of R68.45 (Beringer *et al.* 1978). Since the recombinant plasmid was unable to replicate in *R. meliloti*, selection for Km^r allows the identification of those derivatives in which the *mob* region was inserted into the megaplasmid via homologous recombination. Mobilization of the megaplasmid carrying the *mob*$_{RP4}$ region was effective with several other gram-negative bacteria. The maintenance of the megaplasmid was demonstrated in two other *Rhizobium* species and in *A. tumefaciens*.

Simon (1984) cloned the *mob* region into Tn5 and using transposon mutagenesis techniques, this new transposon (Tn5-Mob) can easily be inserted into the DNA of gram-negative bacteria. The mobilization of *R. meliloti* plasmids was also achieved by this method.

Recently, cloning vehicles for *Rhizobium* were developed from various broad-host range plasmids (Ditta *et al.* 1980; Kiss and Kalman 1982; Simon *et al.* 1983; David *et al.* 1983). These vehicles allowed the establishment of gene banks in *E. coli* and recombinant plasmids were mobilized into *R. meliloti* (see next Section) where these plasmids were stably maintained. In this way, complementation analyses can now be performed in *R. meliloti*, as long as recombination between the homologous regions can be ruled out. A recombination deficient mutant of Rm41 was recently reported (Olasz *et al.* 1983). Recently the *rec*A gene and its Tn5 mutant derivative were isolated from Rm102F34 (Better and Helinski 1983), allowing the easy construction of recombination deficient *R. meliloti* derivatives as was already shown for Rm1021 (Finan *et al.* 1984).

10.2.2 Transduction

Generalized transduction was described for several *R. meliloti* strains, RmL530 (Kowalski 1967), RmGR4 (Casadesús and Olivares 1979), Rm41 (Sik *et al.* 1980), and Rm1021 (Finan *et al.* 1984; Martin and Long 1984). Specialized transduction was demonstrated in Rm41 (Svab *et al.* 1978). The usefulness of transduction systems for fine-linkage mapping in these strains have, however, been somewhat limited.

10.2.3 Transformation

Early reports on transformation in rhizobia including *R. meliloti* were reviewed by Balassa (1963) and by Gabor-Hotchkiss (1972). There has been practically no progress since these early experiments on transformation of chromosomal markers of *R. meliloti*. Transfection of *R. meliloti* by purified phage DNA was achieved only at barely detectable frequencies (Sik and Orosz 1971; Staniewski and Rugala 1975; Kondorosi *et al.* 1974).

Recently, transformation of *R. meliloti* (Selvaraj and Iyer 1981; Kiss and Kalman 1982) and *R. leguminosarum* (Bullerjahn and Benzinger 1982) with low-molecular-weight plasmid DNA has been reported. The transformation frequencies were, however, rather low. Further increases in the transformation efficiency would be necessary in order to establish gene banks directly in *R. meliloti*.

10.3 CLONING SYSTEMS

Various cloning vehicles suitable for *R. meliloti* have been developed from broad-host-range plasmids. The original plasmid vehicles were too big or had no suitable restriction sites, etc., although the use of RP4 was reported (Julliot and Boistard 1979; Vincze *et al.* 1981). Later cloning vehicles carried the broad-host-range replication region; the *mob* region, resistance marker(s) as well as suitable restriction sites.

Since the efficiency of transformation in *R. meliloti* was still rather low, in most cases recombinant plasmids were introduced into *E. coli*, then the plasmids were mobilized back into *R. meliloti* by a helper plasmid providing transfer functions. Plasmid pRK290 (Ditta *et al.* 1980) is a shortened derivative of the P-1 type plasmid RK2, which is identical to RP4 or R68 (Burkardt *et al.* 1979). It can be mobilized at high frequency by the helper plasmid pRK2013 which contains the transfer genes of RK2 and has the narrow host range *col*E1 replicon. A cosmid derivative of pRK290, pLAFR1 was constructed by inserting the λ *cos* region into it (Friedman *et al.* 1982).

Gene fusion vehicles for the analysis of gene expression in *R. meliloti* were also developed from RK2 (Kahn and Timblin 1984). Some of these contain the *cos* site of bacteriophage P4 and these plasmids are efficiently packaged into phage P2 capsids. Surprisingly, *R. meliloti* strain 104A14 can be infected with P2, therefore these plasmids can be introduced into this particular strain by transduction.

Plasmid pGV1106, a cloning vehicle construced from the W-incompatibility group S-a (Leemans *et al.* 1982), can also be used for *R. meliloti* (Kiss and Kalman 1982). When this vehicle is mobilized by the P-1 type plasmid pJB3JI, however, insertion of a 1.3 kb fragment from pJB3JI, results in the recombinant plasmid pKK2. This can be more efficiently mobilized by pJB3JI than the original vehicle.

Plasmid RSF1010, belonging to the Q incompatibility group, also replicates in *R. meliloti*. Several derivatives for cloning purposes were constructed (Bagdasarian *et al.* 1981) which can be mobilized into *R.*

meliloti (David *et al.* 1983). It is advantageous if the helper plasmid is not transferred into the recipient. *E. coli* strains, carrying RP4 integrated into the chromosome, and consequently unable to transfer into the recipient, but still able to mobilize other plasmids, have been constructed (Simon *et al.* 1983).

Bacteriophage 16-3 may be a potential cloning vector for *R. meliloti*. A correlated genetic-physical map is now available for this phage (Orosz *et al.* 1973; Dallmann *et al.* 1980) and it is strikingly similar to phage λ of *E. coli*.

10.4 ISOLATION OF SYMBIOTIC MUTANTS

The development of symbiotic nitrogen-fixing nodules is a complicated process, in which both partners interact at each stage (Vincent 1980; Vance 1983; Verma and Long 1983). It is very likely that numerous bacterial genes are involved in the developmental, as well as in the nitrogen fixation processes. Therefore, isolation of symbiotic mutants defective either in nodulation (Nod⁻) or fixation (Fix⁻) may greatly aid understanding of the genetic control of nodule formation and function.

Symbiotic mutants of *R. meliloti* have been isolated in various ways. First, numerous mutant phenotypes which were observable in free-living bacteria (such as resistance to certain antibiotics, auxotrophy, changes in exopolysaccharide production, etc.) were correlated with changes in the symbiotic properties (see, e.g., Dénarié *et al.* 1976). However, the genes mutated in these derivatives are not specifically required for symbiosis. Nevertheless, biochemical analysis of such mutants may elucidate the metabolic events in the bacterial partner during the establishment of symbiotic association. This mutant class has recently been extensively reviewed (Schwinghamer 1977; Kuykendall 1981). Another approach is to mutagenize bacteria randomly, to plate for single colonies, and then to screen each colony for altered on symbiotic phenotype. In principle, this technique allows the identification of practically all symbiotic genes, including those which express exclusively during symbiosis. This latter class of symbiotic genes control the crucial steps of symbiotic development and nitrogen fixation.

Random mutagenesis of *R. meliloti* with either chemical agents or with transposons has been reported (Paau *et al.* 1981; Meade *et al.* 1982; Forrai *et al.* 1983; Ali *et al.* 1984). Since some chemical mutagens, such as N-methyl-N'-nitro-N-nitrosoguanidine (NTG) induce mainly reversible point mutations, while transposon insertions have strong polar effects, the two sets of mutants serve different purposes. Nevertheless, transposon mutagenesis is particularly useful in generating symbiotic mutations, since the symbiotic genes are marked both genetically and physically. By following the antibiotic resistance marker of the transposon, the inactivated gene can easily be mapped and cloned.

A general method was developed for transposition mutagenesis of gram-negative bacteria by Van Vliet *et al.* (1978). In the initial experiment a broad-host-range plasmid carrying both bacteriophage Mu and transposon Tn7 was transferred from *E. coli* to *A. tumefaciens*. Since in *Agrobacterium* (and in the majority of gram-negative bacteria) Mu-containing replicons are not maintained, transposition derivatives of the recipient can be directly selected. A similar 'suicide' plasmid (pJB4JI) carrying Tn5 was constructed by Beringer *et al.* (1978). After transferring pJB4JI into *R. leguminosarum*, kanamycin resistant derivatives were obtained. About 0.5 per cent of these clones were auxotrophs. The auxtrophy and Km^r markers showed 100 per cent linkage in genetic mapping experiments. The 'suicide' plasmid pJB4JI was used to isolate symbiotic mutants of *R. leguminosarum* (Buchanan-Wollaston *et al.* 1980), of *R. trifolii* (Rolfe *et al.* 1980), and also of *R. meliloti* (Meade *et al.* 1982; Forrai *et al.* 1983). It was found that not every *Rhizobium* strain yielded Tn5-containing derivatives (Rolfe *et al.* 1980; Meade *et al.* 1982) or in other strains parts of pJB4JI were retained (Banfalvi *et al.* 1981; Meade *et al.* 1982). In some derivatives, Mu sequences transposed together with Tn5, which severely decreased the applicability of these insertion mutants for either genetic or physical studies (Meade *et al.* 1982; Forrai *et al.* 1983). Another vehicle for Tn5 mutagenesis, p^{1011}, was constructed by Simon *et al.* (1983). It carries a *col*E1 replicon, the *mob* region of RP4, allowing its mobilization into *Rhizobium* and Tn5. Since this plasmid cannot replicate in *Rhizobium*, Km^r derivatives are transposition derivatives. Suicide plasmid vehicles for insertion mutagenesis in *R. meliloti* were constructed by Selvaraj and Iyer (1983). The vectors contain the *mob* region of N type plasmids which is efficient in conjugations between *E. coli* and *R. meliloti*, but these plasmids do not replicate in *Rhizobium*. Suicide plasmids for Tn1, Tn7, and Tn9 mutagenesis were also reported (Selvaraj and Iyer 1983; Bolton *et al.* 1984). Tn7 was shown to insert specifically into two unique sites on the *R. meliloti* megaplasmid (Bolton *et al.* 1984).

It was observed that after insertion of Tn5 into the *R. meliloti* genome, Tn5 did not transpose to new locations (Meade *et al.* 1982; Forrai *et al.* 1983). If one copy of Tn5 is already present in *E. coli*, transposition of Tn5 is also reduced (Biek and Roth 1980) but only to a certain extent. In *R. leguminosarum*, Tn5 can still jump to new sites (Buchanan-Wollaston *et al.* 1980). There are several possible explanations for this behaviour. For example, the transposition–inhibition activity of protein 2 coded by Tn5 (Johnson *et al.* 1982) is more active in *R. meliloti* than in other gram-negative bacteria or an early step of transposition is missing in *R. meliloti*, etc. There are other variations in the expression of Tn5 functions in different bacterial species and strains. It was shown that Tn5 codes for streptomycin resistance; this gene is expressed in *R. meliloti* but not in *E. coli* (Putnoky *et al.* 1983). Recently, other laboratories for other *Rhizobium* strains confirmed these findings (Selvaraj and Iyer 1984; O'Neill *et al.* 1984; DeVos *et al.* 1984).

After NTG (Paau *et al.* 1981; Forrai *et al.* 1983) or Tn5 mutagenesis (Meade *et al.* 1982; Forrai *et al.* 1983) both Nod$^-$ and Fix$^-$ mutants of *R. meliloti* have been isolated. The frequency of Fix$^-$ mutants was about the same as that of auxotrophs, indicating that numerous bacterial genes code for functions required in symbiotic nitrogen fixation. The number of Nod$^-$ mutants were 5–10 times lower. In earlier screenings of NTG-mutagenized populations of *R. japonicum* (Maier and Brill 1976) and of *R. leguminosarum* (Beringer *et al.* 1977) such differences were also detected. It is possible that only a few Nod$^+$ revertants, among a Nod$^-$ bacterial population will induce nodules (Brewin *et al.* 1980; Long *et al.* 1982). In contrast to this, Fix$^+$ revertants have no strong selective advantage over the Fix$^-$ mutants. Moreover, the number of *nod*-genes is definitely lower (see below).

A serious difficulty in the use of random mutagenesis procedures arises from the fact that indigenous insertion elements have been detected in some *R. meliloti* strains. It was found for Rm1021 that out of 50 symbiotic mutants carrying Tn5, only four were 'true' Tn5-induced mutants. At least in 33 mutants the symbiotic mutations were due to the insertion of a 1.4 kb insertion element (ISRm1) into the *nif* region (Ruvkun *et al.* 1982*a*). This IS element was present in at least ten copies in Rm1021, but was not detectable in the other *R. meliloti* strains or *Rhizobium* species tested. A 2.6 kb IS element (ISRm2) was found in Rm41 (Kondorosi *et al.* 1984*a*). Both ISRm1 and ISRm2 seem to show a certain degree of specificity for the *nif* region. It is not clear whether the presence of such actively translocating IS elements has any role in controlling symbiotic nitrogen fixation.

Due to the presence of IS elements in these strains it was necessary to demonstrate that mutants obtained after Tn5 mutagenesis were 'true' Tn5-induced mutants. The simplest way to do this, is to test for the 100 per cent linkage of Tn5 with the symbiotic mutation (Meade *et al.* 1982; Forrai *et al.* 1983). The resolution of genetic mapping, however, does not exclude the possibility of two, very close, independent mutations. Ruvkun *et al.* (1982*a*, *b*) have reported a more reliable method. Fragments carrying the insertion were cloned and then transferred back into the wild type *R. meliloti*. After replacing the normal genomic fragment with the fragment carrying the insertion, the phenotype of the resulting strain was tested. If the symbiotically defective phenotype of the original mutant was due to the insertion, the resulting strains should also be symbiotically defective.

A site-directed mutagenesis technique allowing efficient mutagenesis of previously cloned *Rhizobium* genes was developed by Ruvkun and

Ausubel (1981). *R. meliloti* DNA fragments cloned into multicopy plasmids in *E. coli* were mutagenized with Tn*5*, then recloned into a P-1 type Mob⁺ Tra⁻ vector, pRK290 (Ditta *et al.* 1980), and mobilized into the wild type *R. meliloti*. The wild-type DNA fragment was replaced by the cloned sequences carrying Tn5 via recombination between the homologous regions. Conjugation of a second P-1 type plasmid (incompatible with pRK290) into the *R. meliloti* strain carrying the recombinant pRK290 plasmid resulted in the loss of the pRK290 derivative. In those transconjugants which inherited the second P-1 plasmid and the Km^r marker of Tn5, the double cross-over event at the homologous region had occurred. It was shown that Tn5 was located exactly at the same position in the *R. meliloti* genome as in the cloned fragment.

The directed Tn5 mutagenesis technique was very powerful in the construction of a correlated physical-genetic map of the *nif* and *nod* regions (see below). Random mutagenesis is useful in detecting DNA regions carrying symbiotic genes, while directed Tn5 mutagenesis, is a suitable method for the fine-structure analysis of cloned genes, or in searching for other symbiotic genes in the vicinity of the cloned genes.

Fix⁻ and Nod⁻ deletion mutants have also been isolated in *R. meliloti*. Using a plasmid elimination procedure (Kiss *et al.* 1980) a series of Nod⁻ and Fix⁻ derivatives of Rm41 were isolated. It was shown that deletions of varying size occurred in the megaplasmid (Banfalvi *et al.* 1981). Several Nod⁻ mutations have simultaneously lost the *nif* structural genes. Nod⁻ mutants with *nif* deletions also spontaneously appeared (Banfalvi *et al.* 1981; Rosenberg *et al.* 1981).

10.5 MAPPING OF SYMBIOTIC MUTATIONS

The chromosomal linkage maps of *R. meliloti* strains (Kondorosi *et al.* 1977*a*; Meade and Signer 1977; Casadesús and Olivares 1979) contain auxotrophic and resistance markers which provide a framework for the localization of any chromosomal mutations. Using P-1 plasmid-mediated recombination, it is relatively simple to locate a non-selectable mutation on the linkage map. During plasmid promoted chromosome mobilization in *Rhizobium*, very large chromosomal segments are transferred and incorporated into the recipient DNA. The occurrence of multiple cross-over events on the incorporated region is relatively infrequent (Kondorosi and Johnston 1981).

In Rm41, the entire chromosome can be transferred in eight fragments (Kondorosi *et al.* 1980). By testing these eight recombinant classes for symbiotic phenotype, the symbiotic mutation can be localized on these chromosomal segments. In this way five *fix* mutations were localized in three different chromosomal regions (Fig. 10.1). Two of these mapped mutations were caused by Tn5 insertion. Mapping of

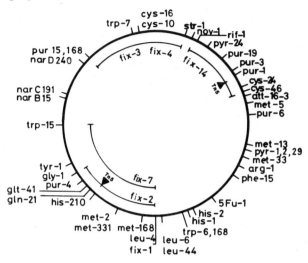

FIG. 10.1 Locations of symbiotic mutations on the *R. meliloti* 41 chromosome. The linkage map was constructed by means of R68.45-mediated recombination (Kondorosi *et al.* 1977*a*, 1980; Svab *et al.* 1978; Kiss *et al.* 1979; Vincze *et al.* 1981). Mapping of symbiotic mutations was described by Forrai *et al.* (1983).

the Km[r] marker of Tn5 allowed the precise localization of the symbiotic gene. The majority of *fix* mutations, however, and all *nod* mutations tested did not map onto the chromosome (Banfalvi *et al.* 1981; Forrai *et al.* 1983). In Rm1021, two Tn5-induced symbiotic mutations were localized on the chromosome, but the majority of the mutations were extrachromosomal (Buikema *et al.* 1983).

In five Tn5-induced symbiotic mutants where Tn5 did not map onto the chromosome, Tn5 was localized on the megaplasmid. The indigenous plasmids of Rm41 were separated on agarose gel and a ^{32}P-labelled Tn5 probe hybridized to the megaplasmid band (Forrai *et al.* 1983).

Cloned fragments of the *nod-nif* region of the megaplasmid are now available. Thus, localization of Tn5- or other IS-induced mutations on this DNA region can be performed by hybridizing the cloned fragments to total DNAs from the mutants (Ruvkun *et al.* 1982*a*; Kondorosi *et al.* 1983*c*).

10.6 INDIGENOUS PLASMIDS IN *R. meliloti*

10.6.1 Medium-size plasmids

Large plasmids have been detected in all *R. meliloti* strains so far tested. Various analytical and preparative procedures were developed to isolate plasmids in CCC DNA form with a size up to 300 Mdal (see review by Dénarié *et al.* 1981). The majority of the strains tested possess at least one plasmid with a molecular weight of 60 to 300 Mdal (Olivares *et al.* 1977; Bechet and Guillaume 1978; Casse *et al.* 1979; Spitzbarth *et al.* 1979; Prakash *et al.* 1981; Kuchko *et al.* 1983; Adachi *et al.* 1983). Plasmids isolated from different strains share a significant homology (Jouanin *et al.* 1981). There is no indication, however, that any of these plasmids code for symbiotic properties. Some of these plasmids share a considerable degree of homology with plasmids from *R. leguminosarum* (Prakash *et al.* 1981) and from *A. tumefaciens* (Jouanin *et al.* 1983; Koncz, C., unpublished). The DNA region coding for plasmid replication seems to be conserved on these plasmids.

Self-transmissibility of several plasmids has been observed (J. Dénarié, personal communication; Adachi *et al.* 1983). Other functions coded by these plasmids have not been identified so far. Thus, their biological role is still obscure.

10.6.2 Megaplasmids

Using a very gentle lysis and agarose gel electrophoresis procedure (Eckhardt 1978), plasmids larger than 300 Mdal (megaplasmids) were detected in all *R. meliloti* strains tested so far (Banfalvi *et al.* 1981; Rosenberg *et al.* 1981). Recent screening of more than 200 *R. meliloti* isolates indicates that the presence of a megaplasmid is a characteristic feature of the *R. meliloti* species (W. J. Broughton, personal communication). Using electron-microscopic techniques the length of the megaplasmid from *R. meliloti* strain MVII/1 was found to be about 1000 Mdal (Burkardt and Burkardt 1984). These plasmids may represent at least 20–30 per cent of the total DNA of *R. meliloti*. Moreover, in Rm41 two megaplasmids (pRme41*b* and pRme41*c*) are present (Banfalvi *et al.* 1981; Simon *et al.* 1983). There is both genetic and physical evidence that one of the megaplasmids (pRme41*b* in Rm41) carries *nod, fix*, and *nif* genes. The first indication came from the analysis of symbiotic mutants in which part of the megaplasmid was deleted (Banfalvi *et al.* 1981; Rosenberg *et al.* 1981). Banfalvi *et al.* (1981) have isolated numerous Nod⁻ and Fix⁻ mutants using a heat-treatment procedure, originally worked out for the elimination of R-plasmids from *R. meliloti* (Kiss *et al.* 1980). In some Nod⁻ mutants the deletion of megaplasmid segments was noticed on 'Eckhardt' gels; in others hybridization of cloned *nif*-genes to the total DNA isolates of the mutant revealed loss of megaplasmid regions. In the latter case the majority of Nod⁻ mutants lost the *nif*-genes as well (Banfalvi *et al.* 1981). Spontaneous Nod⁻ mutants also lacked the *nif* structural genes (Banfalvi *et al.* 1981; Rosenberg *et al.* 1981). These results suggested that some *nod-, fix-,* and *nif*-genes are linked on the megaplasmid. The megaplasmid was made susceptible to mobilization by inserting the mobilization region of RP4 into it (Kondorosi *et al.* 1982). The megaplasmid pRme41*b* was introduced into *nod-nif* deletion mutants where both Nod$^+$ and Fix$^+$ phenotypes were restored. After its transfer into two different *Rhizobium* species and into *A. tumefaciens*, the transconjugants of all three species induced nodule formation on alfalfa, indicating that the megaplasmid carried all essential genes controlling the early steps in nodulation.

These nodules, however, were small and electronmicroscopic investigations revealed that they developed only to a certain extent (Wong *et al.* 1983). These nodule-like structures were composed of meristematic cells but were not invaded by bacteria. Bacteria were only found in a few sub-epidermal cells of the outer root cortex, within infection threads and within intercellular spaces of the outermost cells. The growth of infection threads into the inner part of the nodules was prevented. Based on these observations, it was suspected that bacteria degenerate and disappear during cellular differentiation. One may speculate that genes not present on this megaplasmid might be involved in the later stages of nodule development, for instance to overcome the plant defense mechanism against bacterial invasion. These functions cannot be provided by other *Rhizobium* species tested. It is possible that chromosomal *fix*-genes code for this bacterium-plant species specific function.

Recently, the symbiotic (*psym*) megaplasmid of Rm2011 was mobilized by plasmid RP4 into *A. tumefaciens* and the resulting transconjugants induced again nodule-like structures but infection threads were not detected (Truchet *et al.* 1984). Using suitable selection markers, Banfalvi *et al.* (1985) demonstrated a low level of self-transmissibility for both pRme41*b* and pRme41*c*.

Plasmid control of symbiotic genes has been demonstrated for other fast-growing *Rhizobium* species, such as *R. leguminosarum* (Johnston *et al.* 1978*a*), *R. phaseoli* (Beynon *et al.* 1980), *R. trifolii* (Zurkowski and Lorkiewicz 1979), and a broad-host-range *Rhizobium* sp. NGR234 (Pankhurst *et al.* 1983). Thus, the existence of so-called p*sym*-plasmids in fast-growing rhizobia is widespread. Transfer of pRme41*b* allowed the demonstration of other plasmid-coded functions. In this way, it was shown that the octopine utilization ability of Rm41 is coded by pRme41*b* (Kondorosi *et al.* 1983*b*).

Nodules formed on alfalfa roots, inoculated with RmL5-30, contain compounds which are specific growth substrates for RmL5-30 (Tempé *et al.* 1982). By analogy to the opines in crown gall tumours induced by *Agrobacterium*, it was suggested that these compounds might serve as C or N sources for the bacterium during symbiosis. Preliminary experiments suggest that genes coding for the utilization of these compounds are carried by the megaplasmid (J. Tempé, P. Murphy, and A. Kondorosi, unpublished).

The physical and genetic analysis of the megaplasmid is difficult, because of its very high molecular weight and lack of a procedure for its isolation. One possibility to overcome these problems was to construct R-primes carrying various sections of the megaplasmid. Since studies on the megaplasmid deletion mutants indicated that some *nod* and *nif* genes are fairly closely located in one area of the plasmid (Banfalvi *et al.* 1981; Rosenberg *et al.* 1981), it was likely that R-primes carrying both *nod*- and *nif*-genes could be obtained. Indeed, a set of pRme41*b* R-primes carrying the *nif* structural genes were shown to contain *nod* genes (Banfalvi *et al.* 1983).

Transfer of these R-primes into different Nod⁻ point insertion or deletion mutants restored the Nod⁺ phenotype. When they were introduced into *A. tumefaciens*, the transconjugants formed small nodules on alfalfa. These nodule-like structures were very much like the nodules induced by *A. tumefaciens* carrying pRme41*b*, suggesting that all *nod* genes of pRme41*b* are present on the R-primes. These R-primes carried 100–250 kb megaplasmid DNA segments, indicating again that the *nod*- and *nif*-genes are fairly closely linked.

A *nod-nif* R-prime of about 250 kb in size, has been isolated from Rm2011 (Dusha *et al.* 1983). This R-prime restored Nod⁺, in all Nod⁻ Rm41 mutants tested. In some Fix⁻ mutants the mutations were suppressed by this plasmid. Several of these mutations were not located in the vicinity of the *nif*-genes (Dusha, I., personal communication). R-primes carrying other megaplasmid segments have also been isolated (Z. Banfalvi, unpublished). Symbiotic genes on these plasmids have not yet been identified.

Physical maps of the *nod-nif* regions (Fig. 10.2) have been established for Rm1021 (Buikema *et al.* 1983) and for Rm41 (Kondorosi *et al.* 1983*c*). For Rm1021 a total *Rhizobium* gene bank was constructed in *E. coli*, using the cosmid vehicle pHC79 (Hohn and Collins 1980). A cloned DNA fragment carrying the *nif* structural genes was used to identify hybridizing clones. Overlapping clones were then identified. The cosmids selected were partially digested with *Eco*RI, the large DNA fragments were selected and re-ligated. From these overlapping clones, an *Eco*RI restriction map of about 90 kb size was established (Fig. 10.2; Buikema *et al.* 1983).

For Rm41 a gene library was constructed in *E. coli*, using the cosmid vehicle pJB8 (Ish-Horowitz and Burke 1981). Recombinant cosmids, hybridizing with a *nif* probe and with a *nod-nif* R-prime were selected and a restriction map of about 130 kb size with the help of four restriction endonucleases was constructed (Fig. 10.2; Kondorosi *et al.* 1983*b*, 1984*b*).

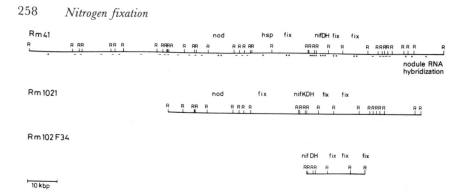

Fig. 10.2 Physical maps of the *nod–nif* regions for different *R. meliloti* strains. Maps for Rm41 (Kondorosi *et al.* 1983*c*), for Rm1021 (Buikema *et al.* 1983), and for Rm102F34 (Corbin *et al.* 1982) show only the *Eco*RI restriction sites. Hybridization of cloned fragments of Rm41 was performed with total nodule RNA as a hybridization probe (Kondorosi *et al.* 1983*b*).
+ = hybridization; − = no hybridization above background.

The structure and organization of symbiotic genes on this megaplasmid region is discussed below.

10.7 ANALYSIS OF SYMBIOTIC GENES

10.7.1 The *nif*-genes

The DNA sequences of the nitrogenase structural genes are highly conserved among various nitrogen-fixing organisms (Nuti *et al.* 1979; Ruvkun and Ausubel 1980; Mazur *et al.* 1980). On this basis, using plasmid pSA30 carrying the *K. pneumoniae nif* structural genes (Cannon *et al.* 1979) as a hybridization probe, it was possible to identify and clone the *nif* structural genes from other diazotrophs, including *R. meliloti*. Fig. 10.3 shows the cloned *nif* regions from Rm41 (Banfalvi *et al.* 1981), from Rm1021 (Ruvkun and Ausubel 1980), and from Rm102F34 (Ditta *et al.* 1980). The restriction maps are fairly similar; only minor differences exist.

Hybridization of the cloned DNA fragments with each of *K. pneumoniae nifH, D*, and *K* genes from pSA30 revealed that *nifH*- and *D*-genes displayed a fairly strong interspecies homology (Ruvkun and Ausubel 1981; Banfalvi *et al.* 1981). Homology for the *nifK* gene was not detected in these studies. In further experiments, Ruvkun *et al.* (1982*b*) found that using another probe, containing the *K. pneumoniae nifK-, Y-, E*-genes, a 1.7 kb Rm1021 fragment, adjacent to the *nifH, D* fragment, hybridized to the probe. It is likely that this fragment also contains *nifK*. Hybridization of other cloned *K. pneumoniae nif* genes to the cloned *R. meliloti nif* region was not observed (Ruvkun *et al.* 1982*b*).

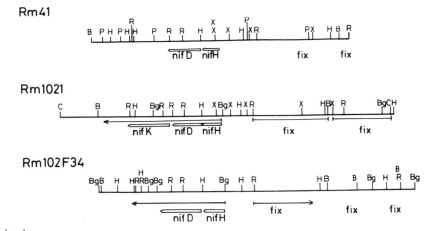

FIG. 10.3 The *nif* regions in three *R. meliloti* strains. Data for Rm41 (Banfalvi *et al.* 1981, 1983; Török and Kondorosi 1981; Kondorosi *et al.* 1983*c*), for Rm1021 (Ruvkun and Ausubel 1981; Ruvkun *et al.* 1982*b*), and for Rm102F34 (Corbin *et al.* 1982; 1983) show the location of *nif* structural genes as determined from hybridizations with the respective *K. pneumoniae nif* probes and the location and orientation of transcription units. For Rm41 the transcripts have not been reported, although genetic data indicate that their locations are the same as for Rm1021 and Rm102F34.

The nucleotide sequence data for the DNA region showing homology with the *K. pneumoniae nifH-* and *D*-genes provided further evidence that this region indeed codes for *nifH* and *D*. The entire nucleotide sequence of *nifH* and a partial-sequence of *nifD* were determined for Rm41 (Fig. 10.4; Török and Kondorosi 1981). Partial *nifH* and *D* sequences for Rm1021 were also reported (Ruvkun *et al.* 1982*b*). Since entire or partial amino-acid sequences of the *nifH* and *D* products (the nitrogenase reductase and one subunit of dinitrogenase), have been reported for *K. pneumoniae* (Sundaresan and Ausubel 1981; Scott *et al.* 1981) for *Anabaena* 7120 (Mevarech *et al.* 1980), for *Clostridium pasteurianum* (Tanaka *et al.* 1977; Hase *et al.* 1981) and for *Azotobacter vinelandii* (Lundell and Howard 1978), the amino acid sequences of the *nifH* (Fig. 10.5) and *nifD* (Fig. 10.6) gene products can be deduced. Rm41 nitrogenase reductase protein consists of 297 amino acid residues, has a molecular weight of 32 740 dal and contains five cysteine residues (Török and Kondorosi 1981). It shares 70 per cent, 67 per cent, and 60 per cent amino acid homology with nitrogenase reductases from *Anabaena* 7120, *K. pneumoniae*, and *C. pasteurianum*, respectively. At a base pair level, the homology is 63 per cent, and 66 per cent with *nifH*

```
                    -140              -120              -100              -80            -60
    ──────▶      ◀───┬──────       ──────┬──────▶     ◀───┬──────        -80┬          -80┬
    ACCATCTGGTCAATTCCAGACTAACTATCTGAAAGAAAAGCCGAGTAGTTTTATTTCAGACGGCTGGCACGACTTTGCACGATCAGCCCTGGCGCGCATGCTGTTGC

              -40┬              -20┬              -1┬ ┬              20┬              40┬
    GCATTCATGTGTCCGACCAACCGAAATAGCTTAAACAACAAAGGAAGCAAG ATG GCA GCT CTG CGT CAG ATC GCG TTC TAC GGT AAG GGG GGT

                   60┬                        80┬                       100┬                      120┬
    ATC GGC AAG TCC ACG ACC TCC CAA AAT ACA CTC GCC GCG CTT GTC GAC CTG GGG CAA AAG ATC CTT ATT GTC GGC TGC GAT

                  140┬                       160┬                       180┬                      200┬
    CCG AAA GCG GAC TCC ACG CGC CTC ATC CTG AAC GCA AAG GCA CAG GAC ACC GTA CTC CAT CTT GCG GCA ACC GAA GGT TCG

                  220┬                       240┬                       260┬                      280┬
    GTC GAA GAT CTC GAG CTC GAG GAC GTC CTC AAA GTC GGT TAC AGA GGC ATC AAG TGC GTG GAG TCC GGT GGC CCA GAG CCG

                  300┬                       320┬                       340┬                      360┬
    GGC GTC GGC TGC GCC GGA CGC GGC GTT ATC ACC TCG ATC AAC TTC CTG GAA GAG AAC CGC GCT TAC AAC GAT GTC GAT TAC

                  380┬                       400┬                       420┬                      440┬
    GTC TCA TAC GAC GTC CTA GGG GAC GTA GTA TGC CGC GGC TTT GCG ATG CCT ATT CGC GAA AAC AAG GCT CAG GAA ATC TAC

                  460┬                       480┬                       500┬                      520┬
    ATC GTC ATG TCC GGT GAC ATG ATG GCG CTC TAT GCC GCC AAC AAC ATC GCG AAG GGT ATC CTG AAG TAC GCC CAT GCG GGC

                  540┬                       560┬                       580┬                      600┬
    GGC GTC CGC CTG GGG GGG TTG ATT TGC AAC GAG CGC CAC ACC GAT CGG GAG CTC GAC CTC GCC GAG GCA CTT GCC GCC CGC

                  620┬                       640┬                       660┬                      680┬
    CTC AAT TCC AAG CTC ATC CAC TTC GTG CCG CGC GAC AAT ATC GTT CAG CAC GCA GAG CTC AGA AAG ATG ACA GTG ATC CAA

                  700┬                       720┬                       740┬                      760┬
    TAT GCG CCG AAC TCT AAG CAA GCC GGG GAA TAT CGC GCC CTG GCT GAA AAG ATC CAT GCA AAT TCC GGC CGA GGC ACC GTC

                  780┬                       800┬                       820┬                      840┬
    CCT ACA CCG ATC ACT ATG GAG GAA CTG GAG GAC ATG CTG CTC GAC TTT GGA ATC ATG AAG AGC GAC GAG CAG ATG CTT GCC

                  860┬                       880┬                       900┬                      920┬          940┬
    GAA CTC CAC GCC AAG GAA GCC AAG GTA ATA GCC CCC CAC TGA ATCGCCGCCGAGAGGGTGGCGCAGCTGGACGCGGCCGTGCCATTCCACAAAC

                  960┬                       980┬                       1000┬              1020┬
    GGCGCCATTGATGAGGTGTCACCTAGATCGTGAAATGGGCAGGCCCAA       ATG AGC CTC GAC TAT GAG AAT GAC AAT GCT TTG CAC GAG AAG

            1040┬                    1060┬                    1080┬                    1100┬
    CTT ATC GAA GAG GCT ATC GCA CTA TCC AGA CAA GGG GCG AAG CGC CGT AAA AAG CAC CTG AGT GTC GCA AAG AAC AAG CAG

            1120┬                    1140┬                    1160┬
    GAG ACC GCC GAG GAA GGA CAG GTC GTT TCC GAG TGC GAC GTA AAG TCG AAC ATC A--
```

FIG. 10.4 Nucleotide sequence of the Rm41 *nifH*-gene with the two flanking regions. The sequence (Török and Kondorosi 1981) is identical with the *nifH* mRNA and the triplets correspond to the amino acid sequences shown in Figs. 10.4 and 10.5. The putative Shine–Dalgarno sequence is boxed. The start-point for transcription was determined for the Rm1021 *nifH* promoter (Sundaresan *et al.* 1983) but, due to the nearly complete homology of the two sequences, it is very likely that the site indicated by the arrow is the transcription initiation startpoint in Rm41. The −35 and −10 sequences, homologous to the corresponding sequences of *K. pneumoniae nifH* promoter, are underlined. Horizontal arrows above the sequences indicate two sequences with a partial two-fold symmetry.

of *Anabaena* 7120 and of *K. pneumoniae* respectively. At the triplet codon level however, it shares only 27 per cent and 34 per cent homology with the above species, indicating that the amino-acid sequences are more conserved than the nucleotide sequences. This structural conservation of nitrogenase reductase is most probably related to function, and not

```
Rm:   H₂N-Met-Ala-[Ala]-Leu-Arg-Gln-Ile-Ala-Phe-Tyr-Gly-Lys-Gly-Gly-Ile-Gly-Lys-Ser-Thr-Thr-Ser-Gln-Asn-
Kp:        H₂N-Thr-[Met]-Arg-Gln-Cys-Ala-Ile-Ala-Phe-Tyr-Gly-Lys-Gly-Gly-Ile-Gly-Lys-Ser-Thr-Thr-Thr-Gln-Asn-
An:   H₂N-Met-Thr-Asp-Glu-Asn-[Ile]-Arg-Gln-Ile-Ala-Phe-Tyr-Gly-Lys-Gly-Gly-Ile-Gly-Lys-Ser-Thr-Thr-Ser-Gln-Asn-
Cp:        H₂N-Met-Arg-Gln-[Val]-Ala-Ile-Tyr-Gly-Lys-Gly-Gly-Ile-Gly-Lys-Ser-Thr-Thr-Gln-Asn-
Av:        H₂N-[Ala]-Met-Arg-Gln-Cys-Ala-Ile-Tyr-Gly-Lys-Gly-Gly-Ile-Gly-Lys-Ser-Thr-Thr-Thr-Gln-Asn-

Rm:   Thr-Leu-Ala-Ala-Leu-[Val-Asp-Leu]-Gly-Gln-Lys-Ile-[Leu]-Ile-Val-Gly-[Cys]-Asp-Pro-Lys-Ala-Asp-Ser-Thr-Arg-
Kp:   Leu-Val-Ala-Ala-Ala-Leu-Gly-Gln-Lys-Val-Met-Ile-Val-Gly-[Cys]-Asp-Pro-Lys-Ala-Asp-Ser-Thr-Arg-
An:   Thr-Leu-Ala-Ala-[Met]-Ala-Glu-Met-Gly-Gln-Arg-Ile-Met-Ile-Val-Gly-[Cys]-Asp-Pro-Lys-Ala-Asp-Ser-Thr-Arg-
Cp:   Leu-[Thr-Ser-Gly-[Leu]-His-Ala-Met-Lys-Lys]-Thr-Ile-Met-[Val]-Val-Gly-[Cys]-Asp-Pro-Lys-A.a-Asp-Ser-Thr-Arg-
Av:   Leu-Val-Ile-                                                Val-Met-Ile-Val-Gly-[Cys]-Asp-Pro-Lys-

Rm:   Leu-Ile-Leu-[Asn]-Ala-Lys-Ala-Gln-[Asp]-Thr-[Val-Leu-His-Leu-Ala-Ala]-Thr-Glu-Gly-Ser-Val-Glu-Asp-Leu-Glu-
Kp:   Leu-Ile-Leu-His-Ala-Lys-Ala-Gln-Asn-Thr-[Ile-Met-Glu-Met]-Ala-Ala-Glu-Val-Gly-Ser-Val-Glu-Asp-Leu-Glu-
An:   Leu-[Met]-Leu-His-[Ser]-Lys-Ala-Gln-Thr-Thr-Thr-Val-Leu-His-Leu-Ala-Ala-Glu-[Arg]-Gly-[Ala]-Val-Glu-Asp-Leu-Glu-
Cp:   Leu-Leu-Leu-[Gly-Gly-Leu]-Ala-Gln-Lys-Ser-Val-Leu-Asp-Thr-Leu-Arg-Glu-Glu-Gly------Gly-Asp-[Val]-Glu-

Rm:   Leu-Glu-Asp-Val-Leu-Lys-Val-Gly-Tyr-Arg-Gly-Ile-Lys-[Cys]-Val-Glu-Ser-Gly-Gly-Pro-Glu-Pro-Gly-Val-Gly-
Kp:   Leu-Glu-Asp-Val-Leu-Gln-Ile-Gly-Tyr-Gly-Asp-Val-Arg-Cys-Ala-Glu-Ser-Gly-Gly-Pro-Glu-Pro-Gly-Val-Gly-
An:   Leu-[His-Glu-Val]-Met-Leu-Thr-Gly-[Phe]-Arg-Gly-Val-Lys-[Cys]-Val-Glu-Ser-Gly-Gly-Pro-Glu-Pro-Gly-Val-Gly-
Cp:   Leu-Asp-Ser-Ile-Leu-Lys-Glu-Gly-Tyr-Gly-Gly-Ile-Arg-Cys-Val-Glu-Ser-Gly-Gly-Pro-Glu-Pro-Gly-Val-Gly-
Av:                                              [Cys]-Val-Glu-Ser-Gly-Gly-Pro-Glu-Pro-Gly-Val-Gly-

Rm:   [Cys]-Ala-Gly-Arg-Gly-Val-Ile-Thr-Ser-Ile-Asn-Phe-Leu-Glu-Glu-Asn-Gly-Ala-Tyr-
Kp:   [Cys]-Ala-Gly-Arg-Gly-Val-Ile-Thr-Ala-Ile-Asn-Phe-Leu-Glu-Glu-[Glu]-Gly-Ala-Tyr-Glu-Asp-Asp-Leu-Asp-Phe-
An:   [Cys]-Ala-Gly-Arg-Gly-Ile-Ile-Thr-Ala-Ile-Asn-Phe-Leu-Glu-Glu-Asn-Gly-Ala-Tyr-Glu-Asp-Asp-[---]-Leu-Asp-Phe-
Cp:   Ala-Gly-Arg-Gly-Ile-Ile-Thr-Ser-Ile-Asn-[Met]-Leu-Glu-Glu-Gly-Ile-Leu-Gly-Ala-Tyr-Thr-Asp-Asp-Leu-Asp-Tyr-
Av:   [Cys]-Ala-Gly-Arg-

Rm:   Val-Ser-Tyr-Asp-Val-Leu-Gly-Asp-Val-Val-[Cys]-Gly-Gly-Phe-Ala-Met-Pro-Ile-Arg-Glu-Asn-Lys-Ala-Gln-Glu-
Kr:   Val-Phe-Tyr-Asp-Val-Leu-Gly-Asp-Val-Val-[Cys]-Gly-Gly-Phe-Ala-Met-Pro-Ile-Arg-Glu-Asn-Lys-Ala-Gln-Glu-
An:   Val-Ser-Tyr-Asp-Val-Leu-Gly-Asp-Val-Val-[Cys]-Gly-Gly-Phe-Ala-Met-Pro-Ile-Arg-Glu-Gly-Lys-Ala-Gln-Glu-
Cp:   Val-Phe-Tyr-Asp-Val-Leu-Gly-Asp-Val-Val-[Cys]-Gly-Gly-Phe-Ala-Met-Pro-Ile-Arg-Glu-Gly-Lys-Ala-Gln-Glu-
Av:                        Asp-Val-Val-[Cys]-Gly-Gly-Phe-Ala-Met-Pro-Ile-Arg-

Rm:   Ile-Tyr-Ile-Val-[Met]-Ser-Gly-Glu-Met-Met-Ala-Leu-Tyr-Ala-Ala-Asn-Asn-Ile-Ala-Lys-Gly-Ile-Leu-Lys-Tyr-
Kp:   Ile-Tyr-Ile-Val-Cys-Ser-Gly-Glu-Met-Met-Ala-Met-Tyr-Ala-Ala-Asn-Asn-Ile-Ser-Lys-Gly-Ile-[Val]-Lys-Tyr-
An:   Ile-Tyr-Ile-Val-[Thr]-Ser-Gly-Glu-Met-Met-Ala-Met-Tyr-Ala-Ala-Asn-Asn-Ile-Ser-Lys-Gly-Ile-Leu-Lys-Tyr-
Cp:   Ile-Tyr-Ile-Val-[Ala]-Ser-Gly-Glu-Met-Met-Ala-Leu-Tyr-Ala-Ala-Asn-Asn-Ile-Ser-Lys-Gly-Ile-[Gln]-Lys-Tyr-
Av:   Ile-Tyr-Ile-Val-Cys-Ser-Gly-Glu-Glu-

Rm:   Ala-His-[Ala]-Gly-Gly-Val-Arg-Leu-Gly-Gly-Leu-Ile-[Cys]-[Asn]-Glu-Arg-His-Thr-Asp-Arg-Glu-Leu-Asp-[Leu]-Ala-
Kp:   Ala-Lys-Ser-Gly-[Lys]-Val-Arg-Leu-Gly-Gly-Leu-Ile-[Cys]-Asn-Ser-Arg-[Gln]-Thr-Asp-Arg-Glu-Asp-Glu-Leu-Ile-
An:   Ala-His-Ser-Gly-[Gly]-Val-Arg-Leu-Gly-Gly-Leu-Ile-Cys-Asn-Ser-Arg-Lys-Val-Asp-Gly-Asp-[Arg]-Glu-Leu-Ile-
Cp:   Ala-Lys-Ser-Gly-Gly-Val-Arg-Leu-Gly-Gly-Ile-Ile-Cys-Asn-Ser-Arg-Lys-Val-[Ala-Asn]-Glu-Tyr-Glu-Leu-Leu-
Av:                Leu-Gly-Gly-Leu-Ile-Cys-Asn-Ser-Arg-

Rm:   Glu-[Ala-Leu-Ala]-Ala-[Arg]-Leu-Asn-Ser-[Lys]-Leu-Ile-His-Phe-Val-Pro-Arg-Asp-Asn-Ile-Val-Gln-His-Ala-Glu-
Kp:   Ile-[Ala]-Glu-Ala-Glu-[Lys]-Leu-Gly-Thr-Gln-Met-Ile-His-Phe-Val-Pro-Arg-Asp-Asn-Ile-Val-Gln-[Arg]-Ala-Glu-
An:   Met-[Asn]-Leu-Ala-Glu-Arg-Leu-Gly-Thr-Gln-Ile-His-Phe-Val-Pro-Arg-Asp-Asn-Ile-Val-Gln-His-Ala-Glu-
Cp:   Asp-[Ala]-Phe-Ala-Lys-Glu-Leu-Gly-Ser-Gln-Ile-His-Phe-Val-Pro-Arg-Ser-Pro-Met-Val-Thr-Lys-Ala-Glu-

Rm:   Leu-Arg-Lys-Met-Thr-Val-Ile-[Gln]-Tyr-Ala-Pro-[Asn]-Ser-Lys-Gln-Ala-[Gly]-Glu-Tyr-Arg-Ala-Leu-Ala-Glu-Lys-
Kp:   Ile-Arg-Arg-Met-Thr-Val-Ile-Glu-Tyr-Asp-Pro-Ala-Cys-Lys-Gln-Ala-Asn-Glu-Tyr-Arg-[Thr]-Leu-Ala-Gln-Lys-
An:   Leu-Arg-Arg-Met-Thr-Val-Ile-Glu-Tyr-Asp-Pro-Asp-[Asn]-Ser-Gln-Ala-[Gly]-Glu-Tyr-Arg-Ala-Leu-Ala-Lys-Lys-
Cp:   Ile-[Asn]-Lys-[Gln]-Thr-Val-Ile-Glu-Tyr-Asp-Pro-Thr-[Cys]-Glu-Gln-Ala-Glu-Glu-Tyr-Arg-[Glu]-Leu-Ala-Arg-Lys-
Av:                    Tyr-Asp-Pro-Lys-Ala-Lys-Gln-Ala-Asp-Glu-

Rm:   [Ile]-His-[Ala-Asn]-Ser-Gly-Arg-Gly-[Thr-Val-Pro-Thr-Pro]-Ile-[Thr]-Met-Glu-Glu-Leu-Glu-Asp-Met-[Leu]-Leu-Asp-
Kp:   Ile-[Val]-Asn-Asn-Thr-Met-Lys-----Val-Val-Pro-Thr-Pro-[Cys]-Thr-Met-Asp-Glu-Leu-Glu-Glu-Ser-Leu-Leu-Met-Glu-
An:   Ile-[Asn----]-Asn-Asp-Lys-Leu-----Thr-Ile-Pro-Thr-Pro-Met-[Glu]-Met-Glu-Leu-Glu-Glu-Leu-Glu-[Ala]-Leu-Leu-Glu-
Cp:   Val-Asp-[Ala-Asn]-Glu-Leu-Phe-----Val-Ile-Pro-[Lys]-Pro-Met-Thr-[Gln]-Glu-Arg-Leu-Glu-Glu-[Ile]-Leu-Met-[Gln]-
Av:                                     Glu-Leu-Glu-Leu-Leu-Leu-Glu-

Rm:   Phe-Gly-Ile-Met-Lys-Ser-[Asp]-Glu-Gln-Met-Leu-Ala-Glu-Leu-His-Ala-Lys-----[Gly-Ala]-Lys-Val-Ile-[Ala]-Pro-
Kp:   Phe-Gly-Ile-Met-Glu-[Glu]-Glu-Glu-Asp-[Asp]----Ile-Ile-Gly-Lys-Thr-Ala-[Ala]-Glu-[Glu]-Asn-[Ala-Ala]-COOH
An:   Tyr-Gly-Leu-[Leu]-Asp-Asp-Asp-[Thr]-Lys-His-Ser-Glu-Ile-Ile-`ly-Lys-[Pro]-Ala-Glu-Ala-Thr-[Asn]-Arg-Ser-Cys-
Cp:   Tyr-Gly-Leu-Met-Asp-Leu-COOH
Av:   Phe-Gly-Ile-Met-Glu-Val------Glu-Ser-[Ile]-Val-Gly-Lys-Thr-Ala-Glu-Glu-Val-COOH

Rm:   His-COOH
An:   Arg-Asn-COOH
```

FIG. 10.5 Amino acid sequences of the nitrogenase reductase from different nitrogen-fixing organisms. *R. meliloti*: Rm (Török and Kondorosi, 1981); *K. pneumoniae*: Kp (Scott *et al.* 1981; Sundaresan and Ausubel 1981); *Anabaena* 7120: An (Mevarech *et al.* 1980); *C. pasteurianum*: Cp (Tanaka *et al.* 1977); *Azotobacter vinelandii*: Av (Hausinger and Howard 1980). Those amino acid residues which occur more than once at a given position are enclosed in boxes. The five conserved cysteine residues are hatched.

merely to the late evolution of the *nifH*-genes, as suggested by Postgate (1974). When the *nifH*-gene from Rm41 was placed behind a strong *E. coli* promoter, a polypeptide of about 33 kdal was detected in *E. coli* minicells (Z. Svab and A. Kondorosi, unpublished). This size is in fairly good agreement with that of calculated from the sequence data.

```
Rm:
Kp:
An: H₂N(Met-Arg-Ser-Asn-Ser-Ser-Ser-His-Ser-Pro-Phe-Pro-Thr-Pro-His-Ser-Pro-Lys-Tyr-Thr-Ser-Ile-Pro-Pro-Phe-
Cp:
Av:

Rm:                                                           H₂N-Met-Ser┤Leu├Asp-Tyr┤Glu-Asn├Asp-Asn┤Ala-Leu┤His-
Kp:                                               H₂N-Met-Met┤Thr-Asn┤Aln-Thr-Gly┤Glu┤Arg-Asn-Leu┤Ala-Leu-Ile-
An: Val-Arg-Val-Thr-Glu-Ala-Asp)Met-Thr-Pro-Pro-Glu-Asn-Lys┤Asn-Leu┤Val-Asp┤Glu-Asn┤Lys┤Glu┤---┤Leu-Ile-
Cp:                                                                             Ser┤Glu┤Asn┤Leu┤Lys-
Av:                                                           H₂N┤Thr┤Gly-Met-Ser-Arg┤Glu┤Glu-Val┤Glu┤Ser┤Leu-Ile-

Rm: ---┤Glu┤Lys┤Leu┤Ile┤Glu┤Glu-Ala-Ala-Leu-Leu-Ser-Arg-Gln-Gly-Ala┤Lys┤Arg┤Arg┤Lys┤Lys-His┤Leu-Ser┤Val┤
Kp: Gln-Glu-Val-Leu┤----┤Glu-Val┤Thr----┤Pro┤Glu┤Thr┤--------Ala-Arg┤Lys┤Glu┤Arg┤Arg┤Lys┤His┤Met-Met┤Val┤
An: Gln-Glu-Val-Leu┤
Cp: Asp┤Glu┤Ile┤Leu┤---┤Glu┤Lys┤Tyr┤Ile┤Pro┤Lys┤Thr┤-----------Lys┤Lys┤Thr┤Arg┤Ser-Gly┤His┤Ile----┤Val┤
Av: Gln-Glu-Val-Leu┤---┤Glu-Val-Tyr┤

Rm: Ala┤Lys┤Asn┤Lys┤Gln┤Glu-Glu-Thr┤Ala-Glu-Glu-Gly-Gln-Val┤Val-Ser┤Glu-Cys-Asp----┤Val┤Lys-Ser-Asn┤Ile┤
Kp: Ser┤Asp-Pro┤Lys┤Met-Val-Val-Ser┤Val-Gly-Lys-Cys-Ile┤Ile┤----┤Ser-Asn┤Arg┤Lys┤Ser-Gln┤Pro-Gly┤Val-Met┤
An:
Cp: Ile┤Lys┤Thr-----Glu┤Glu-Glu-Thr┤Pro-Asn-Pro-----Glu┤Ile-Val┤Ala┤Asn┤Thr┤Lys┤Thr┤Val-Pro-Gly┤Ile┤Ile┤
Av:
```

FIG. 10.6 Partial amino-acid sequences for the *nifD*-gene product. The symbols are the same as in Fig. 10.5. The Rm sequence was deduced from data for Rm41 (Török and Kondorosi, 1981); Kp (Scott *et al.* 1981; Sundaresan and Ausubel 1981); An (Mevarech *et al.* 1980); Cp (Hase *et al.* 1981); Av (Lundell and Howard 1978). (After Raschke and Eberhard, thesis, Erlangen)

From the sequence data, a potential Shine–Dalgarno ribosome binding sequence (Shine and Dalgarno 1975) was found at position -6 to -10 from the translation initiation codon. Although the 3'-end of the *R. meliloti* 16S RNA has not been determined, it is very likely that the ribosome binding-sequences for *E. coli* and *R. meliloti* are fairly similar.

Combined genetic and physical analysis of the *nifH-, D-,* and *K*-genes revealed that the organization of these genes is essentially the same as in *K. pneumoniae*. It was shown that in *R. meliloti* the *nifHDK*-genes also form one transcriptional unit (Ruvkun *et al.* 1982*b* Corbin *et al.* 1983). Using site-directed Tn5 mutagenesis, it was found that insertions in a 6.3 kb region comprising the *nifHDK*-genes all resulted in Fix⁻ phenotype. Complementation analysis between genomic and pRK290-carried *nif*::Tn5 insertions indicated that these genes belong to one transcriptional unit, where transcription starts from the right of the *nifH*-gene. This 6.3 kb DNA region may code for another protein of about 30 kdal in size. In *K. pneumoniae*, four genes—*nifHDK* and *Y*—form one transcriptional unit, coding for polypeptides of 35, 60, 60, and 24 kdal (Pühler and Klipp 1981).

Weber and Pühler (1982) used the *E. coli* minicell system to correlate the physical map with the protein map of this DNA region. They applied the technique developed for *K. pneumoniae nif* genes for the same purpose (Pühler and Klipp 1981). Various *R. meliloti* 1021 DNA fragments carrying the *nif*-genes were placed after strong promoters of different *E. coli* vector plasmids. The proteins synthesized in *E. coli* minicells were radioactively labelled and then detected via autoradiography on protein gels. The *nifH*-gene product was identified and mapped by Tn5 mutagenesis. These data fitted well with the DNA-sequencing results discussed above. In contrast, the *nifD-* and *K*-gene products were not produced from the same transcription unit, indicating that the expression of the *R. meliloti nifHDK* operon is somewhat different in *E. coli*.

The structure and control of the *nifH* promoter region of *R. meliloti* and of *K. pneumoniae* show striking similarities. The promoter region of Rm1021 *nifH* was identified by mapping the *in vivo* start points of transcription (Sundaresan *et al.* 1983*a*). Based on genetic data, it was already clear that the DNA region upstream from the N-terminal coding region of *nifH* should contain the promoter sequences. Indeed, when this DNA was probed with RNA isolated from bacteroids, followed by S_1 digestion, a *nif*-specific transcript starting 69 nucleotides upstream from the ATG start codon was identified.

The nucleotide sequence determined upstream from the ATG translational start point is the same in Rm1021 (Sundaresan *et al.* 1983*a*) as in Rm41 (Török and Kondorosi 1981). Consequently, the controlling sequences should also be the same (Fig. 10.4).

The promoter region of the *nifH*-gene did not exhibit strong homology with consensus promoter sequences of *E. coli*. Using *E. coli* RNA polymerase and purified DNA fragments as templates, no significant transcription could be detected (I. Török, unpublished; Sundaresan *et al.* 1983*a*). The same was found for the *K. pneumoniae* promoter region (Sundaresan *et al.* 1983*a*). When the *nifH* sequences upstream from the transcriptional start site in the two species were compared, considerable homology was observed. There is about 50 per cent homology for a 40 bp region upstream from the transcription start point. Certain regions show exact homology (Fig. 10.4). The different *K. pneumoniae* *nif* promoters have a similar structure, characterized by the consensus sequence 5'-TTGCA-3' at the -15 to -10 region (Ow *et al.* 1983; Sundaresan *et al.* 1983*b*) and the sequence 5'-CTGG-3' at -26 to -23 (Beynon *et al.* 1983) and this is conserved in the *R. meliloti nif* promoters as well (Sundaresan *et al.* 1983*b*; Better *et al.* 1983). Interestingly, the *xyl*ABC operon of *Pseudomonas putida*, coding for the degradation of toluene has promoter sequences nearly identical with these *nif* promoters (Inouye *et al.* 1984). As suggested by Johnston and Downie (1984) the *nif* regulon system might be one member of a larger family of regulons in gram-negative bacteria. Since both the *nif* and *xyl*ABC are positively regulated, it is possible that these consensus promoter sequences are characteristic for certain positively controlled regulons. It is not known which sequences, if any, are characteristic only for the *nif* promoters.

The regulation of *R. meliloti nifH* promoter is also similar to that of *K. pneumoniae*. Fusions of each of the two promoters to the *lacZ*-gene from *E. coli* were constructed and both promoters were shown to be activated by the *K. pneumoniae nifA* gene (Sundaresan *et al.* 1983*a*).

Moreover, a multicopy plasmid, containing the *R. meliloti nifH* promoter inhibits N_2-fixation in *K. pneumoniae*, as do multicopy plasmids containing the *K. pneumoniae nifH* promoter (Riedel *et al.* 1979). It was suggested that due to binding of the *nifA* products to the *nifH* promoters from either *K. pneumoniae* or from *R. meliloti*, the higher copy number of *nifH* promoter titrates the *nifA* products (Sundaresan *et al.* 1983*a*). In contrast to the *K. pneumoniae nif*H promoter, the *R. meliloti nif*H promoter, however, could also be activated by the product of the *E. coli ntrC* gene which positively controls the expression of a variety of nitrogen assimilatory genes (Sundaresan *et al.* 1983*a*; Szeto *et al.* 1984). Based on the above results, Sundaresan *et al.* (1983*a*) suggested that a regulatory protein similar to *K. pneumoniae nifA* should be present in *R. meliloti*. Since *R. meliloti* does not derepress its *nif* structural genes *ex planta*, this regulatory protein must be under some kind of symbiotic control.

The *R. meliloti nifA*-type gene was recently found on a 1.8 kb region located 5.5 kb upstream of the *nifH* promoter (Zimmerman *et al.* 1983; Szeto *et al.* 1984). Fix⁻ Tn5 insertion mutants mapping in this region induced such nodules which did not accumulate *nifHDK* mRNA or the nitrogenase polypeptides. Another 4.0 kb symbiotic gene region, located 1.8 kb from the *nifH* promoter and containing the consensus *nif* promoter (P2 promoter, Corbin *et al.* 1983), was also not transcribed in these nodules (Szeto *et al.* 1984). Szeto *et al.* (1984) demonstrated in a series of elegant experiments that this *nif* regulatory gene of *R. meliloti* is related both in structure and function to the *K. pneumoniae nifA* and to the *E. coli ntrC* genes. The *R. meliloti nif* regulatory gene hybridized to *E. coli ntrC* DNA and somewhat weaker hybridization with the *K. pneumoniae nifA* DNA was also detected. The *nifH* and the P2 promoters could be activated in the free-living *R. meliloti* carrying such plasmids from which either the *K. pneumoniae nifA* or the *E. coli ntrC gene products were produced constitutively. Szeto et al.* (1984) suggested that the *R. meliloti* regulatory gene is more like *nifA* rather than *ntrC*, since this gene is located on the symbiotic megaplasmid, closely linked to the *nifHDK* operon. Moreover, the regulatory mutant was not impaired in the utilization of other nitrogenous compounds; (these metabolic pathways are controlled by the *ntrC* gene in enterobacteria and probably also in *Rhizobium*).

In Rm102F34, a 5–6 kb transcript for the *nif* structural genes was identified by Northern blot analysis (Corbin *et al.* 1983). As supported by S_1 nuclease mapping it has the same polarity as in Rm1021. It is interesting to note that Ruvkun *et al.* (1982*b*) and Corbin *et al.* (1983) have found Tn5 insertions between *nifH* and *nifD* with a Fix⁺ phenotype. It is suggested that this might be due to either a weak secondary promoter in the intergenic region or that transcription from the termini of Tn5 insertions may hve occurred (Corbin *et al.* 1983).

The organization of nitrogen fixation genes in *R. meliloti* shows similarities to those found for *K. pneumoniae*. In this latter organism, 17 *nif*-genes are highly clustered in seven or eight operons and reside on a 24 kb DNA segment (MacNeil *et al.* 1978; Cannon *et al.* 1979; Merrick *et al.* 1980). In *R. meliloti, fix*-genes have been localized in the vicinity of the *nif* structural genes (Corbin *et al.* 1982; Ruvkun *et al.* 1982*b*; Banfalvi *et al.* 1983). Performing directed Tn5 mutagenesis on a 18 kb region adjacent to, and surrounding the *nif* structural genes of Rm1021, a total of 31 Tn5 insertions were constructed and assayed for Fix phenotype. Fix⁻ mutations were found in two clusters: one is the 6.3 kb *nifHDK* operon and another of at least 5 kb in size. They are separated by a 1.6 kb region, on which no symbiotic genes were identified. Complementation analysis showed that this second cluster forms two transcriptional units (Ruvkun *et al.* 1982*b*) (Fig. 10.3). Three *fix*-genes forming one transcriptional unit code for proteins of 30 000, 37 000, and 43 000 kdal, respectively, as demonstrated in experiments using *E. coli* minicells (Pühler *et al.* 1984). The other transcriptional unit contains *nifA*.

In Rm102F34, a 35 kb DNA region, including the *nif* structural genes, was subjected to Tn5 mutagenesis. The *fix* (probably *nif*) genes were localized within a 14–15 kb region comprising the *nif* structural genes. No symbiotic genes were detected on the neighbouring 10 kb region on both sides of the 14 kb *nif* gene cluster. An inessential 1.9 kb region to the right of the *nifH*-gene was also detected. It seems that in *R. meliloti* the number of *nif*-genes is either lower than in *K. pneumoniae* or some *nif*-genes are located in another DNA region.

As discussed above, transcription of the *nifHDK* genes proceeds from *nifH* to the left (Fig. 10.3). The direction of transcription for another transcript with opposite direction was also identified in Rm102F34 (Corbin *et al.* 1983). Apart from the *nifHDK*- and the *nifA* genes, the functions of the other *nif* genes have not yet been identified.

These *nif*-genes are specifically expressed in the nodules and do not express in the free-living bacterium (Corbin *et al.* 1982; Kondorosi *et al.* 1983*a*). When RNA isolated from alfalfa nodules was hybridized to this cloned DNA region, strong hybridization of the DNA fragments carrying *nif*-genes was obtained. In contrast, no hybridization with RNA isolated from vegetatively grown bacteria was detected. The neighbouring DNA regions where no *fix*-genes were found did not hybridize with nodule RNA (Kondorosi *et al.* 1983*c*).

The relative rate of transcription of the *nif* structural genes in *R. meliloti* during symbiotic development was studied by Paau and Brill (1982). They found that *nif*-specific transcripts first appear at a very early stage of symbiosis, when bacteria are still in the infection thread. As expected, the transcripts are most abundant, however, in the mature bacteroids of young nodules. In senescent root-nodules their level is much lower.

There is no indication that the nitrogenase genes are amplified or rearranged during symbiosis (Paau and Brill 1982). Although bacteroids have a higher DNA content (Paau *et al.* 1979), this is probably not due to specific rearrangement of symbiotic genes. The higher DNA content of bacteroids can be explained by their lack of cell division.

Nitrogen fixation in *K. pneumoniae* is positively controlled (either directly or indirectly) by the accumulation of guanosine tetraphosphate, ppGpp (Riesenberg *et al.* 1982). In *R. meliloti*, however, ppGpp is not accumulated during amino acid starvation (Belitsky and Kari 1982). Moreover, RNA accumulation stops when protein synthesis is halted and this process is not regulated by ppGpp. It seems that the metabolism of ppGpp is different from that observed in enteric bacteria (Belitsky and Kari 1982). It will be interesting to find out whether this unusual ppGpp metabolism has any bearing to the inability of *R. meliloti* to fix nitrogen in the free-living state.

10.7.2 The *nod*-genes

The process of nodulation of leguminous plants by rhizobia consists of several steps, such as colonization of root hairs, adhesion of bacteria to the root hairs, curling and branching of root hairs, infection thread formation, development of plant meristematic cells, bacterial release, conversion of bacteria to bacteroids, etc. (see reviews by Broughton 1978; Vincent 1980). It is likely that several bacterial genes are involved in nodule development. Here the operational umbrella term '*nod*-genes' for those genes which are involved in nodule formation will be used. In contrast, '*fix*-genes' will be referred to as those contributing to the late development of N_2-fixing symbioses and to the functioning and maintenance of nitrogen fixation.

Our knowledge about the *nod*-genes is still very scanty, emerging mainly from studies on *Rhizobium* mutants defective in nodule formation (Nod⁻ mutants).

As discussed in Section 10.4, *R. meliloti* mutants defective in nodulation of alfalfa have been reported by several laboratories. The characterization of these mutants was mostly limited to light or electron-microscopic analyses, however. Some of the Nod⁻ mutants characterized were unable to evoke root-hair curling (Hac⁻) or were 'non-reactive' mutants (Hirsch *et al.* 1982; Forrai *et al.* 1983), not penetrating the plant cells. In contrast, the 'reactive' mutants induce root-hair curling and enter the epidermal cells although no infection threads may be observed. These Nod⁻ mutants caused proliferation and hypertrophization of root hairs. A limited number of ineffective nodules were also induced. Paau *et al.* (1981) reported that this mutant type did not attach to root hairs and did not bind to purified alfalfa agglutinin. Hirsch *et al.* (1982) found that the mode of entry of these bacteria into plant cells was different from that of the wild-type *R. meliloti* and necrotic symptoms on the infected root tissue were observed.

Nodulation genes defined by Tn5-induced Hac⁻ mutations were cloned by direct complementation of Nod⁻ mutants (Long *et al.* 1982). Since the host plant is able to select rare Nod⁺ *R. meliloti* cells from many Nod⁻ bacteria, the Nod⁺ phenotype restored by the cloned wild-type *nod* allele could be identified (Brewin *et al.* 1980). A clone bank of Rm1021 was prepared by the use of the wide-host-range vector pLAFR1, and a population of the recombinant plasmids was mass-conjugated into Nod⁻ mutants. *Medicago* seedlings were inoculated with batches of 200–300 transconjugants. Some batches induced nodules. Bacteria were re-isolated from the nodules and in this way a recombinant plasmid (pRmSL26) carrying the wild allele of the mutation complemented was identified. Plasmid pRmSL26 restored the ability of two Nod⁻ mutants to curl root-hairs and form nodules. Both Nod⁻ mutants had an insertion in an 8.7 kb fragment which mapped 20 kb from the *nif* structural genes (Fig. 10.2).

In Rm41 the same *nod* region was identified by deletion mapping. The extent of deletions in various *Nod⁻* and *Fix⁻* deletion mutants was determined by the use of cosmid clones overlapping a 135 kb region of pRme41b shown in Fig. 10.2 (Kondorosi *et al.* 1983b, 1984b). The cosmid clones were labelled with ³²P and hybridized with total DNA isolated from numerous mutants. It was found that all deletions started from the left end fragment or from outside of the mapped region but the extent of deletion towards right were different. In all Fix⁻ deletion mutants, the deletion ended at different points before an 8.5 kb *Eco*RI fragment. Comparison of the physical maps of the *nod-nif* regions from Rm1021 and Rm41 indicated that this 8.5 kb *Eco*RI fragment of Rm41 is the equivalent to the 8.7 kb *Eco*RI *nod* fragment of Rm1021 (Fig. 10.2). In some mutants the end point was in its left neighbouring *Eco*RI fragment. In all Nod⁻ deletion mutants, the 8.5 kb fragment was missing. The shortest deletions extended about 15 kb further, (AK635), indicating that at least one *nod* gene was present on this DNA region. This region was recloned into pRK290 (pKSK5), transferred into Nod⁻ recipients and it was found that the 8.5 kb *Eco*RI fragment restored the Nod⁺ Fix⁺ phenotype to the mutant with the shortest deletion. Three Nod⁻ point mutations were also suppressed by this recombinant plasmid (Kondorosi *et al.* 1983c; and unpublished results from the author's laboratory). Banfalvi *et al.* (1981) reported that these three Nod⁻ point mutations and the Nod⁻ phenotype of AK635 were suppressed upon the introduction of pJB5JI, a *sym*-plasmid of *R. leguminosarcum* (Beringer *et al.* 1978).

The fact that *nod* genes on the 8.5 kb fragment can be substituted by *R. leguminosarum nod*-genes, indicates that these genes code for non-plant specific nodulation functions. It was found that these Nod⁻ mutants are complementable upon introduction of the symbiotic plasmid of *R. trifolii* (Kondorosi *et al.* 1984*b*) and of nodulation plasmids from other *Rhizobium* species as well (Kondorosi *et al.* 1985*a*). Therefore, the *nod*-genes on the 8.5 kb *Eco*RI fragments were designated as 'common' *nod*-genes (Kondorosi et al. 1984*b*).

The complemented deletion mutants and two point mutants are Hac⁻ 'non-reactive' types (Forrai *et al.* 1983). It seems that this nodulation function is common for various rhizobia. The 8.5 kb fragment was shown to hybridize to DNA from several other *Rhizobium* species (Pankhurst *et al.* 1983; Kondorosi *et al.* 1984*a*; E. Kondorosi, A. Kondorosi, A. Szalay, and R. Hadley, unpublished). In this way, nodulation genes from these *Rhizobium* species can be identified and cloned. Recently results from other laboratories supported the existence of 'common' *nod*-genes (Djordjevic *et al.* 1983; Pühler *et al.* 1984; Long *et al.* 1985).

In the Nod⁻ mutant ZB157, the delection was about 3 kb larger than in AK635: it ended in a 6.8 kb *Eco*RI fragment. When its Nod⁻ phenotype was suppressed either by the 8.5 kb fragment or by pJB5JI, the nodules were ineffective. It is suggested that in this mutant a *fix* gene was also deleted (Kondorosi *et al.* 1984*b*).

In Nod⁻ mutants with deletions extending up to the *nif* structural genes, the Nod⁻ phenotype was not suppressed by pJB5JI. In contrast, a *nod–nif* R-prime pGR3 (Banfalvi *et al.* 1983) containing the region from the last-but-one left fragment adjacent to the fragment carrying *nifH*, restored the Nod⁺ phenotype in Nod⁻ mutants with very large deletions. Moreover, *A. tumefaciens* carrying pGR3-induced nodules on alfalfa, indicating that all essential *nod*-genes coding for early nodulation functions are present on this 80 kb region (Banfalvi *et al.* 1983). From this data a second *nod* region, coding also for host-specificity of nodulation, may be postulated. This gene)s) may reside somewhere between the 8.5 kb fragment and the *nif*-genes. This is in line with the recent finding that a Nod⁻ mutant obtained after random mutagenesis, mapped on a 6.8 kb *Eco*RI fragment.

Recently, directed Tn5 mutagenesis of the 8.5 and 6.8 kb *nod* fragments was applied to obtain a large collection of Nod⁻ mutants (Kondorosi *et al.* 1983*c*, 1984*b*). First, based on the analysis of 17 Tn5 insertions on the 8.5 kb *Eco*RI fragment, a rough correlated physical-genetic map was established (Kondorosi *et al.* 1984*b*); then by mapping 61 new Tn5 insertions, a contiguous, 3.0 kb long *nod* region was delimited (Török *et al.* 1984). At the same time an about 1 kb *fix* region was also revealed on the 8.5 kb fragment. On the 6.8 kb *Eco*RI fragment the symbiotic phenotype of 20 different Tn5 insertions was determined, which resulted in the delimitation of a 2.0–2.5 and a 1.0 kb long *nod* regions (Kondorosi *et al.* 1984*b*). More recent studies indicate the existence of at least one more *nod* region (E. Kondorosi and B. Horvath, personal communication).

The protein coding regions of the 8.5 kb *Eco*RI fragment were determined using *E.coli* minicells and an *in vitro* transcription/translation system from *E. coli* (Schmidt *et al.* 1984). Since proteins coded by the *nod* gene region did not express from their own promoter, subfragments from the 8.5 kb fragment were cloned after strong *E. coli* promoters and the protein coding regions were mapped by analysing deletion and Tn5 insertions mutants. In this way three polypeptide chains (23 000, 28 500, and 44 000 kdal) were localized on the 'common' *nod* region. Recently an *in vitro* transcription/translation system was worked out with *R. meliloti* extracts. Using this system the results were the same as obtained with the heterologous *E.coli* system (Dusha *et al.* in preparation). The nucleotide sequence of this DNA segment was determined by Török *et al.* (1984). The sequence data indicated three large open reading frames with the same polarity, coding for three proteins of 196, 217, and 402 (or 426, depending on the first ATG) amino acid residues, respectively. The location, polarity, and size of these proteins were in agreement with the values obtained in the expression studies of Schmidt *et al.* (1984). On this basis the three 'common' *nod* genes were designated as *nod*A, *B*, and *C*.

The nucleotide sequence of three *nod* genes from *R. leguminosarum* was determined by Rossen *et al.* (1984). Comparison of the *R. meliloti* *nod*ABC nucleotide and amino acid sequences shows 69–72 per cent homology, confirming the high degree of conservation of 'common' *nod* genes in these *Rhizobium* species. The nucleotide sequence of the *nod*C gene of Rm1021 was also determined (S. Long and T. Jacobs, personal communication) and, except for two base pair changes, the sequences are identical for the two strains. The *nod*C gene nucleotide sequence from a broad-host-range *Rhizobium* MPIK3030 again shares about 70 per cent homology with that of Rm41 (Bachem *et al.* 1985).

As mentioned above, the 6.8 kb *Eco*RI fragment contains at least three *nod* genes. Mutations in two *nod* regions result in clear Nod⁻ phenotype, although sparse curling of root hairs was observed and after 6–8 weeks infrequent nodulation occurred (Kondorosi *et al.* 1985*b*). Mutations in the third region caused 1–3 weeks delay of nodulation.

Two types of Nod⁻ mutants mapped on the 6.8 kb fragment were not complementable to Nod⁺ (tested on alfalfa) upon introduction of the *R. leguminosarum nod* plasmid (Kondorosi *et al.* 1984*a,b*). Moreover, using the cloned *nod* region of *R. leguminosarum* as hybridization probe, no homology was detected with the 1 kb region and weak homology was observed with the other region (Kondorosi *et al.* 1984*a*).

When plasmid pEK10 was introduced into some other *Rhizobium* species, the transconjugants induced nodule formation on *Medicago sativa* at low frequency. These results indicated that the 6.8 kb fragment carries genes coding for host specificity of nodulation (*hsn* genes; Kondorosi *et al.* 1984*b*).

Recently a recombinant plasmid (pPP346), made in pLAFR1 carrying a 24 kb megaplasmid insert, comprising the 8.5 kb and the 6.8 kb fragments was identified. Upon introduction of pPP346 into Nod⁻ mutants with very large megaplasmid deletions the Nod⁺ phenotype was restored (Kondorosi *et al.* 1984*a*). It was shown that *A. tumefaciens* carrying pPP346 was able to nodulate *Medicago* as efficiently as *A. tumefaciens* containing the *nod* R-prime pGR3, indicating that genes determining the essential *nod* genes from Rm41 are located on pPP346 (Kondorosi *et al.* 1985*d*). Nodulation by *Agrobacterium* transconjugants carrying either pKSK5 or pEK10 were not observed at detectable frequency. For strain 1021 somewhat different results were obtained by Hirsch *et al.* (1984) and Long *et al.* (1985). *Agrobacterium* transconjugants carrying pRmSL26 formed nodules on alfalfa and low frequency of nodulation was observed with transconjugants carrying the 8.7 kb*nod* fragment. On the other hand, the most efficient nodulation was found when the transconjugants contained not only the 'common' *nod* region but also a DNA region between the 'common' *nod* and *nif* genes which may correspond to the *hsn* region of Rm41 (Long *et al.* 1985).

The data discussed above indicate that the essential early nodulation genes on the *R. meliloti* megaplasmid are organized in two regions: the 'common' *nod* and the *hsn* region. Preliminary results suggests that a megaplasmid region, 25 kb away from the *nif* structural genes and opposite to the 'common' and *hsn nod* genes, is also needed for efficient nodulation (Kondorosi *et al.* 1985*c*).

The control of expression of nodulation genes is still not understood and efforts to find conditions where the *nod* genes are derepressed have failed (Long *et al.* 1985, Kondorosi *et al.* 1985*a*). The biochemical function of the *nod* gene products are still to be determined.

Little is known about genes involved in the later steps of nodule development, such as growth of infection threads into the inner part of the nodule, bacterial release and conversion of bacteria to bacteroids. Our techniques for the study of these processes are even more limited. Since mutations in genes controlling these processes give rise to Nod⁺ Fix⁻ phenotypes, such mutants are at present classified as Fix⁻. Recently, *R. meliloti* mutants inducing abnormal nodules were isolated

(E. Signer *et al.*; A. Pühler *et al.*, personal communications). In these nodules the meristematic regions were normal, however, no infection threads and shepherd's crook form were observed (A. Hirsch, personal communication). These nodules seem to be similar to those induced by *A. tumefaciens* transconjugants carrying *R. meliloti nod* genes (Wong *et al.* 1983; Truchet *et al.* 1984; Hirsch *et al.* 1984).

The mutants were deficient in exopolysaccharide production, had an altered phage sensitivity, and antibody reactivity (Finan *et al.* 1985). This again suggests, that there are genes outside the p*sym* megaplasmid which are involved in the process of nodule development.

An important property of rhizobia in nodulation is the ability to compete with other *Rhizobium* strains. For instance, the success of the inoculant strains to nodulate the host plant depends on its competitiveness against indigenous rhizobia as well as other micro-organisms already present in the soil. Behavioural mutants of *R. meliloti* altered in their mobility have been isolated (Ames *et al.* 1980; Ames and Bergman 1981; Soby and Bergman 1983). The motile strains had selective advantage in nodule formation over nonmotile mutants. Mobility seems to be a useful property of *Rhizobium* strains as inoculants.

10.7.3 The *fix*-genes

Genetic analyses of *R. meliloti* suggest that the number of genes controlling the ability of rhizobia to *fix* nitrogen within a nodule (*fix*-genes) is much higher than the number of *nod*-genes. With the exception of *nif*-gencs, however, it has not been proven whether most *fix*-genes are specifically involved in symbiotic nitrogen fixation or whether some of these genes may also express in free-living bacteria. In this section those *fix*-genes are discussed which are located outside the *nif*-gene cluster.

Several Fix⁻ mutants were isolated which grew on poor carbon and nitrogen sources without added amino acids or bases and mapped onto the chromosome (Forrai *et al.* 1983). Although these mutants did not exhibit an altered phenotype in free-living conditions, they may still have defects in certain metabolic pathways.

Fix⁻ mutations have been located on the megaplasmid outside the *nif* gene cluster (Kondorosi *et al.* 1983*c*; Z. Banfalvi unpublished; P. Boistard, personal communication) but these mutants have not yet been characterized.

Auxotrophic and antibiotic resistant mutants exerting a pleiotropic Fix⁻ phenotype have been described (recently reviewed by Kuykendall (1981)). Such mutants may help to understand biochemical processes occuring during symbiotic development. A block in the release of bacteria from the infection threads into plant cells in nodules infected by a Leu⁻ mutant was observed by Truchet *et al.* (1980). This could be overcome by the addition of L-leucine. For most mutants however, reversion studies have not been performed to prove the pleiotropy.

Mutants defective in haem biosynthesis are particlularly interesting, since haem is a constituent of leghaemoglobin, essential for symbiotic nitrogen fixation. Mutants defective in δ-aminolaevulinic acid synthetase, the first specific enzyme in the haem pathway have been isolated (Leong *et al.* 1982). The wild-type gene was cloned and this gene weakly suppressed an *E. coli haemA* mutation. Full restoration of Haem$^+$ phenotype was observed when this fragment was sub-cloned and was under *E. coli* promoter control. A temperature sensitive mutant with temperature sensitive enzyme activity *in vitro* was isolated. Since this mutation was in the cloned gene, this gene is the structural gene of δ-aminolaevulinic acid synthetase. Using site-directed mutagenesis, the exact location of the *haemA* gene was determined on the cloned fragment. The Haem$^-$ mutants were able to nodulate alfalfa only poorly, and the nodules were white and did not *fix* N$_2$. Upon addition of δ-aminolaevulinic acid, a partial complementation was achieved. This suggests that the plant does not provide δ-aminolaevulinic acid to the bacteroids; this is in line with earlier suggestions (Godfrey and Dilworth 1971).

Biochemical mutants defective in ammonia assimilation (Kondorosi *et al.* 1977*b*; Osburne 1982), in carbohydrate metabolism (Arias *et al.* 1979; Duncan and Fraenkel 1979; Gardiol *et al.* 1980, 1982), in nitrate reduction (Kondorosi *et al.* 1973; Kiss *et al.* 1979) in glutamine synthetase (Somerville and Kahn 1983; De Bruijn *et al.* 1984) are now available. Their study may shed light on the role of these pathways during symbiosis.

10.8 CONCLUSIONS

In the past few years there has been great interest in the molecular genetics of nitrogen fixation and the advances in this field have been dramatic. In particular, investigation into the *R. meliloti* system have contributed considerably to this progress. The development of sophisticated techniques in various fields of molecular genetics and biochemistry have undoubtedly been a great asset.

It has been established that the majority of fast-growing rhizobia contain plasmids of various sizes some of which code for information, important in the symbiosis. In *R. meliloti*, for example, not only *nif-* but also *nod-* and *fix-*genes have been identified, all mapping in a relatively small region of this extremely large plasmid. The other regions of the plasmid as well as the functions carried by the other indigenous plasmids remained largely obscure until now (1984).

Although it may be expected that most of the genes involved in symbiosis will be genetically and physically mapped and characterized in the near future, the products of these genes and their metabolic functions will need extensive study.

The genetic components contributed by the plant and their role during symbiosis is as yet unclear. Consequently, this side of the interaction will have to be the focus of more detailed research in the future.

ACKNOWLEDGEMENTS

My thanks are due to C. Bachem and G. B. Kiss for correcting the manuscript.

REFERENCES

ADACHI, T., HOOPER, I., and IYER, V. N. (1983). *Can. J. Microbiol.* **29**, 1601-6.

ALI, H., BECHET, M., NIEL, C., GUILLAUME, J.-B. (1984). *Can. J. Microbiol.* **30**, 507-11.

AMES, P., SCHLUEDERBERG, S. A., and BERGMAN, K. (1980). *J. Bact.* **141**, 722-7.

— and BERGMAN, K. (1981). *J. Bact.* **148**, 728-9.

BACHEM, C., KONDOROSI, E., BANFALVI, Z., HORVATH, B., KONDOROSI, A., and SCHELL, J. (1985). *Molec. & gen. Genet.* **199**, 271-8.

BAGDASARIAN, M., LURZ, R., RÜCKERT, B., FRANKLIN, F. C. H., BAGDASARIAN, M. M., FREY, J., and TIMMIS, K. N. (1981). *Gene* **16**, 237-47.

BALASSA, G. (1963). *Bact. Rev.* **27**, 228-41.

BANFALVI, Z., SAKANYAN, V., KONCZ, C., KISS, A., DUSHA, I., and KONDOROSI, A. (1981). *Molec. & gen. Genet.* **184**, 318-25.

— RANDHAWA, G. S., KONDOROSI, E., KISS, A., and KONDOROSI, A. (1983). *Molec. & gen. Genet.* **189**, 129-35.

—— KONDOROSI, E., and KONDOROSI, A. (1985). *Plasmid* **13**, 129-38.

BECHET, M. and GUILLAUME, J. B. (1978). *Can. J. Microbiol.* **24**, 960-6.

BEDMAR, E. J. and OLIVARES, J. (1980). *Molec. & gen. Genet.* **177**, 329-31.

BELITSKY, B. and KARI, Cs. (1982). *J. biol. Chem.* **257**, 4677-9.

BERINGER, J. E. (1974). *J. gen. Microbiol.* **84**, 188-98.

— JOHNSTON, A.W.B., and WELLS, B. (1977). *J. gen. Microbiol.* **98**, 339-3.

— BEYNON, J. L., BUCHANAN-WOLLASTON, A.V., and JOHNSTON, A.W. B. (1978). *Nature, Lond.* **276**, 633-4.

— BREWIN, N. J., and JOHNSTON, A.W. B. (1980). *Heredity* **45**, 161-86.

BETTER, M. and HELINSKI, D. R. (1983). *J. Bact.* **155**, 311-16.

— LEWIS, B., CORBIN, D., DITTA, G., and HELINSKI, D. R. (1983). *Cell* **35**, 479-85.

BEYNON, J. L., BERINGER, J. E., and JOHNSTON, A.W. B. (1980). *J. gen. Microbiol.* **120**, 421-9.

— CANNON, M., BUCHANAN-WOLLASTON, V., and CANNON, F. (1983). *Cell* **34**, 665–71.

BIEK, D. and ROTH, J. R. (1980). *Proc. natn. Acad. Sci. U.S.A.* **77**, 6047–51.

BOLTON, E., GLYNN, P., and O'GARA, F. (1984). *Molec. & gen. Genet.* **193**, 153–7.

BREWIN, N. J., BERINGER, J. E., BUCHANAN-WOLLASTON, A.V., JOHNSTON, A.W. B., and HIRSCH, P. R. (1980). *J. gen. Microbiol.* **116**, 261–70.

BROUGHTON, W. J. (1978). *J. appl. Bact.* **45**, 165–94.

BUCHANAN-WOLLASTON, A.V., BERINGER, J. E., BREWIN, N. J., HIRSCH, P. R., and JOHNSTON, A.W. B. (1980). *Molec. & gen. Genet.* **178**, 185–90.

BUIKEMA, W. B., LONG, S. R., BROWN, S. E., VAN DEN BOS, R. C., EARL, C., and AUSUBEL, F. M. (1983). *J. molec. appl. Genet.* **2**, 249–60.

BULLERJAHN, G. S. and BENZINGER R. H. (1982). *J. Bact.* **150**, 421–4.

BURKARDT, B. and BURKARDT, H.-J. (1984). *J. molec. Biol.* **175**, 213–18.

BURKARDT, H. J., RIESS, G., and PÜHLER, A. (1979). *J. gen. Microbiol.* **114**, 341–8.

CANNON, F. C., RIEDEL, G. E., and AUSUBEL, F. M. (1979). *Molec. & gen. Genet.* **174**, 59–66.

CASADESÚS, J. and OLIVARES, (1979). *Molec. & gen. Genet.* **174**, 203–9.

CASSE, F., BOUCHER, C., JULLIOT, J. S., MICHEL, M., and DÉNARIÉ, J. (1979). *J. gen. Microbiol.* **113**, 229–42.

CORBIN, D., DITTA, G., and HELINSKI, D. R. (1982). *J. Bact.* **149**, 221–8.

— BARRAN, L., and DITTA, G. (1983). *Proc. natn. Acad. Sci. U.S.A.*, **80**, 3005–9.

DALLMAN, G., OLASZ, F., and OROSZ, L. (1980). *Molec. & gen. Genet* **178**, 443–6.

DATTA, N., HEDGES, R. W., SHAW, E. J., SYKES, R. B., and RICHMOND, M. H. (1971). *J. Bact.* **108**, 1244–49.

DAVID, M., VIELMA, M., and JULLIOT, J. S. (1983). *FEMS Lett.*, 335–41.

DeBRUIJN, P., SUNDARESAN, V., SZETO, W. W., OW, D. W., and AUSUBEL, F. M. (1984). In *Advances in nitrogen fixation research* (ed. C. Veeger and W. E. Newton), pp. 627–33. M. Nijhoff, The Hague.

DÉNARIÉ, J., TRUCHET, G., and BERGERSON, B. (1976). In *Symbiotic nitrogen fixation in plants* (ed. P. S. Nutman), pp. 47–61. Cambridge University Press.

— BOISTARD, P., CASSE-DELBART, F., ATHERLY, A. G., BERRY, J. O., and RUSSELL, P. (1981). *Int. Rev. Cytol., Suppl.* 13, 225–46.

DITTA, G., STANFIELD, S., CORBIN, D., and HELINSKI, D. R. (1980). *Proc. natn. Acad. Sci. U.S.A.* **77**, 7347–51.

DIXON, R. A., CANNON, F. C., and KONDOROSI, A. (1976). *Nature, Lond.* **260**, 268–71.

DJORDJEVIC, M. A., ZURKOWSKI, W., SHINE, J., and ROLFE, B. G. (1983). *J. Bact.* **156**, 1035–45.

DUSHA, I., RENALIER, M. H., TERZAGHI, B., GARNERONE, A. M., BOISTARD, P., and JULLIOT, J.-S. (1983). In *Molecular genetics of bacteria plant interaction* (ed. A. Pühler), pp. 46–54. Springer, v. Berlin, Heidelberg.

ECKHARDT, T. (1978). *Plasmid* **1**, 584–8.

ELKAN, G. H. (1981). *Int. Rev. Cytol. Suppl.* 13, 1–14.

FINAN, T. M., HARTWIEG, E., LeMIEUX, K., BERGMAN, K., WALKER, G. C., and SIGNER, E. R. (1984). *J. Bact.* **159**, 120–4.

— JOHANSEN, E., LEIGH, J. A. WALKER, G. C., SIGNER, E., HIRSCH, A., KULDAU, G., and DEEGAN, S. (1985). In *Advances in the molecular genetics of bacteria-plant interaction* (ed. A. A. Szalay and R. P. Legocki) pp. 99–101. Cornell Univ., Ithaca, N.Y.

FORRAI, T., VINCZE, E., BANFALVI, Z., KISS, G. B., RANDHAWA, G. S., and KONDOROSI, A. (1983). *J. Bact.* **153**, 635–43.

FRIEDMAN, A. M., LONG, S. R., BROWN, S. E., BUIKEMA, W. J., and AUSUBEL, F. M. (1982). *Gene* **18**, 289–96.

GABOR-HOTCHKISS, M. (1972). In *Uptake of informative molecules by living cells* (ed. L. Ledoux), pp. 212–34. North-Holland, Amsterdam.

GODFREY, C. A. and DILWORTH, M. J. (1971). *J. gen. Microbiol.* **69**, 385–90.

GRAHAM, P. H. (1964). *J. gen. Microbiol.* **35**, 511–17.

HAAS, D. and HOLLOWAY, B. W. (1976). *Molec. & gen. Genet.* **144**, 243–51.

HASE, T., NAKANO, T., MATSUBARA, H. and ZUMFT, W. G. (1981). *J. Biochem.* **90**, 295–8.

HIRSCH, A. M., LONG, S. R., BANG, M., HASKINS, N., and AUSUBEL, F. M. (1982). *J. Bact.* **151**, 411–19.

HIRSCH, A. M., WILSON, K. J., JONES, J. D. G., BANG, M., WALKER, V. V., and AUSUBEL, F. M. (1984). *J. Bacteriol.* **158**, 1133–43.

HOHN, B., and COLLINS, J. (1978). *Gene* **11**, 291–8.

HOOYKAAS, P. J. J., DEN DULK-RAS, H., and SCHILPEROORT, R. A. (1982). *J. Bact.* **150**, 395–7.

— PEERBOLTE, R., REGENSBURG-TUINK, A. J. G., DE VRIES, P., and SCHILPEROORT, R. A. (1982). *Molec. & gen. Genet.* **188**, 12–17.

INOUYE, S., EBINA, Y., NAKAZAWA, A., and NAKAZAWA, T. (1984). *Proc. natn. Acad. Sci. U.S.A.* **81**, 1688–91.

ISH-HOROWITZ, D. and Burke, J. F. (1981). *Nucl. Acids Res.* **9**, 2989–98.

JOHNSON, R. C., YIN, J. C., and REZNIKOFF, W. S. (1982). *Cell* **30**, 873–82.

JOHNSTON, A.W. B., BEYNON, J. L., BUCHANAN-WOLLASTON, A.V., SETCHELL, S. M., HIRSCH, P. R., and BERINGER, J. E. (1978a). *Nature, Lond.* **276**, 635–6.

— BIBB, M. J., and BERINGER, J. E. (1978b). *Molec. & gen. Genet.* **165**, 323–30.

—— and DOWNIE, J. A. (1984). *Trends biochem. Sci.* **9**, 367–8.

JOUANIN, L., DE LAJUDRE, D., BAZETOUX, S., and HUGUET, T. (1981). *Molec. & gen. Genet.* **182**, 189–95.

JULLIOT, J. S. and BOISTARD, P. (1979). *Molec. & gen. Genet.* **173**, 289–98.

— DUSHA, I., RENALIER, M. H., TERZAGHI, B., GARNERONE, A. M., and BOISTARD, P. (1984). *Molec. & gen. Genet.* **193**, 17–26.

KAHN, M. L. and TIMBLIN, C. R. (1984). *J. Bact.* **158**, 1070–7.

KISS, G. B., VINCZE, E., KALMAN, Z., FORRAI, T., and KONDOROSI, A. (1979). *J. gen. Microbiol.* **133**, 105–18.

— DOBO, K., DUSHA, I., BREZNOVITS, A., OROSZ, L., VINCZE, E., and KONDOROSI, A. (1980). *J. Bact.* **141**, 121–218.

— and KALMAN, Z. (1982). *J. Bact.* **150**, 465–70.

KONDOROSI, A., BARABAS, I., SVAB, Z., OROSZ, L., SIK, T., and HOTCHKISS, R. D. (1973). *Nature, Lond.* (*New Biol.*) **246**, 153–4.

— OROSZ, L., SVAB, Z., and SIK, T. (1974). *Molec. & gen. Genet.* **132**, 153–63.

— Kiss, G. B., Forrai, T., Vincze, E., and Banfalvi, Z. (1977*a*). *Nature, Lond.* **268**, 525–7.

— Svab, Z., Kiss, G. B., and Dixon, R. A. (1977*b*). *Molec. & gen. Genet.* **151**, 221–6.

— Vincze, E., Johnston, A.W. B., and Beringer, J. E. (1980). *Molec. & gen. Genet.* **178**, 403–8.

— and Johnston, A.W. B. (1981). *Int. Rev. Cytol., Suppl.* 13, 191–224.

— Kondorosi, E., Pankhurst, C. E., Broughton, W. J., and Banfalvi, Z. (1982). *Molec. & gen. Genet.* **188**, 433–39.

—— Banfalvi, Z., Dusha, I., Putnoky, P., Toth, J., and Bachem, C. (1984*a*). In *Genetics: new frontiers, Vol. II, Proc. of the XV. Int. Congr. of Genetics*, pp. 205–16, Oxford and IBH Publ. Co., New Delhi.

——— Torok, I., Schmidt, J., and John, M. (1985*a*). In *Nitrogen fixation and CO₂ metabolism* (ed. P. W. Ludden and J. E. Burris), pp. 91–8. Elsevier, New York.

————— Stepkowski, T., Schmidt, J., John, M. (1985*b*). In *Natural Products Chemistry* (ed. R. I. Zalewski and J. J. Skolik). pp. 643–54. Elsevier Amsterdam.

—— Torok, I., Putnoky, P., Banfalvi, Z. (1985*c*). In *Proc. of the 16th FEBS Congress* (ed. Yu. Ovchinnikov). part A, pp. 397–403. VNU Science Press, Utrecht (in press).

— Banfalvi, Z., Broughton, W. J., Forrai, T., Kiss, G. B., Kondorosi, E., Pankhurst, C. E., Randhawa, G..S., Svab, Z., and Vincze, E. (1983*a*). In *Structure and function of plant genomes* (ed. O. Ciferri and L. Dure), pp. 247–52. Plenum Publ. Co., New York and London.

— Kondorosi, E., Banfalvi, Z., Broughton, W. J., Pankhurst, C. E., Randhawa, G. S., Wong, C. H., and Schell, J. (1983*b*). In *Molecular genetics of bactera–plant interaction* (ed. A. Pühler), pp. 55–63. Springer V., Berlin, Heidelberg.

Kondorosi, E., Banfalvi, Z., Slaska-Kiss, C., and Kondorosi, A. (1983*c*). In *UCLA Symp. Molec. & Cell. Biol.* new series, Vol. 12. *Plant molecular biology* (ed. R. B. Goldberg) pp. 259–75. A. R. Liss, Inc., New York.

—— and Kondorosi, A. (1984*b*). *Molec. & gen. Genet.* **193**, 445–52.

— Putnoky, P., Torok, I., Banfalvi, Z., Schmidt, J., John, M., and Kondorosi, A. (1985*d*). In *Advances in the molecular genetics of bacteria–plant interaction* (ed. A. Szalay and R. P. Legocki) pp. 49–52. Cornell Univ. Ithaca, N.Y.

Kowalski, M. (1967). *Acta microbiol. pol.* Ser. A. **16**, 7–12.

Kuchko, V. V., Zaretskaya, A. N., Golubkov, V. I., Fokina, I. G., and Simarov, B. V. (1983). *Mikrobiologija* **52**, 986–90.

Kuykendall, L. D. (1981). *Int. Rev. Cytol. Suppl.* 13, 299–309.

Leemaans, J., Langenakens, J., De Greve, H., Deblaere, R., Van Montagu, M., and Schell, J. (1982). *Gene* **19**, 361–4.

Leong, S. A., Ditta, G. S., and Helinski, D. R. (1982). *J. biol. Chem.* **257**, 8724–30.

Long, S. R., Buikema, W. J., and Ausubel , F. M. (1982). *Nature, Lond.* **298**, 485–8.

— JACOBS, T. W., EGELHOFF, T. T., TU, J., and FISHER, R. (1985). In *Nitrogen fixation and CO$_2$ metabolism* (ed. P. W. Ludden and J. E. Burris), pp. 75–82. Elsevier, New York.

LUDWIG, R. A. and JOHANSEN, E. (1980). *Plasmid* **3**, 359–61.

LUNDELL, D. and HOWARD, J. B. (1978). *J. biol. Chem.* **253**, 3422–6.

MAIER, R. J. and BRILL, W. J. (1976). *J. Bact.* **127**, 763–9.

MARTIN M. Q. and LONG, S. R. (1984). *J. Bact.* **159**, 125–9.

MAZUR, B. J., RICE, D., and HASELKORN, B. (1980). *Proc. natn. Acad. Sci. U.S.A.* **77**, 186–90.

MEADE, H. M. and SIGNER, E. R. (1977). *Proc. natn. Acad. Sci. U.S.A.* **74**, 2076–8.

— LONG, S. R., RUVKUN, G. B., BROWN, S. E., and AUSUBEL, F. M. (1982). *J. Bact.* **149**, 114–22.

MERRICK, M., FILSER. M., DIXON, R. A., ELMERICH, C., SIBOLD, L., and HOUMARD, J. (1980). *J. gen. Microbiol.* **117**, 509–20.

MEVARECH, M., RICE, D., and HASELKORN, R. (1980). *Proc. natn. Acad. Sci. U.S.A.* **77**, 6476–80.

MacNEIL, T., MacNEIL, D., ROBERTS, G. P., SUPIANO, M. A., and BRILL, W. J. (1978). *J. Bact.* **136**, 253–66.

NUTI, M. P., LEPIDI, A. A., PRAKASH, R. K., SCHILPEROORT, R. A., and CANNON, F. C. (1979). *Nature, Lond.* **282**, 533–5.

OLASZ, F., DORGAI, L., PAY, A., and OROSZ, L. (1983). *Molec. & gen. Genet.* **191**, 393–6.

OLIVARES, J., MONTOYA, E., and PALOMARES, A. (1977). In *Recent developments in nitrogen fixation* (ed. W. E. Newton, J. R. Postgate, and C. Rodriguez-Barrueco), pp. 375–85. Academic Press, New York.

O'NEILL, E. A., KIELY, G. M., and BENDER, R. A. (1984). *J. Bact.* **159**, 388–9.

OROSZ, L., SVAB, Z., KONDOROSI, A., and SIK, T. (1973). *Molec. & gen. Genet.* **125**, 341–50.

OW, D. W., SUNDARESAN, V., ROTHSTEIN, D. M., BROWN, S. E., and AUSUBEL, F. M. (1983). *Proc. natn. Acad. Sci. U.S.A.* **80**, 2524–8.

PAAU, A. S., LEPS, W. T., and BRILL, W. J. (1981). *Science* **213**, 1513–15.

— and BRILL, W. J. (1982). *Can. J. Microbiol.* **28**, 1330–9.

PANKHURST, C. E., BROUGHTON, W. J., BACHEM, C., KONDOROSI, E., and KONDOROSI, A. (1983). In *Molecular genetics of bacteria–plant interaction* (ed. A. Pühler), pp. 169–76. Springer V., Heidelberg.

POSTGATE, J. R. (1974). *Symp. Soc. gen. Microbiol.* **24**, 263–92.

PRAKASH, R. K., SCHILPEROORT, R. A., and NUTI, M. P. (1981). *J. Bact.* **145**, 1129–36.

PÜHLER, A. and KLIPP, W. (1981). In *Biology of inorganic nitrogen and sulfur* (ed. H. Bothe and A. Trebst) pp. 276–86. Springer V., Berlin.

— AQUILAR, M. O., HYNES, M., MULLER, P., KLIPP, W., PRIEFER, U., SIMON, R., and WEBER, G. (1984). In *Advances in nitrogen fixation research* (ed. C. Veeger and W. F. Newton), pp. 609–19. M. Nijhoff, W. Junk Publ., The Hague.

PUTNOKY, P., KISS, G. B., OTT, I., and KONDOROSI, A. (1983). *Molec. & gen. Genet.* **191**, 288–94.

RIEDEL, G. E., Ausubel, F. M., and CANNON, F. C. (1979). *Proc. natn. Acad. Sci. U.S.A.* **76**, 2866–70.

RIESENBERG, D., ERDEI, S., KONDOROSI, E., and KARI, C. (1982). *Molec. & gen. Genet.* **185**, 198–204.

ROLFE, B. G., GRESSHOFF, P. M., and SHINE, J. (1980). *Pl. Sci. Lett.* **19**, 277–84.

ROSENBERG, C., BOISTARD, P., DÉNARIÉ, J., and CASSE-DELBART, F. (1981). *Molec. & gen. Genet.* **184**, 326–33.

ROSSEN, L., JOHNSTON, A. W. B., and DOWNIE, J. A. (1984). *Nucl. Acids Res.* **12**, 9497–508.

RUVKUN, G. B. and AUSUBEL, F. M. (1980). *Proc. natn. Acad. Sci. U.S.A.* **11**, 191–5.

—— (1981). *Nature, Lond.* **289**, 85–8.

— LONG, S. R., MEADE, H. M., VAN DEN BOS, R. C., and AUSUBEL, F. M. (1982*a*). *J. molec. appl. Genet.* **1**, 405–18.

— SUNDARESAN, V., and AUSUBEL, F. M. (1982*b*). *Cell* **29**, 551–9.

SCHMIDT, J., JOHN, M., KONDOROSI, E., KONDOROSI, A., WIENECKE, U., SCHRODER, G., SCHRODER, J., and SCHELL, J. (1984). *Eur. molec. and Biol. Orgn J.* **8**, 1705–11.

SCHWINGHAMER, E. A. (1977). In *A treatise on dinitrogen fixation* (ed. R. W. F. Hardy and W. S. Silver) Sect. III, pp. 577–622. Wiley-Interscience, New York.

SCOTT, K. F., ROLFE, B. G., and SHINE, J. (1981). *J. molec. appl. Genet.* **1**, 71–81.

SELVARAJ, G. and IYER, V. N. (1981). *Gene* **15**, 279–83.

—— (1983). *J. Bact.* **156**, 1292–300.

—— 1984). *J. Bact.* **158**, 580–9.

SIK, T. and OROSZ, L. (1971). *Pl. Soil*, Special Vol., pp. 57–62.

— HORVATH, J., and CHATTERJEE, S. (1980). *Molec. & gen. Genet.* **178**, 511–16.

SIMON, R., PRIEFER, U., and PÜHLER, A. (1983). In *Molecular genetics of bacteria–plant interaction* (ed. A. Pühler), pp. 98–106. Springer V., Berlin, Heidelberg.

—(1984). *Molec. & gen. Genet.* **196**, 413–20.

SHINE, J. and DALGARNO, L. (1975). *Nature, Lond.* **254**, 34–8.

SOBY, S. and BERGMAN, K. (1983). *Appl. & env. Microbiol.* **46**, 995–8.

SOMERVILLE, J. E. and KAHN, M. L. (1983). *J. Bact.* **156**, 168–76.

SPITZBARTH, M., PÜHLER, A., and HEUMANN, W. (1979). *Arch. Microbiol.* **121**, 1–7.

STANIEWSKI, R. and RUGALA, A. (1975). *Acta microbiol. pol. Ser. A.* **8**, 151–60.

SUNDARESAN, V. and AUSUBEL, F. M. (1981). *J. biol. Chem.* **256**, 2808–12.

— JONES, J. D. G., OW, P. W., and AUSUBEL, F. M. (1983*a*). *Nature, Lond.* **301**, 728–32.

— OW, D. W., and AUSUBEL, F. M. (1983*b*). *Proc. natn. Acad. Sci. U.S.A.* **80**, 4030–4.

SVAB, Z., KONDOROSI, A., and OROSZ, L. (1978). *J. gen. Microbiol.* **106**, 321–7.

SZETO, W. W., ZIMMERMAN, J. L., SUNDARESAN, V., and AUSUBEL, F. M. (1984). *Cell* **6**, 1035–43.

TANAKA, M., HANIN, M., YASUNOBU, K. T., and MORTENSON, L. E. (1977). *J. biol. Chem.* **252**, 7013-100.

TEMPÉ, J., PETIT, A., and BANNEROT, H. (1982). *CR Acad. Sci (Paris) derIII* **295**, 413-16.

TÖRÖK, I. and KONDOROSI, A. (1981). *Nucl. Acids Res.* **9**, 5711-23.

— KONDOROSI, E., STEPKOWSKI, T., POSFAI, J., and KONDOROSI, A. (1984). *Nucl. Acids. Res.* **12**, 9509-24.

TRUCHET, G., MICHEL, M., and DÉNARIÉ, J. (1980). *Differentiation* **16**, 163-73.

— ROSENBERG, C., VASSE, J., JULLIOT, J.-S., and CAMUT, S., DÉNARIÉ, J. (1984). *J. Bact.* **157**, 134-42.

VANCE, C. P. (1983). *A. Rev. Microbiol.* **37**, 399-424.

VAN VLIET, F., SILVA, B., VAN MONTAGU, M., and SCHELL, J. (1978). *Plasmid* **1**, 446-55.

VERMA, D. P. S. and LONG, S. (1983). *Int. Rev. Cytol. Suppl.* **14**, 211-45.

VINCENT, J. M. (1980). In *Nitrogen fixation* (ed. W. E. Newton and W. H. Orme-Johnson). Vol. 2, pp. 103-29. University Park Press, Baltimore.

VINCZE, E., KONCZ, Cs., and KONDOROSI, A. (1981). *Acta Biol. Acad. Sci. hung.* **32**, 3-4, 195-204.

WEBER, G. and PÜHLER, A. (1982). *Plant molec. Biol.*, **1**, 305-20.

WONG, C. H., PANKHURST, C. E., KONDOROSI, A., and BROUGHTON, W. J. (1983). *J. Cell Biol.* **97**, 787-94.

ZIMMERMAN, J. L., SZETO, W. W., and AUSUBEL, F. M. (1983). *J. Bacteriol.* **156**, 1025-34.

ZURKOWSKI, W. and LORKIEWICZ, Z. (1979). *Arch. Microbiol.* **123**, 195-201.

11 Host-specific gene expression in legume root nodules

T. Bisseling, R. C. van den Bos, and A. van Kammen

11.1 INTRODUCTION

Nitrogen fixation in root nodules of legumes results from specific interactions between the plants and *Rhizobium* capable of invading the roots. Nodules develop from the root cortex after infection by rhizobia (for review see, e.g., Dart 1977; Bergersen 1982). *Rhizobium* penetrates through an infection thread into the cortex and induces dedifferentiation of cortical cells resulting in the formation of meristematic tissue from which the root nodule grows. Cells of this tissue are then invaded by rhizobia released from the infection threads by endocytosis. When they are filled with bacteria, the plant cells differentiate into non-dividing cells and the synthesis of a number of different proteins is induced. Amongst these proteins the best known are leghaemoglobins (lHb) which are produced in large amounts and give the root nodules a characteristic red-brownish colour. Concurrently, the rhizobia within the plant cells undergo drastic changes. They stop dividing and often enlarge greatly to form pleomorphic bacteroids with characteristic shapes, which differ in nodules of different legumes. When the bacteria are endocytotically taken into the cytoplasm of the plant cells they are surrounded by host plasma membranes, and the intracellular development of the bacteria into bacteroids proceeds within the membrane envelopes (the peribacteroid membrane).

In the bacteroid forms of *Rhizobium*, the genes for nitrogen fixation are expressed. N_2 is reduced to ammonia, which is excreted by the bacteroids into the cytoplasm of the plant cell where the ammonia is assimilated into glutamate + glutamine which are subsequently converted into translocatable products (Bergersen 1982). The nitrogen-fixing root-nodule then becomes a distinct organ supplying the plant with nitrogen compounds for growth. It is obvious that the formation of root nodules involves the specific expression of both plant and bacterial genes. A complex series of interactions between plant and rhizobia regulates the physiology of the root nodules so that productive nitrogen fixation is effected (Bergersen 1982). Sometimes nodules fail to fix nitrogen and are termed ineffective. Since an effective nodule is the result of a series of interactions between plant and bacteria, the inability of ineffective nodules to fix nitrogen may be due to defects in either the rhizobia or the plant at various stages in nodule development (for review see, e.g., Meijer and Broughton 1982).

Since the role of bacterial determinants in the symbiosis is discussed in detail in Chapters 9 and 10 of this volume, we shall confine ourselves to a review of the specific changes in the expression of host-plant genes during nodule formation.

Hereditary host factors effecting nodulation and nitrogen fixation have been described by Nutman (1981). There is evidence that several plant genes control the development of an effective root nodule but physiological functions have yet to be assigned to any of these genes excepting the genes coding for lHb. Recently, immunological techniques have been used to detect the occurrence of nodule specific polypeptides synthesized during nodule development (Legocki and Verma 1980; Bisseling *et al.* 1983). Leghaemoglobin is of course one of these proteins but many other proteins have also been detected.

Analysis of mRNA isolated from root nodules has demonstrated the presence of different classes of nodule specific mRNAs not detected in uninfected root tissue (Auger *et al.* 1981) indicating the induction and expression of nodule-specific host genes during root nodule development. By synthesizing complementary DNA (cDNA) upon mRNA from effective root nodules and molecular cloning of the duplex cDNA it has become possible to prepare specific cDNA probes, which can be used to locate on the chromosomal DNA of the plant the genes encoding the nodule-specific proteins. These methods are now being used to gain insight into the number of plant genes concerned, their organization on the plant genome and their expression under the influence of infection by *Rhizobium*. While these studies are still in their infancy and there is a long way to go before all questions are properly answered, we shall review the development of these studies and their potential.

The nodule-specific host proteins have been termed nodulins by Legocki and Verma (1980) and we shall adopt this definition. Among the plant genes the expression of which is essential for nitrogen fixation in legume root nodules are the genes for lHb, and lHb is in fact a nodulin. Leghaemoglobin stands out at once among the nodulins because it has a defined function in effective root nodules (Bergersen 1982). It is an O_2-binding protein, which acts as a carrier of O_2 in the tissue of the nodule. It is supposed that lHb provides a balanced distribution of O_2 among the bacteroids and the plant cytoplasm. The bacteroids require an efficient supply of O_2 for oxidative phosphory-lation to generate the energy for nitrogen fixation and the plant cells need active synthesis of ATP by respiration for the assimilation of ammonia into translocated products. At the same time lHb might effect a low concentration of free O_2 necessary for optimal activity of the bacteroid nitrogenase, which is a very O_2-sensitive enzyme and functions only at low O_2-concentrations (see Haaker, Chapter 3, this volume).

11.2 STRUCTURE AND ORGANIZATION OF SOYABEAN LEGHAEMOGLOBIN GENES

In soyabean nodules, four major types of lHb (lHba, $lHbc_1$, $lHbc_2$, and $lHbc_3$) were found (Appleby *et al.* 1975; Fuchsman and Appleby 1979). In addition, four minor components have been identified that occur as a result of N-terminal modifications of the major components, modifications which are probably post-translational (Whittaker, Moss, and Appleby 1979; Whittaker, Lennox, and Appleby 1981). Amino acid sequence analyses of the major components showed that they are encoded by distinct genes, since they differ by a few amino acids (Whittaker *et al.* 1981; Hurrell, and Leach 1977; Sievers, Huhtala, and Ellfolk 1978); these results are confirmed by sequence analyses of the cloned lHb-genes. In other words, the legheamoglobin components are encoded by a small family of genes.

The structures of the four genes encoding the four major lHb components have been solved completely by the groups of Marcker and Verma. Besides the four lHb-genes, two truncated and one pseudo-lHb-gene have been identified (Marcker *et al.* 1983; Brisson and Verma 1982). Only the soyabean lHb-genes have been sequenced, lHb-genes of other legumes have not been studied so far.

11.2.1 Sequence organization of soyabean lHb genes

The general sequence organization of the lHba, $lHbc_1$-, $lHbc_2$-, and $lHbc_3$-genes is shown in Fig. 11.1. The coding sequences are in all cases interrupted by intervening sequences at three identical positions; the introns are at codon 32 (IVS-I), between codons 68–69 (IVS-2), and between codons 103–104 (IVS-3) (Sullivan *et al.* 1981; Jensen 1981; Hyldig-Nielsen *et al.* 1982; Wilborg *et al.* 1982; Brisson and Verma 1982; Marcker *et al.* 1983). The lengths of the intervening sequences vary between the different lHb genes; especially considerable variations occur for the IVS-2 and IVS-3 sequences.

All the intervening sequences of the four major component lHb genes start with GT and terminate with AG which are the consensus sequences for introns (Breatnach *et al.* 1979). In all lHb-genes sequenced

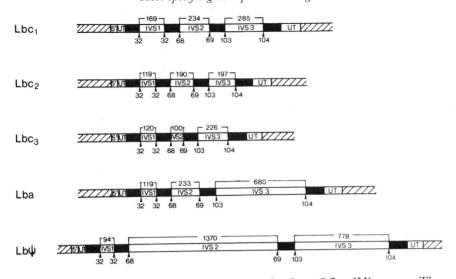

FIG. 11.1. The general DNA sequence organization of five lHb-genes. The introns are denoted IVS-I, IVS-2, and IVS-3, respectively. The exons are shown as solid boxes. UT denotes untranslated regions. The lengths of the introns are indicated above the bars and the location with respect to codon numbers, below the bars. (From Marcker *et al.* 1983.)

so far, the initiation codon ATG is immediately followed by the NH_2-terminal codon (Hyldig-Nielsen *et al.* 1982; Wiborg *et al.* 1982; Brisson and Verma 1982; Marcker *et al.* 1983) indicating that no signal peptide is encoded by the lHb-genes. Therefore it is unlikely that the lHb-polypeptides will be able to pass the peribacteroid membrane. This finding supports the results of Robertson *et al.* (1978), Verma *et al.* (1978), and Bisseling *et al.* (1983) who all showed that lHb is located in the cytoplasm of the host cell and not in the peribacteroid space. The structure of the lHb-genes shows a striking similarity to that of animal globin genes. The positions of the two intervening sequences that interrupt the β-globin genes (codons 30 and 104) coincide with IVS-I and IVS-3 found in the lHb-genes. Go (1981) has defined four regions of the globin polypeptide chain and this analysis revealed that the splicing points in the coding sequence separate these regions from one another except for the central coding sequence which consists of two such regions. Therefore, he suggested that the central exon in globins might be the result of a fusion of two exons with an ancestral splicing point somewhere between amino acid 66 and 71. The location of IVS-2 between codons 68 and 69 in the lHb-genes is in agreement with Go's proposal, and this location supports the notion that lHb genes represent primitive globin-genes.

11.2.2 Regulatory sequences of lHb genes

All four lHb-genes contain regulatory sequences (Fig. 11.2) which are identical or very similar to the corresponding sequences found in other eukaryotic genes (Hyldig-Nielsen *et al.* 1982; Wilborg *et al.* 1982; Brisson and Verma 1982; Marcker *et al.* 1983). A cap addition site (TGCATAA) is found 46–53 nucleotides upstream from the ATG initiation codon and the TATAA box (essential for promoter activity) about 30 nucleotides upstream from the cap addition site; 39–56 nucleotides upstream from the TATAA sequence there is a penta-nucleotide sequence CCAAG which is very similar to the sequence— essential for the initiation of transcription—found in most eukaryotic genes at or close to this position. Grosveld, Rosenthal, and Flavell (1982) have shown that the region immediately preceding the CAT box (in lHb-genes CCAAG sequence) is also important for transcription of the globin-genes. The consensus sequence CACCCTCC can be found in all four lHb-genes at this position (Fig. 11.2).

Analysis of the 3′-non-coding region revealed sequences identical or similar to AATAAA, a consensus sequence for the poly A addition signal (Proudfoot and Brownlee 1976). In all four lHb-genes the hexa-nucleotide GATAAA occurs proximal to the putative poly A addition site. Thus the regulatory signals in lHb-genes are very similar to the corresponding signals in other eukaryotic genes.

11.2.3 The chromosomal arrangement of 1Hb genes

Analysis of overlapping DNA clones, obtained from several soyabean genomic libraries revealed that the normal 1Hb genes are arranged in

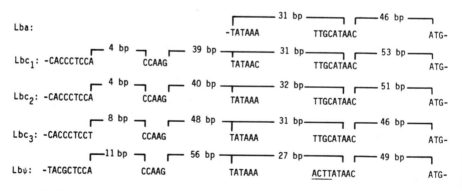

FIG. 11.2. Comparison of putative regulatory sequences on the 5′ flanking region of five lHb-genes. The nucleotide sequences correspond to regulatory sequences cited in the text. Distances are given in base pairs (bp). The underlined nucleotides in the *pseudogene*, lHb ψ, indicate mutations. (From Marcker *et al.* 1983).

two clusters (Fig. 11.3). Three functional genes 1Hba, 1Hbc$_1$, and 1Hbc$_3$ are very closely linked. Between 1Hbc$_1$ and 1Hbc$_3$ a pseudo 1 Hb gene (ψ_1 1 Hb) is located. 1 Hbc$_2$ and ψ_2 1 Hb are located in a second cluster (Lee *et al.* 1983; Marcker *et al.* 1984). Preliminary results of Marcker *et al.* (1983) indicated that both clusters are linked. However, so far no clones connecting both clusters have been identified. Therefore no real evidence exists for linkage of both clusters (Marcker *et al.* 1984). The two truncated 1 Hb genes are located on DNA fragments that are not linked to each other nor to the two clusters of 1 HB genes (Lee *et al.* 1983).

The two 1Hb clusters are flanked on both sides by genes that are expressed in leaves, roots, and root nodules (Lee *et al.* 1983; Marcker *et al.* 1984). These genes are located in the regions A/A′ and C/C′ shown in Fig. 11.3. The sequences of regions A and C show strong cross-hybridization with A′ and C′ respectively. Furthermore a repetitive sequence (10–20 times per genome) is present at the 3′ side of the 1Hb genes in both clusters (Marcker *et al.* 1984). This sequence probably also flanks the two truncated genes (Lee *et al.* 1983).

So far no linkage between 1Hb genes and other nodulin genes has been found. The results described above indicate that if these genes are located on the same chromosome they are interspersed by genes that are expressed in other tissues.

11.2.4 Pseudo and truncated lHb-genes

Besides the structural lHb-genes, two pseudo- and two truncated lHb-genes have been identified (Brisson and Verma 1982; Wiborg *et al.* 1983 Marcker *et al.* 1983; Bojsen *et al.* 1983; Verma *et al.* 1984;

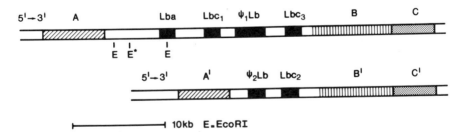

FIG. 11.3. Chromosomal arrangement of six soyabean 1Hb genes. Solid boxes indicate the position of the lHb genes (the three introns present in all lHb genes are not shown). Identical shadings (A and A′, B and B′, C and C′) represent cross-hybridizing non-1Hb regions. The transcriptional polarity of the 1Hb genes is indicated by the two arrows. The *Eco*RI restriction sites proximal to the lHb gene are indicated. E* denotes the EcoRI site which is difficult to cleave in soyabean genomic DNA. (From Marcker *et al.* 1984.)

Marcker *et al.* 1984). The $\psi_1$1Hb is located between lHbc$_1$ and lHbc$_3$ as is shown in Fig. 11.3. This gene has three introns at identical positions to the structural genes but introns 2 and 3 are much larger. For this reason its size is around 3 kb, which is about twice the size of the structural genes. The 3'-end of this gene codes for 6–7 additional amino acids including methionine (Brisson and Verma 1982; Wiborg *et al.* 1983). However, methionine does not occur in soyabean lHb. Furthermore, the sequences essential for transcription, e.g. the CAP addition site and the CACCCTCCT sequence are strongly modified (Fig. 11.2). No RNA molecules transcribed from this gene have been found. These results therefore strongly suggest that this gene is a pseudogene. One of the truncated lHb-genes has been sequenced by Brisson and Verma (1982). Comparison of the sequence of this gene with a functional lHb-gene shows that the homology is limited to the third exon and the non-coding region at the 3'-end. The homology between lHbc$_3$ and this truncated gene ends about three codons from the splicing site.

11.2.5 lHb gene activation and methylation

The DNA of eukaryotes contains a certain amount of 5'-methylcytosine, most often in the sequence CG (Grippo, *et al.* 1968) and most active eukaryotic genes are hypomethylated when compared to the inactive genes (Razin and Riggs 1980). When DNA from the same soyabean tissue, e.g. seedling or nodule DNA is digested with the restriction endonuclease HpaII or MspI no difference in methylation of the lHb-genes is apparent. HpaII and MspI are isoschizomers that recognize the DNA sequence CCGG. Cleavage by these enzymes is inhibited when the cytosines are methylated; methylation of the first cytosine inhibits MspI and methylation of the second inhibits HpaII. However, a HpaII or MspI digest of seedling DNA exhibits a similar difference in the lHb-region when compared to similar digests of nodule DNA (Marcker *et al.* 1983). Thus, in the relevant CCGG sequences in nodule DNA both cytosines are demethylated, while in seedling DNA both are methylated. The sites exhibiting methylation variance have been located 3' to the lHbc$_3$ region (Fig. 11.3).

Although the study of methylation variance is not complete, the results indicate that active lHb-genes are hypomethylated when compared to the corresponding inactive genes and lHb-gene activation appears therefore analogous to other eukaryotic tissue specific gene expression like globin (Waalwijk and Flavell 1978) and ovalbumin gene activation (Mandel and Chambon 1979).

11.3 REGULATION OF lHb-SYNTHESIS

In the following part we will discuss the regulation of lHb-synthesis. Probably the haem-group of lHb is synthesized and excreted by the bacteroid, while apo-lHb is produced by the plant. Arguments in favour of this model will be considered as well as the factors involved in the regulation of haem and apolHb synthesis.

11.3.1 Haem-synthesis by Rhizobium

Soyabean root nodules contain considerable amounts of (up to 0.3 μmole) lHb haem/g fresh weight of nodule (Bergersen and Goodchild 1973). This concentration exceeds by far that normally found in plant tissue or in *R. japonicum* bacteroids (for review see Godfrey *et al.* 1975) and it is obvious that haem synthesis for lHb involves a marked increase in haem biosynthetic activity. The available evidence suggests that to a very large extent the *Rhizobium* bacteroid is responsible for this haem synthesis. *Rhizobium* mutants unable to synthesize haem always produce ineffective (fix⁻) root nodules lacking functional lHb (Nadler 1981; Bisseling *et al.* 1983; Ditta *et al.* 1983). A cooperative effort between the bacteroid and the plant was proposed to be responsible for the lHb-haem synthesis (Godfrey and Dilworth 1971). The results of Cutting and Schulman (1969) on the incorporation of [^{14}C]-ALA (δ-aminolevulinic acid) into different nodule fractions pointed out, however, that the 'particulate fraction', i.e. the bacteroids are the main site of haem synthesis. This contention was confirmed by Nadler and Avissar (1977) who proved that the enzymes involved in haem biosynthesis are present in sufficient quantities in *R. japonicum* bacteroids and correlated with the amounts of lHb produced.

Tetrapyrrole formation in a variety of organisms is limited by the rate of formation of ALA, the first compound committed to haem synthesis. Nadler and Avissar (1977) determined the activity of the enzyme involved in ALA synthesis (ALAS, δ-aminolevulinic acid synthase) from glycine and succinate in bacteroids. It was found that ALAS activity correlates on a fresh weight basis with nodule haem content during nodule development. In young nodule primordia that have no lHb, ALAS activity was not detectable. Thereafter, nodule haem content and bacteroid ALAS activity increase in parallel, both reaching a maximum five weeks after inoculation. As nodules senesce and their haem is degraded, bacteroid ALAS activity declines. In contrast to the nodule ALAS activity, ALAD (δ-aminolevulinic acid dehydrase) activity is found in both bacteroid and plant fractions of effective nodules and therefore the assignment of a specific role of bacteroid or plant ALAD in lHb haem synthesis is more difficult to make (Godfrey and Dilworth, 1971). The major portion of nodule ALAD activity is found in the plant fraction, and is present in both *fix⁺* and *fix⁻* nodules (containing little or no lHb). This might suggest a possible involvement of plant ALAD in haem synthesis; however, as lHb haem accumulates in the developing nodule, the plant ALAD activity decreases sharply although the bacteroid ALAD activity is unchanged or even increases (Fig. 11.4).

A sharp decline in the plant ALAD activity is observed during formation of both *fix*⁺ nodules (formed with *R. japonicum* 3I16-61) and *fix*⁻ (*R. japonicum* 11927) nodules. In contrast, bacteroid ALAD activity is detected only in *fix*⁺ but not in *fix*⁻ nodules and this activity (1–2 mmole PBG, porphobilinogen, formed/hr mg protein) could support as much as 60–120 mmole haem formed/day. g nodule, well in excess of the observed lHb accumulation. Bacteroid ALAD activity, furthermore, is only detectable when lHb is accumulating. The parallel between nodule haem accumulation and bacteroid ALAS as well as ALAD activity suggests a major role for the bacteroid in ALA formation for lHb-haem synthesis, although the possibility that the plant contributes to ALA formation can not be rigorously excluded. It is conceivable that ALA is

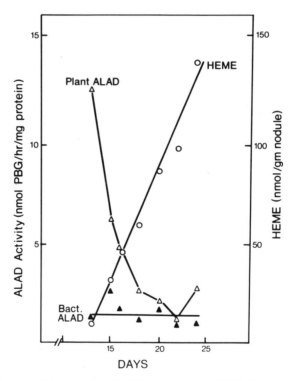

Fɪɢ. 11.4. Plant and bacteroid ALAD activities and lHb haem content of fix⁺ vs. fix⁻ nodules. Extracts were prepared and determinations made as in Nadler and Avissar (1977). Fix⁺-nodulation of *Glycine max* with *R. japonicum* strain 3I16–61; *fix*⁻-nodulation with strain 11927. Bacteroid ALAD activity in *fix*⁻-nodules was below the limits of sensitivity of the assay. Plant ALAD activity (△-△); Lb haem (○-○). (From Nadler and Avissar 1977.)

synthesized in the host plant cell via a five-carbon pathway (Beale and Castelfranco 1974; Beale *et al.* 1975; Meller *et al.* 1975) or that ALA formed in the plant extract *in vitro* is rapidly metabolized to other amino acids (Troxler and Brown 1975). The work of Nadler and Avissar (1977) is at least an indication however, that bacteroids can be the sole site of lHb-haem formation from glycine and succinate.

11.3.2 Enzymes involved in haem synthesis and their regulation

If the synthesis of lHb is a joint effort of both symbiotic partners, regulation of haem production in the *Rhizobium* by exogenous haem and by apolHb is probable. Cutting and Schulman (1972) studied haem synthesis from δ-amino-[4-^{14}C] levulinic acid in *R. japonicum* bacteroids. Their results indicate that within nodule breis, haem synthesis from ALA is subject to negative feedback control as 1.10^{-4} M haem inhibited [^{14}C]-ALA incorporation into haem by 50 per cent. This inhibition was partially reversed by the removal of the exogenous haem. Other tetrapyrroles like protoporphyrin IX and cobalamin, which are present in large amounts within root nodules showed comparable inhibitory effects on haem synthesis from ALA. Addition of haemin also inhibited ALA incorporation into extracellular haem with *R. lupini* bacteroids (Godfrey *et al.* 1975).

Addition of apolHb was reported by Cutting and Schulman (1972) to have a strong stimulatory effect on haem synthesis in *R. japonicum* bacteroids. This stimulation was specific for apolHb and was not observed after addition of holo-lHb. Probably both the haemin inhibition and the protein stimulation observed with soyabean (Cutting and Schulman 1969; 1972) and serradella bacteroids (Godfrey *et al.* 1975) reflect a regulation of the rate of haem release from the bacteroid membrane. In cell-free extracts of *R. japonicum* bacteria (Roessler and Nadler 1982) no influence of haemin (200 μM) on the activity of ALAS or ALAD was noted. Haem synthesis by bacteroids *in vitro* in the presence of apolHb (Cutting and Schulman 1972) suggests that feedback inhibition of haem synthesis may be involved in the coordination of haem and globin synthesis within the nodule. Haem and apolHb can combine *in vitro* under approximately physiological conditions (Ellfolk and Sievers 1965); consequently, the free haem concentration in nodules is probably low, allowing haem to diffuse out of the bacteroids rapidly and therefore minimizing feedback inhibition of haem synthesis in the bacteroids. This line of thought leaves open the question whether the presence of apolHb triggers haem synthesis and excretion by the bacteroids or vice versa. Work by Roessler and Nadler (1982) suggests that iron may play a role in the regulation of haem synthesis in *R. japonicum* bacteroids. Iron-deficient laboratory-grown *R. japonicum* cells excreted protoporphyrin into the growth medium. The activities of the first two enzymes of haem biosynthesis, ALAS and ALAD were diminished in these cells, but were returned to normal levels upon addition of iron to the cultures. It is not clear whether iron (or some metabolite reflecting the iron supply of the cell) acts as a direct modulator of haem synthesis or whether the decreased ALAS and ALAD activities are a result of altered general metabolism in iron-deficient cells. The addition of iron or iron-chelators did not affect ALAS or ALAD activity in cell-free extracts suggesting that iron does not play a direct role in the activity of these enzymes *per se*. These data suggest that the iron supply of the cell affects haem biosynthesis in *R. japonicum* and that high ALAS and ALAD activities are maintained only when sufficient levels of iron are available to the cell. Thus rhizobial haem synthesis in the legume root nodule may be affected to some degree by the release of iron from the host plant to the bacteroids. Another factor that may be responsible for induction or stimulation of haem synthesis by bacteroids is anaerobiosis. Avissar and Nadler (1978) reported that stationary phase *R. japonicum* bacteria when kept under restricted aeration for several days increased their cellular haem content tenfold and released visible amounts of protoporphyrin and coproporphyrin into the culture medium. Concomitant, by a tenfold increase in ALAS and ALAD activity occurred. Reduced oxygen tension may thus play a role in inducing the haem synthesis in *R. japonicum* necessary for lHb synthesis.

In a recent paper, Keithlyt and Nadler (in press) consider the mechanism of haem synthesis by *Rhizobium* under microaerophilic conditions. There are two oxidative steps in this process; the oxidative decarboxylation of coproporphyrinogen III to protophyrinogen IX, catalysed by the enzymes coproporphyrinogen oxidative decarboxylase (COPRO'genase) and protoporphyrinogen oxidase (PROTO'genase) respectively (Jacobs 1974). For this reaction, most aerobic organisms use an aerobic COPRO'genase which has an absolute requirement for molecular oxygen as the electron acceptor and requires no cofactors. However, *R. japonicum* bacteroids appear to have an anaerobic COPRO'genase which transfers electrons to NADP$^+$ and has an allosteric requirement for S-adenosyl-l-methionine. Keithlyt and Nadler observe COPRO'genase activity in extracts of *R. japonicum* bacteria as well as bacteroids under anaerobic conditions. This is the first report of an anaerobic COPRO'genase in an obligate aerobe and shows that an anaerobic enzyme system may have a role in haem synthesis by *Rhizobium* bacteroids. These data lead us to propose tentatively that the following series of events takes place during nodule formation. Rhizobia in the plant divide and consume most of the available oxygen, thereby creating a low-oxygen environment. This induces haem production y the bacteroid, which stimulates apolHb synthesis by the plant; other, yet unknown factors, probably play also a role in this process, as *Rhizobium* mutants disturbed in haem synthesis do induce apolHb synthesis in the plant (Bisseling *et al.* 1983). The resulting lHb then regulates the necessary oxygen supply to the bacteroids and buffers the free oxygen concentration at an optimal value.

11.3.3 ApolHb synthesis in root nodules

It is proven now that lHb protein is encoded on the chromosomal DNA of the legume. cDNA of mRNA coding for lHb has been cloned and used to isolate genomic clones of lHb-coding sequences from soyabean (Brisson *et al.* 1982; Verma *et al.* 1981; Verma *et al.* 1980; Marcker *et al.* 1983) and pea (Bisseling *et al.*, unpublished). The expression of the lHb-genes has most extensively been studied in soyabeans and peas. The apolHb in soyabean is synthesized on free polysomes in the host-cytoplasm (Verma and Bal 1976; Verma *et al.* 1978). Electrophoresis separates the lHbs in a fast-migrating lHbF (corresponding to the lHbc-gene products) and a slow-migrating lHbS-fraction comprising the lHba-component. The relative rates of biosynthesis of lHbS and lHbF change during root-nodule development; lHbF synthesis predominates in young nodules (<12 days) while lHbS synthesis predominates in mature nodules (Verma *et al.* 1974; Verma, 1976; Verma *et al.* 1979; Nash and Schulman 1976). However, between these two major forms of lHb in soyabean root nodules, lHbF always exists in a larger quantity than lHbS (ratio of $>1.5{:}1$) in nodules of different ages (Fuchsman *et al.* 1976). By DEAE-cellulose chromatography (Verma *et al.* 1979) lHbF could be separated into two components, lHbF(c1) and lHbF(c2) of approximately equal amounts. The different rates of lHbF and lHbS synthesis reflect different mRNA concentrations. Table 11.1 shows the rates of synthesis of the two lHb components and their total accumulation (lHbF/lHbS) in soyabean root nodules.

A detailed study of the rates of transcription of the different 1Hb-genes during nodule development in soyabeans was recently published by Marcker *et al.* (1984). Hybridization probes specific for each 1Hb gene were constructed to monitor the appearance of the separate 1Hb mRNAs. The 1Hb genes appear to be activated sequentially in the opposite order to which they are arranged in the soyabean genome. At a specific stage during nodule development transcription of the genes encoding 1Hbc$_1$, 1Hb$_3$ and 1Hba increases sharply, while the transcription of 1Hbc$_2$ is not amplified to a similar extent. This increase in 1Hb gene transcription precedes the activation of the bacterial nitrogenase genes.

Pea root nodules contain two major lHb components (lHbI and lHbIV); a correlation between these two and the soyabean lHbs has

TABLE 11.1

Ratio of lHbF/lHbS during development of soyabean root nodules

Age of nodules (days after infection)	Rate of synthesis (lHbF/lHbS)		Total accumulation lHbF/lHbS
	in vivo[a]	*in vitro*[b]	
12	1.74	1.9	n.d.
14	1.44		1.75
21	0.95	0.86	1.5
28	0.74		1.46

(From Verma *et al.* 1979.)

[a] Measured by *in vivo* labelling of nodules with [^{3}H]leucine, isolation of lHb and analysis on polyacrylamide gels.

[b] Total polysomes were isolated and translated *in vitro* in a wheat germ translation system. Products were analysed on polyacrylamide gels.

not yet been established. lHbI and lHbIV appear to undergo a shift during nodule development in a similar fashion as the soyabean lHbs (Uheda and Syono 1982a); in early stages, lHbI synthesis is favoured, and later a strong increase in the lHbIV concentrations is noted (Table 11.2). The distribution of lHbI and lHbIV within the nodule appeared in homogeneous. In 20-day nodules the lHbI/lHbIV ratio appeared to be different in the younger and older parts of the bacteroid containing tissue (approximately 1.32 and 0.87 respectively), reflecting the higher content of lHbIV in older cells in the region proximal to the root.

The hypothesis that lHbs with a higher oxygen-binding capacity would be more effective in nitrogen fixation was tested by addition of purified lHb components to *R. japonicum* or *R. leguminosarum* bacteroids isolated from nodules. Uheda and Syono (1982a) measured the oxygen binding capacities of pea lHbs and found oxygen partial pressures for half oxygenation of lHbI and lHbIV of 0.112 and 0.081 mm Hg respectively. Similar measurements on purified lHba and lHbc from soyabean (Uheda and Syono 1982b) gave values of 0.024 and 0.036 mm Hg respectively. These values are smaller than those reported earlier by Appleby (1962) who found 0.040 and 0.068 mm Hg respectively for lHba and lHbc; the reduction system used in these experiments, however, was unstable and lHb degradation may partly be responsible for the difference. It proved that lHbs with the lower oxygen partial pressure for half oxygenation, thus the stronger oxygen-binding, i.e. lHba and lHbIV stimulated respiration as well as nitrogen fixation by isolated bacteroids to a greater extent than lHbc and lHbI respectively. It thus seems that the appearance of nitrogen-fixing activity in nodules is parallelled by an increase in the amount of

TABLE 11.2

Leghaemoglobin contents in pea nodules of different ages

Nodule age (days after infection)	lHb content (nmole/g fresh wt)	Total lHb accumulation lHbI/lHbIV
12	15.48	3.88
14	20.25	2.11
17	23.44	1.28
20	23.58	0.85
23	28.55	0.71
26	25.12	0.53

(From Uheda and Syono 1982*a*.)

Pisum sativum cv. Alaska was inoculated with *R. leguminosarum* strain 128C53. lHb was extracted from nodules at the time indicated and subjected to polyacrylamide gel electrophoresis. Relative lHb-contents were determined from densitometric tracings at 420 mm. Absolute lHb-contents were determined by the pyridine haemochromogen method (Stripf and Werner 1978).

lHb components with higher oxygen-binding affinities. (lHba and lHbIV). These results indicate that nitrogen fixation and oxygen transport are closely correlated. Furthermore, because the relative contents of lHb-components with higher oxygen-binding affinities increase with nitrogen fixation, lHb-heterogeneity may be an evolutionary adaptation for more effective nitrogen fixation.

11.3.4 Factors influencing (apo)lHb synthesis

Nitrogenase activity in root nodules is always associated with the presence of lHb. The question whether lHb synthesis in the nodule is a consequence of the presence of active nitrogen-fixation by the bacteroids or on the other hand is a prerequisite for nitrogen fixation has been approached by several authors. In soyabean root nodules, Bergersen and Goodchild (1973) detected lHb (as haemochromogen) on the second day after nodules appeared and acetylene-reducing activity on day four. Verma *et al.* (1979) determined the appearance of lHb using three different methods differing in sensitivity: (1) direct measurement of haemochromogen; (2) the use of antibodies specific against lHb; (3) analyses of the translation products of nodule polysomes during nodule development. Figure 11.5 shows that lHb measured as haematin appeared on the same day as nitrogenase activity. However, the presence of apo-lHb can be detected three days and the *in vitro* translation product four days prior to nitrogenase activity (at day 10). Of the two lHbs, lHbF appears first and is followed by lHbS (see Fig. 11.4, inset). The accumulated ratio of lHbF/lHbS declines during nodule development

(Table 11.1) suggesting that lHbF turns over with a slower rate than lHbS, resulting in the accumulation of lHbF in mature nodules. Auger *et al.* (1979) quantified the increase in concentration of lHb mRNA in developing nodules by hybridization of lHb-cDNA to polysomal RNA from nodules of different developmental stages. Their results indicate a rapid increase in lHb–mRNA concentration between eight and thirteen days, which parallels the biosynthesis of lHb in this tissue. The conclusions of Verma *et al.* (1979) and Auger *et al.* (1979) confirm earlier reports by Bergersen and Goodchild (1973), where lHb was detected, using the haemochromogen method. Contradictory results of Nash and Schulman (1976) on soyabean and Broughton *et al.* (1978) on bacteroids from *Centrosema pubescens* nodules are probably due to the lower sensitivity at which lHb was detected as compared with the sensitivity of the acetylene reduction assay.

In pea root nodules induced by *R. leguminosarum* PRE lHb was detected at the same day (Bisseling *et al.* 1979) by the Mancini technique (Mancini *et al.* 1965) or two days earlier by a more sensitive radioimmunoassay (Bisseling *et al.* 1980) than nitrogenase activity measured as acetylene reduction.

The above-mentioned reports lead to the conclusion that, by using a sufficiently sensitive assay, lHb is detectable before any nitrogenase activity is present. It thus appears that induction of the lHb-genes does not result from the presence of nitrogen-fixing rhizobia. This is confirmed by experiments of Verma *et al.* (1981) who measured the expression of lHb-genes in fix⁻-nodules induced by respectively a fix⁻ mutant (SM5) or a fix⁻ isolate strain (61A24) of *R. japonicum* in soyabeans. In both cases, lHb-apoprotein was synthesized although in very different concentrations: 30–40 per cent of the wild-type level by SM5 and very low levels by 61A24. *In vitro* translation of polysomes from the nodules followed by immunoprecipitation showed that this result is due to differences in the levels of translatable mRNA rather than degradation of the protein product.

In order to determine if the reduced level of lHb synthesis in fix⁻-nodules was due to a reduction in the level of mRNA or its translatability, the concentration of 1 HB-mRNA in polysomes of fix⁺ and fix⁻ nodules was determined by hybridization of polyA⁺-polysomal RNA to a purified lHb-cDNA probe (Baulcombe and Verma 1978) as well as to the plasmid lHb1 containing a lHb-sequence. The results of these experiments are summarized in Table 11.3. The reduction of lHb-mRNA sequences, as presented per μg RNA, ranges from approximately 5.4-fold in SM-5 induced to about 78-fold in 61A24 induced nodules compared with (wild-type) 61A76 induced nodules. However, even at the lowest level of lHb (as in 61A24 induced nodules) lHb–mRNA sequences are very abundant, representing about 1200 mole-

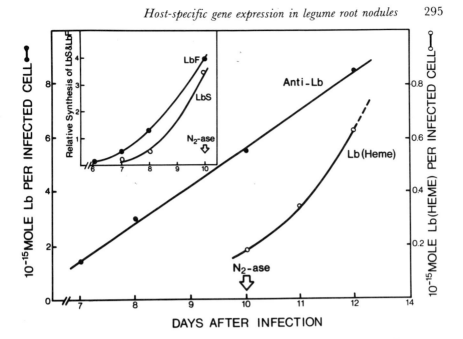

FIG. 11.5. Appearance of lHb in relation to nitrogenase activity during soya-bean root nodule development. Total lHb was determined by haemo-chromogen (lHb-haeme), by rocket immunoclectrophoresis (anti-lHb) and by *in vitro* translation of polysome in wheat germ translation systems (inset). Relative rates of synthesis of lHbS and lHbF were measured by electro-phoresis of the translation products on polyacrylamide gels and densitometric analysis of the fluorographs. Nitrogenase was assayed by reduction of acetylene into ethylene. The data are presented per infected cell which were counted in 10 μm thick sections after staining with haemoxylin. Volume of the nodules was calculated according to Bergersen and Goodchild (1973). Inter-stitial cells were excluded from the infected zone. (From Verma *et al.* 1979.)

cules per infected cell. In soyabean nodules induced by the same *R. japonicum* strains mentioned above lHb–mRNA concentrations were also measured by hybridization to a genomic lHb-clone (pLb1) and similar conclusions were reached (Brisson *et al.* 1982). Since the relative levels of lHb–mRNA in the different nodules correspond roughly to those of their translation products, the primary regulation of lHb genes appears to be at the level of transcription and/or processing of mRNA. Analysis of the *in vitro* translation products under conditions where the two major forms of lHb can be distinguished (Verma *et al.* 1979; Appleby *et al.* 1975) showed that both major forms of lHb (lHbF and lHbS) are being synthesized in SM5-induced nodules, but only a low level of lHbF and no lHbS is induced by strain 61A24. This result suggests that

TABLE 11.3

lHb-mRNA sequences in soyabean nodules induced by fix⁺ and fix⁻
R. japonicum *strains*

R. *japonicum* strain	lHb-mRNA molecules per g nodule	lHb-mRNA molecules per infected cell
61A76 (fix⁺)	1.01×10^{12}	5.49×10^4
SM5 (fix⁻)	1.2×10^{11}	6.52×10^3
	(1.23×10^{11})	(6.68×10^3)
61A24 (fix⁻)	5.88×10^9	1.27×10^3
	(8.82×10^9)	(1.9×10^3)

(From Verma *et al.* 1981.)

lHb-mRNA as determined by [^{3}H]poly(U) hybridization (Verma and Maclachlan 1976), comprised 20 per cent of the poly(A)-RNA in fix⁺-nodules formed by 61A76 (Auger *et al.* 1979); values for other strains were calculated from quantitive hybridizations of RNA to pLb1 (genomic lHb-clone) and lHb-cDNA. Approximate numbers of infected cells per nodule were calculated from observations of ten thick sections of 3-week-old nodules under a microscope (Bergersen and Goodchild 1973; Verma *et al.* 1979), SM5-induced *fix⁻*-nodules have a similar number of infected cells as 61A76-induced, but 61A24-induced nodules have approximately one-quarter as many infected cells on a per g basis.

lHb-genes are regulated independently. SM5 induced nodules are particularly interesting since only the nitrogenase function is affected in this strain of *Rhizobium* (Maier and Brill 1976).

Van den Bos *et al.* (1983) constructed transposon Tn5-insertion mutants of *R. leguminosarum* PRE by recombinant DNA techniques. Due to the construction protocol (Ruvkun and Ausubel 1981) the genotype of these strains can be shown to result from Tn5 insertions in well-defined position of the nitrogenase structural genes. The phenotype of insertions in the *nif*-structural region appeared to be *fix⁻* and the nitrogenase proteins were absent from these mutants as predicted from the site of insertion. lHb-synthesis in nodules induced by these *fix⁻*-mutants was studied; the accumulated lHb at three time-points was estimated; Fig. 11.6 shows that in nodules induced by *fix⁺* as well as *fix⁻* strains lHb is synthesized early in developmental stages. For a continued lHb synthesis, nitrogenase activity appears to be essential; if this activity does not arise lHb is gradually degraded.

Since leghaemoglobin is synthesized in both soyabean (Verma *et al.* 1981) and pea nodules (Van de Bos *et al.* 1983) during the proliferation and differentiation of bacteria prior to nitrogen fixation, it is possible that a reduced level of O_2 in infected cells may in part be responsible for its induction. Such a low pO_2 has been shown to influence the expres-

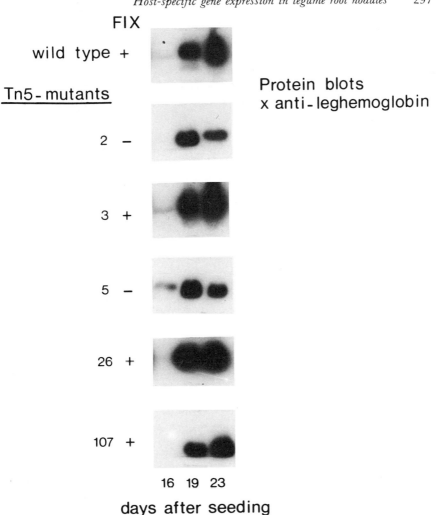

FIX

wild type +

Tn5-mutants

Protein blots
x anti-leghemoglobin

2 −

3 +

5 −

26 +

107 +

16 19 23

days after seeding

Fig. 11.6. Lb concentrations in *fix*⁺ and *fix*⁻ *R. leguminosarum* induced nodules of *Pisum sativum*. Soluble proteins were isolated from nodules at various intervals after seeding; 10 µg of protein was separated by polyacrylamide gel electrophoresis. Proteins were blotted on nitrocellulose and the filter was probed with antiserum specific against lHb. Immune complexes were made visible by incubation with ^{125}I-protein A and autoradiography. (From Van den Bos *et al.* 1983.)

sion of specific *Rhizobium* genes, including the induction of haem excretion and nitrogenase (Avissar and Nadler 1978; Reporter and Hermina 1975). Apart from this low pO$_2$, however, the presence of *Rhizobium* is a prerequisite for lHb-gene induction and for a substantial lHb-synthesis nitrogen fixation by the bacteroids is necessary. We suggest that this low pO_2 directly results from the presence of *Rhizobium* bacteroids. In an early stage of nodule development, actively-dividing rhizobia are responsible for a low pO_2 in the plant cells containing bacteria. This condition stimulates the bacteroids to produce an excess haem, which is excreted into the plant cell and stimulates in these cells the (already) induced synthesis of apolHb. At a later stage, when division of bacteria has ceased, active oxidative phosphorylation is demanded by nitrogen fixation in the bacteroids, which maintains a low pO_2 and thereby the conditions for a continued synthesis of lHb. The amount of lHb present in a nodule has been shown to have a highly significant linear correlation ($r = 0.93$; $P < 0.001$) with nitrogenase activity (Larue and Child 1979) even during nodule senescence. This implies that the rates of lHb-synthesis and/or degradation are influenced by the rate of nitrogenase activity. Chen and Phillips (1977) found that ammonium or nitrate absorption by nodulated pea roots caused an immediate decrease of nitrogenase activity and, after a 2–3 day lag, a similar diminution of the lHb content (measured by the pyridine haemochrome technique). These results were confirmed by Bisseling *et al.* (1978) who showed that apolHb (by radioimmunoassay) as well as haem (by pyridine haemochrome assay) concentrations decrease by the addition of ammonium. Surprisingly, however, the apolHb synthesis was not affected by ammonium. Explanations for this phenomenon may be an increased lHb-degradation by ammonium or an interference of ammonium with the detection of apolHb.

11.4 NODULE-SPECIFIC HOST GENES; NODULINS

In the following paragraphs we will discuss the molecular biological experiments, which have revealed that besides lHb 20–30 root-nodule-specific host genes are expressed in effective root nodules. The identification of these nodulins, their location in the nodule cell and expression in ineffective root nodules will be discussed. For a more detailed review see Bisseling *et al.* (1984)

11.4.1 Expression of nodulin genes as studied by RNA/cDNA hybridization experiments

Fractioned cDNA and single-copy DNA have been sensitive tools in studying tissue specific gene expression (Hastie and Bishop 1979; Ernst, Britten and Davidson 1979; Kamalay and Goldberg 1980). Auger and Verma (1981) have used fractionated cDNA, which was synthesized *in vitro* from nodule polyA(+) RNA, to study the expression of nodule specific host genes, the so-called nodulin genes.

The total mRNA populations of roots and nodules were compared by reassociation kinetics of RNA driven cDNA/RNA hybridizations.

The rate of reassociation of cDNA to its template RNA in vast excess is a pseudo first-order reaction where the Rot (Ro : initial RNA concentration) value at 50 per cent hybridization ($Rot_{1/2}$) is directly related to the base sequence complexity of the RNA. cDNA was prepared from polyA(+) RNA isolated from roots and nodules respectively and hybridized to the homologous or heterologous polyA(+) RNA. These hybridization experiments revealed a marked increase in concentration of the super abundant and moderately abundant sequences *et al.* 1979), which are distinguished by their reassociation kinetics. The super abundant mRNA of nodules consists of lHb sequences and are absent in root tissue. To determine whether in the middle abundant mRNA population nodulin sequences occur, it was essential for Auger and Verma (1981) to fractionate this class of sequences. cDNA prepared from nodule polyA(+) RNA was fractionated, as shown in Fig. 11.7, into a nodule specific (NS-cDNA) fraction and a class which roots and nodules have in common (common M-cDNA). The NS-cDNA and common M-cDNA were used in hybridizations to polyA(+) RNA *et al.* from nodules and uninfected roots (Fig. 11.8). The class of NS-sequences appeared to hybridize to more than 95 per cent with its homologous mRNA ($Rot_{1/2} = 0.24$) and to less than 15 per cent with mRNA from uninfected roots ($Rot_{1/2} = 83$). The hybridization with root mRNA was probably caused by contamination of the NS-cDNA probe with other sequences. The common M-cDNA hybridized to RNA from nodules as well as roots to greater than 90 per cent. However, the difference in kinetics of common M-cDNA to RNA from roots ($Rot_{1/2} = 75$) and nodule tissue ($Rot_{1/2} = 5$) indicates that the relative concentration of these sequences increases significantly during nodule development.

The use of fractionated cDNA probes enabled the quantitative characterization of the NS-cDNA population. Assuming an average molecular weight of 40 000 dal for the nodulins then the hybridization kinetics indicate that about 20 nodulins may be present in soyabean nodules in addition to lHb.

11.4.2 Identification of nodulins

Nodulin 35 (N-35; mw 35 kdal) was the first nodulin detected in soyabean nodules besides lHb (Legocki and Verma 1979). N-35 appears to be induced in parallel with lHb and it represents about 4 per cent of the total soluble plant protein fraction of soyabean nodules. After N-35 was identified, Legocki and Verma (1980) used an immunological assay which allowed the identification of a broader group of nodulins. A similar approach was used by Bisseling *et al.* (1983) who analysed pea nodules. Both groups used nodule specific antiserum preparations. These preparations consisted of antisera raised against total plant

Preparation of M- and NS-cDNA probes

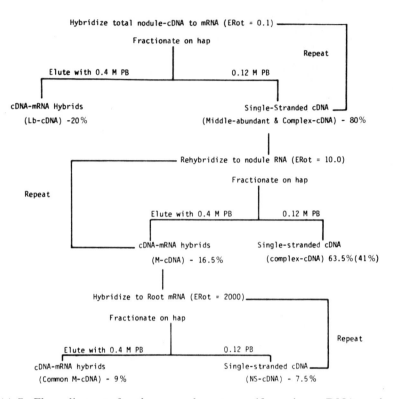

FIG. 11.7. Flow diagram for the procedure to purify various cDNA probes. The approximate mass fraction of total nodule cDNA comprising each moiety is indicated as a percent value. lHb-cDNA—20 per cent, comprises the most abundant nodule mRNA class, M-cDNA—16.5 per cent, the moderately-abundant mRNA population, is sub-divided into the nodule-specific (7.5 per cent) and common-M (9 per cent) moieties. Complex cDNA is only 65 per cent reactive with nodule mRNA, and hence contains some unhybridizable elements. Its corrected mass fraction, 41 per cent (0.65×0.635) is indicated in parentheses. (From Auger and Verma 1981.)

proteins from root nodules, which were made nodule specific by titration with proteins from uninfected roots.

Legocki and Verma (1980) identified soyabean nodulins after *in vitro* translation of (80 S-type) plant polysomes from root nodules. These polysomes have previously been shown to be devoid of bacteroid ribosome contamination (Verma and Bal 1976), providing a suitable system to study the biosynthesis of host proteins. The plant polysomes were translated in a wheat germ system and the nodulins were precipitated

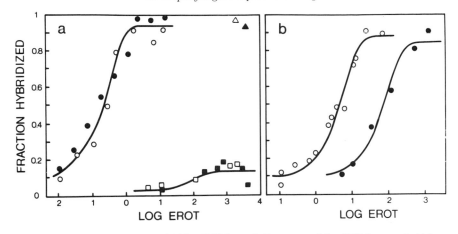

FIG. 11.8. Hybridization of NS–cDNA and Common M–cDNA to poly(A) + RNA from nodules and uninfected root. (a) NS–cDNA was purified from M–cDNA (Fig. 11.7). Open symbols represent a NS–cDNA probe prepared following hybridization with total poly(A) + RNA from uninfected root, assayed by TCA precipitation, and closed symbols represent a different NS–cDNA probe preparation assayed using Whatman DE-81 filters. About 1000 cpm of NS–cDNA were hybridized to 2000-fold excess nodule poly(A) + RNA (●) and 140 000-fold excess uninfected root poly(A) + RNA (■□). At the termination of the hybridization to root-RNA, homologous (nodule) RNA was added, allowed to hybridize to EROT = 10, whereupon NS–cDNA hybridized to greater than 95 per cent with its homologous RNA (▲△), compared to about 15 per cent with uninfected root RNA. (b) Common M–cDNA (see Fig. 11.7), freed of the nodule-specific moiety, hybridized to root-RNA (●), and with nodule RNA (0). From Auger and Verma 1981.)

with a nodulin specific antiserum preparation. The immunoprecipitates were analysed by two-dimensional polyacrylamide gel electrophoresis. Panels A and B of Fig. 11.9 show the total translation products of root and nodule polysomes and panels C and D the corresponding immunoprecipitates. Only one root polypeptide precipitates with the nodule specific antiserum (panel C), while 20 nodulins can still be precipitated from translation products of nodule polysomes (panel D). The nodulins, that were identified, all had a rather low molecular weight (about 20 kdal) (Fig. 11.9D). The reason for this is not clear. It is also unclear why N-35, which is a prominent nodulin in soyabean (Legocki and Verma 1979), is not detectable on these gels. This result illustrates that not all nodulins can be identified by this procedure.

lHb is not detectable in Fig. 11.9 since the translation products were labelled with ^{35}S methionine an amino acid which does not occur in soyabean lHb. Besides the small nodulins shown in Fig. 11.9 some larger ones are present in soyabean nodules. N-35 (Legocki and Verma 1979) is

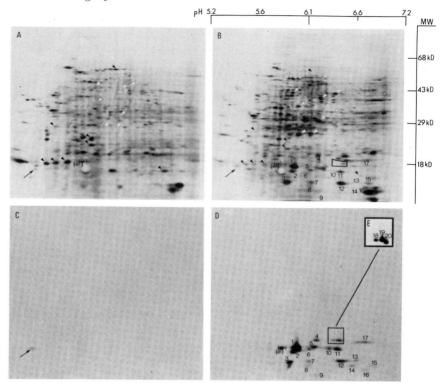

FIG. 11.9. Two-dimensional PAGE (Fluorogram) of ^{35}S-methionine-labelled *in vitro* translation products of total polysomes from uninfected roots (A) and nodules (B) of soyabean. After pre-adsorption with a non-immune serum equal amounts (2.5×10^6 counts/min) of the translation products were reacted with the 'nodule-specific' antiserum and the resulting immunoprecipitates were on separated two-dimensional gels. (C). Immunoprecipitation of proteins from uninfected roots. (D) immunoprecipitation of proteins from root nodules. Isoelectric focusing carried out in a narrower pH range results in a better resolution of polypeptides numbers 18, 19, and 20. (E) twofold magnification. Black arrowheads indicate several root polypeptides which disappear in nodules; white arrowheads show the presence of other 'nodule-specific' proteins which are not immunoprecipitated by this antiserum. Among the nodulins (D) there are two peptides (shown in parentheses) which appear to be in common with the uninfected root (A), but are not immunoprecipitated from the latter (C). The arrow indicates a polypeptide common to uninfected roots and nodules which, however, is not synthesized at this stage of root nodule development. (From Legocki and Verma 1981.)

one example. Furthermore, when nodule specific cDNA clones were used in a hybrid released translation experiment it was shown that the corresponding mRNAs were translated *in vitro* into polypeptides up to

100 kdal (Verma *et al.* 1982; Fuller *et al.* 1983). Bisseling *et al.* (1983) have identified pea nodulins with a nodule-specific antiserum preparation. Plant proteins were separated on a SDS-polyacrylamide gel, blotted onto a nitrocellulose filter and incubated with a nodule specific antiserum preparation. The γ-globulins were visualized with ^{125}I-protein A.

In the pea-*R. leguminosarum* (PRE) symbiosis, the first root nodules become visible on day 10 after inoculation. Nitrogenase activity measured as acetylene reducing activity can be detected on day 12; this activity increases rapidly reaching its maximal rate between three and four weeks after sowing and then decreases (Bisseling *et al.* 1979). Nodulins present at different stages of development were visualized with the nodule specific antiserum as shown in Fig. 11.10. The nodule-specific protein are numbered in order of size; number one (N-1) being

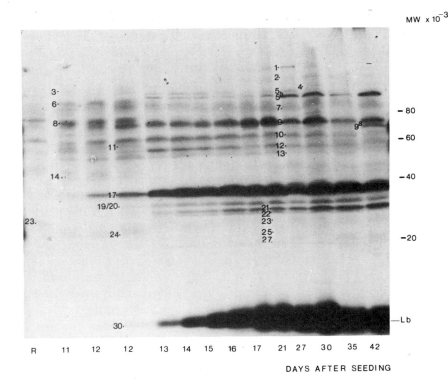

FIG. 11.10. Pea nodule-specific proteins during nodule development: autoradiograph of a protein blot of soluble pea proteins from uninfected roots (R) and root nodules of pea and *R. leguminosalum* (PRE) harvested at different times after sowing, as indicated. The blot was incubated with a nodule-specific antiserum and ^{125}I-protein A. (From Bisseling *et al.* 1983.)

the largest polypeptide detected. The smallest nodule-specific protein N-30, is lHb.

Some nodule-specific proteins, e.g. N-3, N-6, N-8, N-11, N-14, and N-24, were found mainly in nodules at an early stage of root nodule development, i.e. on 11 and 12 days after sowing. Other nodule-specific proteins appeared to accumulate in the course of nodule development as illustrated by lHb, which was first detected at day 12 and the amount of which increased until about day 21. N-17 behaved in a similar way, but the synthesis of N-17 was detectable before that of leghaemoglobin. The antiserum used to detect the nodule-specific proteins was developed against protein from nodules at 21 days after sowing when nitrogen fixation is maximal. It is remarkable that with such an antiserum nodule-specific proteins which are preferentially detectable in early stages of nodule development can be demonstrated. Pea root nodules from 21-day-old plants have a zone of tissue that is still meristematic as well as a zone of senescent tissue in the oldest parts of the nodule (van Brussel 1973; Kijne 1975). Apparently, early nodule-specific proteins still occur in 21-day nodules and are highly antigenic but the concentration of the nodule-specific proteins at day 21 is too low to enable their detection by the blotting procedure.

The polypeptides 10 and 23 were also detected in un-inoculated roots, although the bands were more intense in root nodules. These polypeptides may therefore represent proteins, the synthesis of which is stimulated in nodules, or the bands are composed of polypeptides occurring in un-inoculated roots and nodule-specific polypeptides. Since soyabean and pea nodule-specific proteins are detectable on gels, they have to belong to the population of abundant proteins (Davidson and Britten 1979) and therefore the corresponding mRNA sequences will most probably form part of the moderately abundant mRNA population. The studies of Auger and Verma (1981) on the moderately abundant mRNA population of soyabean nodules revealed that within this population 20–40 sequences are nodule-specific. So these studies and the observations of Legocki and Verma (1980) and Bisseling *et al.* (1983), that 20–30 nodule-specific proteins can be identified, are in good agreement with each other.

The function of most of the identified nodulins is yet unknown, but obviously some of them will be involved in the nitrogen-metabolism. Recent work of Jochimsen and Rasmussen (1982) revealed that probably N-35 of soyabean nodules is the enzyme uricase. This enzyme functions in the catabolism of purines resulting in the formation of ureides which are the most abundant transport form of fixed nitrogen from nodules to shoots (for review see, e.g., Reynolds *et al.* 1982). Furthermore, Cullimore *et al.* (1982) have identified a novel form of glutamine synthetase in root nodules of *Phaseolus vulgaris*, which is different from the root-type enzyme and the two forms of glutamine synthetase found in leaves. Therefore this isoenzyme might also be a nodulin.

11.4.3 Synthesis of nodulins in ineffective root nodules

To study the expression of nodulin genes, several groups have analysed ineffective nodules, which were formed by inoculation with *fix⁻* rhizobia (Legocki and Verma 1981; Noel *et al.* 1982; Jochimsen and Rasmussen 1982; Bisseling *et al.*1983). In principle, mutagenesis of the nodule-specific proteins genes would have been the best approach to study their expression. However, since this technology is not yet applicable to plants, the above-mentioned authors all used *nod⁺ fix⁻ Rhizobium* strains and nodulin synthesis was studied in the corresponding ineffective nodules.

Bisseling *et al.* (1983) used well-characterized Tn*5*-mutated rhizobia and *pop⁻* mutants. The *pop⁻* strains are disturbed in haem synthesis and as a result give rise to ineffective nodules (Nadler *et al.* 1981; Bisseling *et al.* 1983). The Tn*5*-mutants have a transposon inserted in the structural *nif* genes (mutants 116A and 2 in the *nifH* sequence).

The nodule-specific proteins of the ineffective pea-nodules are compared with the corresponding pattern in effective nodules (Fig. 11.11). In all mutants a marked decrease in concentrations of some nodule-specific proteins is observable (N-15 and N-17) while others remain unchanged and some are stimulated (N-22).

Auger and Verma (1981) showed with cDNA/mRNA hybridization experiments that the concentration of NS-sequences decreases in ineffective nodules. With these experiments it is impossible to determine whether the concentration of all nodule-specific mRNAs decreases, or alternatively that some decline while the concentration of others remains unchanged. Legocki and Verma (1981) demonstrated that some nodule-specific mRNAs disappeared, while on the other hand several nodule-specific mRNAs of very low abundance in the effective nodules become easily detectable in the ineffective ones. This stimulation of synthesis of some nodule-specific mRNA is similar to the observations on pea-nodule-specific proteins. All authors who analysed ineffective nodules showed a marked decrease in concentration of some of the nodule-specific proteins. This is a strong indication that at least these nodulins are really involved in nitrogen-fixation and are not the result of, for example, a defence mechanism of the host plant.

11.5 CONCLUDING REMARKS

The studies of Legocki and Verma (1980) and Verma *et al.* (1982) with soyabean and those of Bisseling *et al.* (1983) with pea-root nodules have

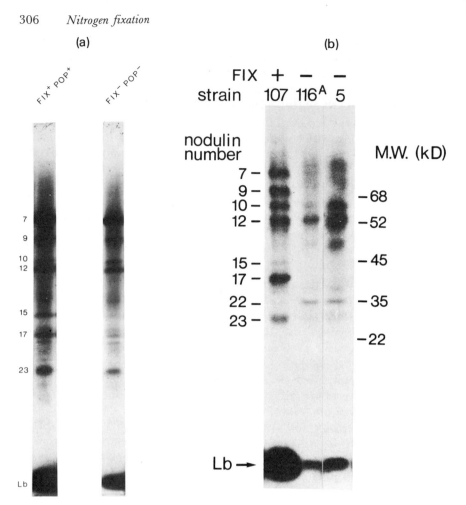

FIG. 11.11. Nodule-specific protein pattern of ineffective nodules formed after infection with a pop⁻ mutant and fix⁺ and fix⁻ PRE::Tn5 strains. Nodule-specific protein were identified by protein blotting and nodulin specific antiserum incubations. A. Nodule-specific protein pattern of nodules of pea and *R. leguminosarum* strain 1062 (fix⁺, pop⁺) and mutant 116 (fix⁻, pop⁻) of this strain and B of pea and *R. leguminosarum* PRE::Tn5 strains 107 (fix⁺), 116A (fix⁻) and 5 (fix⁻). Strain 107 behaves as the wild-type PRE. (From Bisseling *et al.* 1983 and Van den Bos *et al.* 1983.)

demonstrated that besides lHb a considerable number of other polypeptides can be detected, which occur in nodules, but not in uninfected root tissue. Nodulins have been defined as nodule-specific host proteins and it remains to be seen how many of the polypeptides now termed nodulins will in fact meet the requirements of this definition upon further study. It is possible that the group of more than 20 proteins now indicated as nodulins, will appear to be a mixture of proteins unique to effective root-nodules and other proteins. Among the so-called nodulins there may be proteins produced by the plant as a reaction to infection by *Rhizobium* bacteria, as a defence mechanism against invading parasites. Bergersen (1982) pointed out that 'there are some aspects of nodule development which resemble an arrested parasitic infection with the host ultimately elminating the parasite as the nodule senesces (see, e.g., Truchet and Coulomb 1973). The period of N_2-fixation in these terms is an arrested but transient stage and is a special adaptation of general phenomena associated with infection by micro-organisms in plants'. Some nodulins may therefore turn out to be proteins related to infection and not required to accomplish an effective nitrogen-fixing root-nodule. This might specifically apply to the polypeptides detected in later stages of nodule development when part of the tissue containing bacteroids starts to senesce.

Similarly, it appears that there are enzymes and other proteins among the nodulins which occur at the same time, but in much lower concentration, in other tissues of the plant, and which have not yet been detected for that reason. This might apply to proteins characteristic for meristematic tissue and to enzymes involved in the assimilation of NH_3 and/or the carbon metabolism in nodule cells. A number of enzymes may have increased to a large extent because of the specialized function of nodule cells. It should be kept in mind, however, that some of these enzymes like, for example, phosphoenolpyruvate carboxylase or any of the enzymes of the ammonia assimilation pathways, might be isoenzymes like the nodule-specific glutamine synthetase (Cullimore *et al.* 1982), which are expressed from other genes than the enzymes with similar functions in other tissues of the plant, including root tissue.

Furthermore, it remains to be seen whether some of the polypeptides now described as nodule-specific proteins are actually bacterial proteins excreted by the bacteroids. For that matter, special attention should be given to the proteins occurring in the peribacteroid space. The experimental difficulties encountered in studies designed to elucidate this latter type of problem are illustrated by the conflicting results obtained for lHb. Several groups (Robertson *et al.* 1978; Verma *et al.* 1978; Bisseling *et al.*, 1983) have claimed that lHb is confined to the cytoplasm of bacteroids containing nodule cells and does not occur in the peribacteroid space. In contrast, other groups (Dilworth and Kidby 1968; Bergersen and Goodchild 1973; Livanova *et al* 1979; Bergersen and Appleby 1981) maintain that lHb occurs, not only in the cytoplasm, but also in the peribacteroid space and that this specific location of lHb is important in regulating nitrogenase activity. These differences of opinion should be resolved as soon as possible since they implicate different views on the precise functions of, in this case, lHb. lHb is synthesized on polysomes in the plant cytoplasm. Translation of isolated lHb mRNA in an *in vitro* protein synthesing system has shown that lHb is not synthesized as a precursor polypeptide, to be processed into a functional protein (Verma *et al.* 1979). The complete nucleotide sequences of the lHb-genes in soyabean also indicate that there is not a specific lHb with a signal sequence for lHb transport through the membranes of the peribacteroid space as usually found with proteins, which must be transported from their cytoplasmic site of synthesis into a particular cellular compartment, like the mitochondria, the chloroplasts, or through the endoplasmatic reticulum (Blobel and Dobberstein, 1975; Highfield and Ellis, 1978; Neupert and Schatz, 1981). This appears to be a valid argument for the location of lHb in the plant cytoplasm and not in the peribacteroid space as long as the specific transport of lHb through the peribacteroid membrane is not explained.

Bisseling *et al.* (unpublished results), on the other hand, have recently demonstrated that specific proteins may be found in the peribacteroid space surrounding *R. leguminosarum* bacteroids in pea root-nodules, which appear to originate from the *Rhizobium* bacteria. The function of these proteins will be of considerable interest as the peribacteroid space represents the site of closest interaction between the bacteroids and the host cell.

Little is known about the regulation of the processes in the plant cells and the bacteroids which effect a nitrogen-fixing root-nodule. The very existence of lHb of which the protein is encoded on the plant genome and the prosthetic groups is produced by the rhizobia illustrates how closely interlinked these processes are and this is also amply indicated by the studies on the regulation of lHb synthesis summarized in this review. Analysis of the lesions in the symbiotic genes of *Rhizobium* mutants which produce ineffective nodules appear invaluable to gaining further insight in the interaction between rhizobia and the host plant cells. It has been shown already that in ineffective nodules of soyabean (Legocki and Verma 1980) and pea (Bisseling *et al.* 1983) produced by mutants of *R. japonicum* and *R. leguminosarum* respectively, the pattern of nodulins is strikingly different as compared to effective nodules. The possibility, at least with *R. leguminosarum*, to introduce by Tn5 mutations at well-defined positions in the symbiotic genes, as these are largely localized on a plasmid, allow the systemic extension of these studies.

It will not be long before more information becomes available about the genes of legumes involved in the formation of effective root-nodules. The data on the structure of the lHb-genes and their organization on the plant genome are of great interest but do not yet contribute to an insight in nitrogen fixation in root nodules. Soon, however, more

DNA copies of mRNAs for specific nodulins will have been cloned and will be used as probes for isolating the corresponding plant genes. It will then be of considerable interest to compare the structure and organization of the sets of genes for different nodulins and to determine their position on the plant genome. Moreover, it may then become possible to analyse lesions in genes of host plants which are defective in forming an effective nitrogen fixing root nodule and ultimately the molecular basis of the symbiosis between *Rhizobium* bacteria and legumes.

REFERENCES

APPLEBY, C. A. (1962). *Biochim. biophys. Acta* **60**, 226.
— NICOLA, N. A., HURRELL, J. G. R., and LEACH, S. J. (1975). *Biochem.* **14**, 4444.
AUGER, S., BAULCOMBE, D., and VERMA, D. P. S. (1979). *Biochim. biophys. Acta* **563**, 4960.
— and VERMA, D. P. S. (1981). *Biochemistry* **20**, 1300–1306.
AVISSAR, Y. J. and NADLER, K. D. (1978). *J. Bact.* **135**, 782.
BAULCOMBE, D. and VERMA, D. P. S. (1978). *Nucl. Acids Res.* **5**, 4141.
BEALE, S. I. and CASTELFRANCO, P. A. (1974). *Pl. Physiol.* **53**, 291.
— GOUGH, S. P., and GRAMICK, S. (1975). *Proc. natn. Acad. Sci. U.S.A.* **72**, 2719.
BERGERSEN, F. J. and GOODCHILD, D. J. (1973). *Aust. J. biol. Sci.* **26**, 741.
— APPLEBY, C. A. (1981). *Planta* **152**, 534.
— (1982). *Root nodules of legumes: structure and function.* John Wiley, New York.
AVISSAR, Y. J. and NADLER, K. D. (1978). *J. Bact.* **135**, 782.
BAULCOMBE, D. and VERMA, D. P. S. (1978). *Nucl. Acids Res.* **5**, 4141.
BEALE, S. I. and CASTELFRANCO, P. A. (1974). *Pl. Physiol.* **53**, 291.
— GOUGH, S. P., and GRAMICK, S. (1975). *Proc. natn. Acad. Sci. U.S.A.* **72**, 2719.
BERGERSEN, F. J. and GOODCHILD, D. J. (1973). *Aust. J. biol. Sci.* **26**, 741.
— APPLEBY, C. A. (1981). *Planta* **152**, 534.
— (1982). *Root nodules of legumes: structure and function.* John Wiley, New York.
BISSELING, T., VAN DEN BOS, R. C., and VAN KAMMEN, A. (1978). *Biochim. biophys. Acta* **539**, 1.
—— WESTSTRATE, M. W., HAKKAART, M. J. J., and VAN KAMMEN, A. (1979). *Biochim. biophys. Acta* **562**, 515.
— MOEN, A. A., VAN DEN BOS, R. C., and VAN KAMMEN, A. (1980). *J. gen. Microbiol.* **118**, 377.
— BEEN, C., KLUGKIST, J., VAN KAMMEN, A., and NADLER, K. (1983). *Eur. molec. Biol. Orgn. J.* **2**, 961.
BISSELING, T., GOVERS, F., STIEKEMA, W. (1984). In *Oxford Surveys of Plant Molecular and Cell Biology* (ed. B. Y. Miflin| Vol. I, p. 53. Clarendon Press, Oxford.
BLOBEL, G. and DOBBERSTEIN, B. (1975). *J. cell. Biol.* **67**, 835.

BOJSEN, K., ABILDSTEN, D., JENSEN, E. O., PALUDAN, K., and MARCKER, K. A. (1983). *Eur. molec. Biol. Orgn. J.* **2**, 1165.

BRISSON, N., POMBO-GENTILE, A., VERMA, D. P. S. (1982). *Can. J. Biochem.* **60**, 272.

— and VERMA, D. P. S. (1982). *Proc. natn. Acad. Sci. U.S.A.* **79**, 4055–9.

BROUGHTON, W. J., HOH, C. H., BEHM, C. A., and TUNG, H. F. (1978). *Planta* **139**, 183.

VAN BRUSSEL, A. A. M. (1973). Ph.D. Thesis, University of Leiden, the Netherlands.

CHEN, P. C. and PHILLIPS, D. A. (1977). *Pl. Physiol.* **59**, 440.

CULLIMORE, J. V., LARA, M., and MIFLIN, B. J. (1982). In *1st Int. Symp. Molec. Genet. Bacteria–Plant Interactions* (ed. A. Pühler). Springer-Verlag, Heidelberg.

CUTTING, J. A. and SCHULMAN, H. M. (1969). *Biochim. biophys. Acta* **192**, 486.

—— (1972). *Biochim. biophys. Acta* **229**, 58.

DART, P. (1977). In *A treatise on dinitrogen fixation* (ed. R. W. F. Hardy and W. S. Silver), Sect. III, p. 367. John Wiley, New York.

DAVIDSON, E. H. and BRITTEN, R. J. (1979). *Science* **204**, 1052–9.

DILWORTH, M. J. and KIDBY, D. J. (1968). *Exp. Cell Res.* **49**, 148.

DITTA, G., LEONG, S., CORBIN, D., BARRAN, L., STANFIELD, S., and HELINSKI, D. R. (1983). In *Proc. 1st. Int. Symp. Molec. Genet. Bacteria–Plant Interactions*, Springer-Verlag, Heidelberg.

ELLFOLK, N. and SIEVERS, G. (1965). *Acta chem. scand.* **19**, 2409.

ERNST, S. G., BRITTEN, R. J., and DAVIDSON, E. H. (1979). *Proc. natn. Acad. Sci. U.S.A.* **76**, 2209–12.

FUCHSMAN, W. H., BARTON, C. R., STEIN, M. M., THOMPSON, J. T., and WILLETT, R. M. (1976). *Biochem. biophys. Res. Commun.* **68**, 387.

FUCHSMAN, W. H. and APPLEBY, C. A. (1979). *Biochim. biophys. Acta* **579**, 314–24.

FULLER, F., KÜNSTNER, P. W., NGUYEN, T. and VERMA, D. P. S. (1983). *Proc. natn. Acad. Sci. U.S.A.* **80**, 2594.

GO, M. (1981). *Nature, Lond.* **291**, 90–92.

GODFREY, C. A. and DILWORTH, M. J. (1971). *J. gen. Microbiol.* **69**, 385.

— COVENTRY, D. R., and DILWORTH, M. J. (1975). In *Nitrogen fixation by free-living organisms* (ed. W. D. P. Stewart), p. 311. Cambridge University Press.

GOODCHILD, D. J. and BERGERSEN, F. J. (1966). *J. Bact.* **92**, 204.

GRIPPO, P., LACCARIUS, M., PARISI, E., and SCARANO, E. (1968). *J. molec. Biol.* **36**, 195–208.

GROSVELD, G. C., ROSENTHAL, A., and FLAVELL, R. A. (1982). *Nucl. Acids Res.* **10**, 4951–71.

HASTIE, N. D. and BISHOP, J. O. (1976). *Cell* **9**, 761–74.

HIGHFIELD, P. E. and ELLIS, R. J. (1978). *Nature, Lond.* **271**, 420.

HURELL, J. G. R. and LEACH, S. J. (1977). *FEBS Lett.* **80**, 23–6.

HYLDIG-NIELSEN, J. J., JENSEN, E. O., PALUDAN, K., WIBORG, O., GARETT, R., JORGENSEN, P., and MARCKER, K. (1982). *Nucl. Acids Res.* **10**, 689–701.

JACOBS, N. (1974). In *Microbial iron metabolism, a comprehensive treatise* (ed. J. B. Neilands), p. 125, Academic Press, London.

JENSEN, E. O., PALUDAN, K., HYLDIG-NIELSEN, J. J., JORGENSEN, P., and MARCKER, K. A. (1981). *Nature, Lond.* **291**, 677–9.

JOCHIMSEN, B. and RASMUSSEN, O. (1982). In *1st Int. Symp. Molec. Genet. Bacteria–Plant Interaction* (ed. A. Pühler). Abstracts, University of Biclefeld. p. 17.

KAMALAY, J. C. and GOLDBERG, R. B. (1980). *Cell* **19**, 935–46.

LaRUE, T. A. and CHILD, J. J. (1979). *Analyt. Biochem.* **92**, 11.

LEE, J. S., BROWN, G. G., and VERMA, D. P. S. (1983). *Nucl. Acids. Res.* **11**, 5541–53.

LEGOCKI, R. P. and VERMA, D. P. S. (1979). *Science* **205**, 190–3.

—— (1980). *Cell* **20**, 153.

LIVANOVA, G. I., ZISHNEVSKAYA, G. Y. A., and ANDREEVA, I. N. (1979). *Dokl. Akad. nauk. SSSR* **245**, 739–742.

MAIER, R. J. and BRILL, W. J. (1976). *J. Bact.* **127**, 763.

MANDEL, J. L. and CHAMBON, P. (1979). *Nucl. Acids Res.* **7**, 2081–104.

MARCKER, A., LUND, M., JENSEN, E. Ø., and MARCKER, K. A. (1984). *Eur. molec. Biol. Orgn. J.* **3**, 1691–95.

MARCKER, K. A., BOJSEN, K., JENSEN, E. O., PALUDAN, K., and WIBORG, O. (1983). In *Proc. 1st. Int. Symp. Molec. Genet. Bacteria–Plant Interactions* (ed. A. Pühler), p. 149. Springer-Verlag, Heidelberg.

———— (1984). In *Advances in nitrogen fixation research* (ed. C. Veeger and W. E. Newton), pp. 573–8.

MELLER, E., BELKIN, S., and HAREL, E. (1975). *Phytochem.* **14**, 2399.

MEIJER, E. G. M. and BROUGHTON, W. J. (1982). In *Molecular biology of plant tumours* (ed. O. Kahl and J. Setell), p. 107, Academic Press, New York.

NADLER, K. D. (1981). In *Proc. 4th Int. Symp. Nitrogen Fixation* (ed. A. H. Gibson and W. E. Newton), p. 143. Australian Academy of Sciences, Canberra.

— and AVISSAR, Y. (1977). *Pl. Physiol.* **60**, 433.

NALDER, D. (1981). *Pl. Physiol. Suppl.* **67**, 135.

NASH, D. T. and SCHULMAN, H. M. (1976). *Coll. Int. Centre natn. Res. Scient.* **261**, 357.

NEUPERT, W. and SCHATZ, G. (1981). *Trends biochem. Sci.* **6**, 1.

NOEL, K. D., STACEY, G., TANDON, S. R., SILVER, L. E., and BRILL, W. J. (1982). *J. Bact.* **152**, 485–94.

NUTMAN, P. S. (1981). In: *Current perspectives in nitrogen fixation* (ed. A. H. Gibson and W. E. Newton), p. 194. Australian Academy of Science, Canberra.

PROUDFOOT, N. J. and BROWNLEE, G. G. (1976). *Nature, Lond.* **263**, 211–14.

RAZIN, A. and RIGGS, A. D. (1980). *Science* **210**, 604–10.

REPORTER, M. and HERMINA, N. (1975). *Biochem. biophys. Res. Commun.* **64**, 1126.

REYNOLDS, P. H. S., BOLAND, M. J., BLEVINS, D. G., RANDALL, D. D., and SCHUBERT, K. R. (1982). *Trends biochem. Sci.* 366–8.

ROBERTSON, J. G., WARBURTON, M. P., LYTTLETON, P., FORDYCE, A. M. and BULLIVANT, S. (1978). *J. Cell Sci.* **30**, 151–174.

ROESSLER, P. G. and NADLER, K. D. (1982). *J. Bact.* **149**, 1021.

RUVKUN, G. B. and AUSUBEL, F. M. (1981). *Nature, Lond.* **289**, 85.

SIEVERS, G., HUHTALA, M. L., and ELLFOLK, N. (1978). *Acta chem. scand.* Ser. B. **32**, 380–6.

STRIPF, R. and WERNER, D. (1978). *Z. Naturf.* **330**, 373.

SULLIVAN, D., BRISSON, N., GOODCHILD, B., VERMA, D. P. S., and THOMAS, D. Y. (1981) *Nature, Lond.* **289**, 516–18.

TROXLER, R. F. and BROWN, A. S. (1975). *Pl. Physiol.* **55**, 463.

TRUCHET, G. and COULOMB, P. (1973). *J. Ultrastr. Res.* **43**, 36–57.

UHEDA, E. and SYONO, K. (1982*a*). *Pl. Cell Physiol.* **23**, 75.

—— (1982*b*). *Pl. Cell Physiol.* **23**, 85.

VAN DEN BOS, R. C., SCHETGENS, Th,M.P., BISSELING, T., HONTELEZ, J. G. J., and VAN KAMMEN, A. (1983). In *Proc. 1st Int. Symp. Molec. Genet. Bacteria–Plant Interactions* (ed. A. Pühler), p. 121. Springer-Verlag, Heidelberg.

VERMA, D. P. S., NASH, D. T., and SCHULMAN, H. M. (1974). *Nature,* Lond. **251**, 74.

— and BAL, A. K. (1976). *Proc. natn. Acad. Sci. U.S.A.* **73**, 3843.

— and MACLACHLAN, G. A. (1976). *Pl. Physiol.* **58**, 405.

— KAZAZIAN, V., ZOGBI, V., and BAL, A. K. (1978). *J. Cell Biol.* **78**, 919–936.

— BALL, S., GUÉRIN, C., and WANAMAKER, L. (1979). *Biochem.* **18**, 476.

— BRISSON, N., SULLIVAN, D., GOODCHILD, B., THOMAS, D., and HAUGLAND (1980). In *Current perspectives in nitrogen fixation* (ed. A. H. Gibson and W. E. Newton), p. 318. Australian Academy of Science, Canberra.

— HAUGLAND, R., BRISSON, N., LEGOCKI, R. P., and LACROIX, L. (1981). *Biochim. biophys. Acta* **653**, 981.

— FULLER, F, LEE, J., KÜNSTNER, P., BRISSON, N., and NGUYEN, T. (1982). In *Structure and function of plant genomes.* NATO (ed. O. Cifferri).

— LEE, Y., FULLER, F., and BERGMANN, H. (1984). In *Advances in Nitrogen Fixation Research* (eds. C. Veeger and W. E. Newton) p. 557.

WAALWIJK, C. and FLAVELL, R. A. (1978). *Nucl. Acids Res.* **5**, 3231–6.

WHITTAKER, R. G., MOSS, B. A., and APPLEBY, C. A. (1979). *Biochem. biophys. Res. Commun.* **89**, 552–8.

— LENNOX, S., and APPLEBY, C. A. (1981). *Biochem. Int.* **3**, 117–124.

WIBORG, O., HYLDIG-NIELSEN, J. J., JENSEN, E. O., PALUDAN, K., and MARCKER, K. A. (1982). *Nucl. Acids Res.* **10**, 3487–94.

————— (1983). *Eur. molec. Biol. Orgn. J.* **2**, 449–52.

ZABEL, P., MOREMAN, M., VAN STRATEN, F., GOLDBACH, R., and VAN KAMMEN, A. (1982). *J. Virol.* **41**, 1083–8.

Appendix: Contents of Volumes 1-3

VOLUME 1: ECOLOGY

1. Photosynthetic micro-organisms: *P. Fay*
2. Heterotrophic micro-organisms: *V. Jensen*
3. Non-leguminous root-nodules symbioses with actinomycetes and *Rhizobium*: *A. D. L. Akkermans and C. van Dijk*
4. Environmental physiology of the legume-*Rhizobium* symbiosis: *T. A. Lie*
5. Nitrogen fixation in some terrestrial environments: *G. J. Waughman, J. R. J. French, and K. Jones*
6. Nitrogen fixation in waters: *H. W. Paerl, K. L. Webb, J. Baker, and W. J. Wiebe*
7. Paddy Fields: *I. Watanabe and S. Brotonegoro*
8. Forage legumes: *E. F. Henzell*

VOLUME 2: *RHIZOBIUM*

1. Nodulation status of the leguminosae: *G. Lim and J. C. Burton*
2. Ecology: *H. V. A. Bushby*
3. Biology: *M. J. Trinick*
4. Carbohydrate metabolism: *G. H. Elkan and L. D. Kuykendall*
5. Genetics: *J. E. Beringer, N. J. Brewin, and A. W. B. Johnston*
6. Ultrastructure of the free-living cell: *H.-C. Tsien*
7. Surface chemistry: *R. W. Carlson*
8. Serology: *J. M. Vincent*
9. Control of root hair infection: *F. B. Dazzo and D. H. Hubbell*
10. Development of leguminous root nodules: *E. G. M. Meijer*

VOLUME 3: LEGUMES

1. Taxonomy: *H. D. L. Corby, R. M. Polhill, and J. I. Sprent*
2. Mineral nutrition: *A. D. Robson*
3. Agronomy: *P. H. Graham and D. L. Chatel*
4. Physiology and morphology of perennial legumes: *G. H. Heichel and C. P. Vance*

5. Nodule development and senescence: *W. D. Sutton*
6. Nodule metabolism: *D. W. Emmerich, J. E. Lepo, and H. J. Evans*
7. Nitrogen uptake, transport and utilization: *J. S. Pate and C. A. Atkins*
8. Energy relationships: *J. D. Mahon*

Index

acetate 137
  uptake 83
acetyl–CoA reductase 135
Aeschynomene indica 208
Agrobacterium 132, 206, 209, 214 f., 224, 226,
  229, 240, 256
  tumefaciens 202, 225, 223 f., 237, 240,
  247, 256, 271
alanine dehydrogenase 110
alanine transport 200
Alcaligenes 128
  latus 61
alchohol dehydrogenase 62
Alfalfa, see Medicago
alginic acid 131, 137
allantoic acid 64
allantoin 64
allantoinase 64
aminoacid synthesis 62
γ-aminobutyric acid 215
δ-aminolaevulinic acid (ALA) 272, 287, 289
  dehydrogenase (ALAD) 287 ff.
  synthase (ALAS) 287, 289 f.
ammonia 27, 30 f., 109, 130, 175, 189,
  197 f., 212
  assimilation 110 f., 133, 272, 280, 307
  regulation of 110 f.
  transporter 175
  uptake 110
ammonium 24 f., 38, 40 f., 52, 86, 133, 140,
  148, 189, 298
  assimilation 53, 55, 63 ff., 136
  transport 140
Anabaena 50, 168, 170, 173 ff., 259, 261
  cylindrica 1 f., 7
  heterocysts 174 ff.
  map of nif genes of 170
  nif gene organization and regulation in 174 ff.
  nif HDK operon of 180 ff., 188
    aminoacid sequence of 181 f.
  nif H promotor structure 184 ff.
Anthoceros 189, 191
anthranilate 119
Athrobacter variabilis 84 f.
asparagine 62 ff.

synthetase 64
aspartate 24 f., 62, 64
ATP/ADP ratio intracellular 83 ff.
auxins 119 f.
Azolla 50, 189 f.
Azomonas 128, 142
Azomonotrichon macrocytogenes 128
Azorhizophilus paspali 128, 132, 143, 154
Azospirillum 106 ff., 128
  amazonese 106
  in association with plants 118 ff.
  brasilense 106 ff.
  genetic homology with K. pneumoniae nif
    genes 116 ff.
  lipoferum 23 ff., 106 ff., 112 ff.
  mutants 114 ff.
  nitrogen metabolism in 107 ff.
    regulation of 110 f.
  plasmids and phages 112 ff.
Azotobacter 27, 37, 70 f., 73, 75, 82, 84,86 f.,
  89, 127 ff., 170, 204
  beijerinckii 61, 143
  chroococcum 2, 7, 39 f., 128, 132, 134, 145 ff.
  classification 127 f.
  compounds produced by 131 ff.
  cyst formation by 130 f.
  DNA content 142 f.
  electron transport in 138 ff.
  gene transfer into 146 f.
  metabolism in 134 ff.
  molybdenum independent N-fixation in 40
  mutants 147 ff.
    nif 148 ff.
      map of 151 ff.
  nif KDH genes 144 ff.
    promotor of 145
  plasmids 143
  synergistic organisms to 132
  vinelandii 2, 5, 7, 33, 39 ff., 57, 70 ff., 84 ff.,
    108, 128, 130 ff., 142 ff, 203, 259, 261
Azotobacteriaceae 127 ff.

bacteriocin production 230 f.
bacterioferritin 135

315

316   Index

bacteriophages 112 f., 131, 226
  AL-1 112, 114, 249 f.
  λ 114, 249 f.
  M-1 204
  Mu 113, 115
  P2 249
  P4 249
  RL38 237
Bazillus polymyxa 7
beans 234, 236
Beijerinckia 128, 131 f., 134, 142
BH-FeMoco model system 27
Bradyrhizobium 150, 196, 224
  japonicum 7, 39, 56 ff., 77, 155, 194 ff., 252,
      287 ff.
    asymbiotic nitrogen fixation 211 ff.
    bacteroids 205, 287 ff.
    gene transfer 202 ff.
    nif genes 206 ff.
    *nif*H promotor 208
    nod genes 206
    plasmids 204 ff.
    root nodule development by 209 ff.
    symbiotic nutants 197 ff.
  spp 196; see also *Rhizobium*
  taxonomy of 196
Bryophyllum diagremontiansus 62

calcium 130, 140 f.
carbohydrate metabolism 65, 272
Carmichaelia 214
Centrosema pubescens nodules 294
Chromatium vinosum 7, 18, 70, 72
Clostridium pasteurianum 2, 4, 7, 39 ff., 71 ff.,
      89, 259
Clostridium spp 78
clover, *see Trifolium*
cobalamin 289
coproporphyrinogen 290
  oxidative carboxylase 290
Corynebacterium 128
Cowpea rhizobium 2, 52, 154, 194 ff.,
      197 f.
Cowpea nodules 212
Crotonate 141
crown gall tumour 214, 224 f., 256
Cyanobacteria 50, 73, 77 f., 81 f., 84, 168 ff.
  classification of 173
  nitrogen fixation 168 ff.
    aerobically 188 f.
    in heterocysts 174 ff.
    models for 172 f.
    in symbiosis 189 f.
cycloserine 200
cytochrome 135, 139
  oxydase 55, 59 f., 135, 141
cytokinin 119, 131
cytoplasmic membrane 84

Dermocarpa 172
Desulfovibrio 77 f.
Derxia 128, 138, 142
Diazotrophs 1 ff., 70 ff.
dicarboxylate transport system 63
N,N⁺-dicyclohexylcarbodiimide (DCCD) 85
dithionite 70 f., 79

electron carrier reducing enzymes 71, 78 ff.
electron carriers 71 ff.
electron transport to nitrogenase 58 f., 70 ff.,
      138 ff., 169 f.
  models for 86 ff.
Embden–Meyerhoff pathway 137
Entner–Douderoff pathway 134, 137
Erwinia 119
Escherichia coli 4, 42, 54, 76, 78, 83, 95 ff., 110,
      137, 142, 154, 168, 184, 190 f., 208, 225,
      232 ff., 246 f., 253, 257, 272
  lacZ gene 209
  minicells 96, 117, 261 f., 265
exopolysaccharides 209, 212 f., 271

Fe protein 1, 7, 16 ff., 38, 55 f.
  covalent modification of 22 ff.
  electron transfer from 26
  FeS clusters 18 f.
  inhibition by ammonia, glutamine, and
      aspartate 24 f.
  nucleotide binding to 20 ff.
  reactivation of 23
  redox properties of 19 f.
  structure of 17 f.
ferredoxin 18, 70 ff., 87 f., 90, 135, 139, 176
  NAD⁺ oxidoreductase 80 f.
  NADP⁺ oxidoreductase 81 f.
  structure of 71 ff.
flavine mononucleotide (FMN) 76
flavodoxin 18, 37, 71, 75 ff., 87 f., 90, 135,
      139, 169
  hydroquinone 39, 76, 82, 86
  quinone 76, 82
  semiquinone 76, 82, 86
  structure of 76
formate 78 f.
  dehydogenase 79
fructose 63, 187
fumarate 141

galactose 213
galactosidase 209
galacturinic acid 213
giberellin 119, 131
Gloeothece 172 f., 188 f.
  *nif*HDK operon 188 f.
glucohydrolase 135

glucose 63, 106, 129, 134, 137, 148, 213
  dehydrogenase 135, 174
  6-phosphate 87 f., 176
  transport 63, 141
glutamine 1, 175 f., 186, 199 f., 211
  synthetase (GS) 24 f., 53 f., 64, 110, 116,
    135 f., 140 f., 174 f., 186, 189, 199 f.,
    212, 272, 280, 304, 307
  transport 174
glutamate 22, 64, 141, 175, 280
  dehydrogenase (GDH) 53, 64, 110, 190
  synthase (GOGAT) 53, 64, 110 f., 136,
    174 ff.
  transport 174
2-oxo-glutarate 141
glycine 195, 287, 289
Glycine max 51, 59, 61, 196, 198, 205
  nodules 62 ff., 283, 288 ff.
glycolysis 136
glyoxylate cycle 136
goat's rue 214
GS, GOGAT pathway 53
guanosine tetraphosphate (ppGpp) 266

haem biosynthesis 272
hairy root disease 224 f
heterocysts 39, 79, 130, 141, 174 ff.
hexose monophosphate pathway 187
hydrogenase 15, 56 f., 79, 138, 140, 174, 210,
    230 f.
  uptake system (hup) 15, 56 f., 62, 138, 148,
    198 f., 205, 230 f.
hydrogen evolution 28, 138 f.

infection thread 266, 271, 280
indole-acetic-acid 119, 131
indole-lactic-acid 119
insertion elements 252, 254
  ISRm1 252
  ISRm2 252
iron 39, 140 f., 187
  molybdenum cofactor (FeMoco) 5 f., 10 f.,
    14 f., 55 f.
isocitrate 88, 141, 176
  dehydrogenase 135, 174, 176, 187

β-ketoadipate pathway 136
α-ketoglutarate 175 f. 187
  dehydrogenase 174
Klebsiella pneumoniae 2, 5, 7, 17, 28 f., 32 f.,
    36 f., 39 ff., 54, 77 ff., 95 ff., 108, 113,
    116 ff., 144 ff., 168 ff., 206 ff., 238 ff.,
    258 ff.
  electron transport to nitrogenase 169 f.
  glnA 111, 116, 170, 184

nif genes 32 ff., 95 ff., 144, 170 ff, 206 f.,
    238 ff., 258, 265
  map of 96, 171
  regulation of 95 ff., 170 ff.
  products of 32 ff., 207
  stability of 38
ntrA 111, 171
  gene product 172
ntrB 111, 118
ntrC 111, 118, 170 f.
  gene product 54, 264
  see also nif genes

lectin 118
leghaemoglobin 55, 58 ff., 70, 210, 212, 233,
    272, 280 ff.
  apo- 289
    synthesis of 290 ff.
  genes
    activation and methylation of 286
    chromosomal arrangement of 284 f.
    regulatory sequences of 284
    sequence organization in soyabean of
      282 f.
  globin part of 58
    synthesis of 289
  haem-prosthetic group of 58 f.
    synthesis of 287 ff.
  pseudogene 285 f.
  structure of 282 f.
  synthesis of 286 ff.
    regulation of 286 ff.
    relation to nitrogenase activity of 294 ff.
Lotus 196, 200, 204, 210
  lupinus 214
  pedunculatus 210
  rhizobia 200 f., 204 ff.
lucerne, *see Medicago*
lupine nodules 62, 64

Macroptilium atropurpureum 199, 201, 203
maize 120
malate 62, 78 f., 88, 141
  dehydrogenase 62
  export 62
  permease 62
malonate 63, 141
D-mannitol 106, 129, 197
  dehydrogenase 197
mannose 63, 213
Medicago 214, 245
  agglutinin 267
  nodules 256, 265 f.
  sativa 233
Megasphaera elsdenii 71
melanin production (mel) 230, 232
Melilotus 245

membrane potential 84 ff.
4-O-methylglucuronic acid 213
Microcoleus 188
MoFe protein 1 ff., 38, 55 f.
  FeS centres of 10 f.
  inhibitors of 14 f.
  nucleotide binding to 15
  redox centres of 4 f.
  redox properties of 11 ff.
  structures of 2 ff.
  sustrates of 14 f.
molybdenum 39 ff., 138, 140 f., 155 ff., 187
  processing of 42
  regulation of *nif* expression by 39 ff., 155
  storage 41 f., 141, 148, 157
    protein 141
  uptake system 40 f.
Myxobacterium 128

NADH: ferredoxin oxidoreductase 80, 90, 174
NADH/NAD + ratio intracellular 80
NADPH: flavodoxin oxidoreductase 82 f., 90
NADPH/NADP + ratio intracellular 81 ff., 87
nif gene organization in
  Anabaena 174 ff.
  Azospirillum 117 f.
  Azotobacter 144 ff.
  Gloeothece 188 f.
  Klebsiella 32 ff., 95 ff., 170 ff.
  Rhizobium 207 ff., 228 ff., 245 ff.
nif genes
  transcriptional analysis of 95 ff.
*nif*A 33f., 38, 54, 95, 97 f., 100, 116, 118, 145 f., 149 ff., 170 f., 263 ff.
  dependent promotors 97 ff.
  gene product 33 f., 38, 98 ff., 155, 172
*nif*AL operon 97 ff.
  terminator 97 ff.
*nif*B 15, 33 ff., 42, 96 ff., 150, 170
  expression of 97
  gene product 15, 33 f., 98, 155
  promotor 99 f.
  terminator 99 f.
  transcription unit 99 f.
*nif*C 150
*nif*D 2, 4, 9, 33 ff., 54, 100, 118, 144 f., 169 f., 177 ff., 201, 206 ff., 258 f.
  gene product 2, 4, 33 f.
  amino acid sequence of 262
*nif*E 33 f., 36, 38, 42, 100 ff., 116, 150, 170, 206, 258
  gene product 33 f., 36, 38
*nif*F 33 f., 37 f., 77, 79, 96, 103, 169
  gene product 33 f., 37 f.
*nif*H 4, 13, 16 ff., 33 ff., 40, 54, 100, 102, 116 ff., 144 f., 149, 158, 169 f., 177 ff.,

201, 206 ff., 258 f., 305
  gene product 4, 17 f., 33 ff., 158
*nif*HDK operon 54, 101, 116, 143 ff., 180, 206 ff., 238, 258
  promoter 35, 100 f., 145, 208
*nif*HDKY operon 35, 103, 116, 153
*nif*J 33 ff., 38, 77, 79, 103, 150, 169
  gene product 33 ff., 38
*nif*K 2, 4, 7, 33 ff., 54, 100, 102, 117 f., 144 f., 177 ff., 202, 206 ff., 258 f.
  gene product 4, 33 ff.
*nif*L 33 ff., 38, 54, 96 f., 102, 170 f., 183
  gene product 33 f., 38, 98
*nif*M 33 f., 36, 38, 101, 103, 170
  gene product 33 f., 36, 38
*nif*MVSU operon 103
  promoter 103
*nif*N 33 f., 36, 38, 42, 100 ff., 150, 170
  gene product 33 f., 36, 38
*nif*Q 34, 38, 96, 100, 150, 170
  gene product 34, 38, 98
*nif*QB operon 97 f.
  promoter 97 ff.
*nif*S 33 f., 36, 101, 103, 170
  gene product 33 f., 36
*nif*U 34, 101 ff., 170
  gene product 34
*nif*V 15, 28, 33 f., 36, 42, 101, 103, 153, 170
  gene product 15, 28, 33 f., 36
*nif*X 33 f., 96, 101 f., 150
  gene product 33 f.
*nif*XNE operon 102 f.
  promoter 102
*nif*Y 33 ff., 37, 96, 100 ff., 116, 150, 258, 262
  gene product 33 ff., 37
nigericin 84 f.
nitrate 111, 130
  assimilation 111
  dissimilation 111
  reductase 42, 199
  reduction 272
nitrite 130
nitrogen
  assimilation 111
  metabolism 52 ff., 107 ff.
    in *Azospirillum* 107 ff.
    in free living *Rhizobia* 52 ff.
    in *K. pneumoniae* 54
  storage 63
nitrogenase 1 ff., 55, 60, 62, 70 ff., 130, 134 f., 169, 174, 176, 230, 258 ff.
  activity 22 ff., 41 f., 54 ff., 65, 70 ff., 83 ff., 108 f., 111, 119, 155 ff., 194, 199, 209, 211 ff., 238, 293 ff.
  binding of MgATP to 20 ff.
  biosynthesis of 39 f., 108 f., 111, 134, 212
    dependence on iron 39
    dependence on molybdenum 39 ff.
  degradation 65

nitrogenase (*cont.*)
electron transport to 58 ff., 70 ff., 138 ff.
electron carrier reducing enzymes for 78 ff.
electron carriers for 71 ff.
models for 86 ff.
Fe protein of 1, 16 ff., 55 f., 59, 65, 81, 86 f., 89 f., 110, 116, 135, 139, 149 ff., 198 f., 207
induction of 213
inhibition by ammonium 24 f., 41, 86, 138
interaction with hydrogen 28 f.
iron–molybdenum cofactor of 5 f., 10 f., 14 f., 55 f.
model for nitrogen reduction by 29 ff.
MoFe protein of 1 ff., 38, 55 f., 65, 89, 110, 116, 118, 134, 149 ff., 169, 199, 207
structure of 2
protection from inactivation by oxygen 38 f., 134, 136 ff., 187
reductase 178, 259
amino acid sequences of 261
regulation of 22 ff., 83 ff., 110, 136
relation to proton motive force 83 ff.
repression of 109, 212
stimulation by Mg 22 f., 39
substrate specificity of 26 ff.
turnover 25 ff., 56
nod genes 206 f., 230
control of expression of 270
nodulation 266
nodule, *see* root nodule
nodulins 210, 281, 298 ff.
genes for 298 f.
purification of cDNA probes 300 f.
Nostoc 184, 191

oligomycin 83 f.
opines 213 ff., 256
organic acid metabolism 63
Oscillatoria 172 f., 188
oxaloacetate 62, 141
oxydoreductase 169

Parasponia 60, 201
andersonii 194, 210
rhizobium 2
Paspalum 128
notatum 132
rhizobium 7
pea 303
nodules 62, 236, 291, 296 f., 305 f.
pectin 119
Peltigera
aphthosa 189
carita 189
Pennisetum americanum 118
peribacteroid membrane 58 f., 280, 283, 308

peribacteroid space 283, 307 f.
phages, *see* bacteriophages
Phaseolus 214
nodules 64
vulgaris 54, 304
phosphoenolpyruvate (PEP) 62, 137
carboxylase (PEPC) 61 f., 307
6-phosphogluconate 176
dehydrogenase 63, 134 f.
1-1-1-phosphogluconate aldolase 135
phytohormone-production 119 f.
pinitol 63
Pisum sativum 53, 59, 62 f., 206
plasmids 112 f., 224 ff.
colK 191
colE1 249
IncP-group 113, 115 f.
IncP1-group 202, 227, 229, 234, 246, 253
pAB1 117
pAB35 117 f.
pAB36 118
pACYC184 95, 97, 116
pBR322 100, 190, 248
pCK3 118
pEK10 270
pGE10 118
pGR3 268, 270
pGV1106 249
pHC79 257
pJB3JI 115, 248 f.
pJB4JI 114 f., 232 f.
pJB5JI 231, 234 ff., 268
pJB8 257
pKK2 249
pKSK5 270
pKT230 147
pLAFR1 249, 267
pPP346 270
pRD1 113, 202 f.
pRi 229
pRK290 115 ff., 249, 253, 267
pRK2013 115, 249
pRL1JI 228, 233, 236
pRL6JI 230, 236
pRL10JI 236
pRlel001a 239 f.
map of 239 f.
pRme41a 247
pRme41b 248, 255 ff., 267
pRme41c 256
pRmeSL26 267, 270
pSA30 117,. 206, 258
pSUP2021 115
pSym 225 ff., 247 f.
incompatibility 235 ff
pSym1 234 f., 238 f.
pSym5 228 f., 234, 238 f.
pSym9 234, 238
pTi 214, 225, 229, 234 f. 239 f.

plasmids (*cont.*)
  pVK100 115
  pWK22 95
  R68 249
  R68.45 113, 115, 202 f., 227, 230, 246 ff.,
    254
  R772 235
  R1822 202
  RK2 115 f., 249
  RP4 113, 115, 147, 154, 191, 202, 230,
    246 f., 249f., 256
    cointegrate formation with pRme41a 247
  RP41 147
  RSF1010 115, 147, 249
Plectonema 170, 172 f., 187
  boryanum 40, 42, 84 f.
polygalacturonase 204
polygalacturonic acid 119
poly-3-hydroxybutyric acid (PHB) 58, 60 f.,
  136 f., 141
proton motive force 83 ff., 89
  relation to nitrogenase activity 83 ff.
Protoporphyrin IX 289
protoporphyrinogen 290
  oxidase 290
Pseudoanabaena 172 f.
Pseudomonas
  aeroginosa 115
  putida 112, 263
purine
  biosynthesis 64
  oxidases 64
pyridine nucleotides 86 f.
pyruvate 78 f., 90, 141, 176, 187
  dehydrogenase 135, 137, 174
  :ferredoxin oxidoreductase 77 ff., 90, 174,
    187
  :flavodoxin oxidoreductase 78 f., 90
  kinase 62, 135, 137

rhenium 156
Rhizobiaceae 51, 53, 225, 239, 245
Rhizobium 50 ff., 56, 58, 70, 86 f., 113, 120,
  128, 151, 154, 172, 280 f., 287
  bacteroids 51, 55 f., 58 ff., 70, 280 f., 287
  common nod genes 206 f., 266 ff.
    control of expression of 270
  fredii 196
  gene transfer 202 ff., 246 ff.
  haem synthesis 287 ff.
  japonicum (*Bradyrizobium*) 7, 39, 56 ff., 77,
    155, 194 ff., 252, 287 ff.
    asymbiotic nitrogen fixation 211 ff.
    bacteroids 205, 287 ff.
    gene transfer 202 ff.
    nif genes 206 ff.
    *nif*H promotor 208
    nod genes 206

    root nodule development 209 ff.
    plasmids 204 ff.
    symbiotic mutants of 197 ff.
  leguminosarum 53 f., 60, 63, 154, 194 f., 197,
    202 ff., 224 ff., 247, 249, 252, 269 f.,
    292, 294, 296 f., 303, 308
    chromosome 226 ff.
    nod genes 268
    plasmids 224 ff.
  loti 154, 196, 224
  lupini 7, 61, 63, 194 ff., 224, 227, 233, 237,
    245 ff.
  meliloti 2, 7, 54, 116, 144, 153 f., 194 f.,
    197, 201 ff., 214, 224 f., 227, 233, 237,
    245 ff.
    cloning systems 249 f.
    fix genes 271 f.
    gene bank 257, 267
    gene transfer systems 246 ff.
    mega plasmids 245, 255 ff.
    *nif*HDK operon 54, 206 ff., 262 ff.
      promotor of 208, 260, 263 f.
        regulation of 263 f.
    *nif*H
      nucleotide sequence of 260
      Shine–Dalgarno sequence 260, 262
    nif genes 258 ff.
      map of 258 f.
    nod genes 258, 266 ff.
      map of 258
    symbiotic mutants of 250 ff.
    sym plasmid 214
  nitrogen metabolism 52 ff.
  phaseoli 54, 63, 84, 144, 194 f., 206, 224 ff.
    plasmids 224 ff.
  spp 52, 60, 194 ff.
    asymbiotic nitorgen fixation 211 ff.
    bacteroids 212
    gene transfer systems 202 ff.
    nif genes 206 ff.
    nod genes 206
    plasmids 204 ff.
    root nodule development 209 ff.
    symbiotic mutants of 197 ff.
    taxonomy of 195 f.
  symbiotic nitrogen fixation 50 ff.
    biochemical physiology of 50 ff.
    in free living condition 52 ff., 211
    genes for 206 ff., 224 ff.
    syntrophic growth 57 f.
  sym plasmids 204, 224 ff.
    expression of 230 ff.
    mapping on 238 ff.
  taxonomy of 195 f., 227
  transduction 237 f.
  transformation 237
  trifolii 2, 7, 63, 154 f., 194 f., 197, 201, 203,
    205, 210, 224 ff., 268
    nif genes 234
    plasmids 224 ff.

Rhizopogen rubescens 133
Rhodopseudomonas 186
   capsulata 7, 22 ff., 39
   palustris 22, 25
   sphaeroides 22 f., 25, 84 ff.
   vanelli 22
Rhodospirillum 170
   rubrum 7, 22 ff., 78, 80, 108
ribulose-PHkinase 174
rice 120
root hair curling (Hac) 266 f.
root nodule 51, 55, 280 ff.
   ammonia assimilation in 63 f.
   biochemical physiology of 58 ff.
   cells 61 f.
      fermentative pathways in 58, 61 ff.
   development 209 ff.
   electron transport in 55 f., 58
   general physiology of 58
   nitrogene storage 63 f.
   nodulins 281
   oxygen tension 55
   specific mRNA 281
RUBP carboxylase 174, 176 f.

Saccharomyces cerevisiae 133
Salmonella typhimurium 54
Sesbania rostrata 52, 209
sorbitol 106
soyabean, *see Glycine max*
spinach 81
   ferredoxin 87
sunstrates for nitrogenase 26 ff.
succinate 141, 211, 287, 289
   dehydrogenase 63
sucrose 63, 129
sulfite 71
superoxide dismutase 134, 139
Synechococcus 172 f., 188
syntrophism 57 f.

TCA-cycle 134, 137, 174
T-DNA 214, 240
tetraphyenylphosphonium (TPP) 86
Ti-plasmid 214, 225, 229, 234 f., 239 f.
transacetylase pathway 136
transposon mutagenesis 96, 114 f., 146, 201,
     228 f., 232 f., 248, 250 ff., 262, 265, 267,
     296, 305
transposons
   Tn5 96, 100, 114 f., 146, 201, 228 ff.,
     232 f., 248, 251 ff., 262, 264 f., 267 f.,
     296, 305
   Tn7 146, 251
   Tn9 251
Trifolium 214 f., 233, 236
   nodules 234
   repens 232
Trigonella 245

ubiquinone 139
uncoupler 83, 85
ureide 62 f.
uronic acids 141

valinomycin 84 f.
vanadim 156
Vicia 214, 233
   faba 214, 232
   nodules 62, 234
Vigna 214
   radiata 199
   sinensis 210

wheat 120

xanthine dehydrogenase 64
Xanthobacter 128
   autotrophicus 7